Studies in Systems, Decision and Control

Volume 315

Series Editor

Janusz Kacprzyk, Systems Research Institute, Polish Academy of Sciences, Warsaw, Poland

The series "Studies in Systems, Decision and Control" (SSDC) covers both new developments and advances, as well as the state of the art, in the various areas of broadly perceived systems, decision making and control–quickly, up to date and with a high quality. The intent is to cover the theory, applications, and perspectives on the state of the art and future developments relevant to systems, decision making, control, complex processes and related areas, as embedded in the fields of engineering, computer science, physics, economics, social and life sciences, as well as the paradigms and methodologies behind them. The series contains monographs, textbooks, lecture notes and edited volumes in systems, decision making and control spanning the areas of Cyber-Physical Systems, Autonomous Systems, Sensor Networks, Control Systems, Energy Systems, Automotive Systems, Biological Systems, Vehicular Networking and Connected Vehicles, Aerospace Systems, Automation, Manufacturing, Smart Grids, Nonlinear Systems, Power Systems, Robotics, Social Systems, Economic Systems and other. Of particular value to both the contributors and the readership are the short publication timeframe and the world-wide distribution and exposure which enable both a wide and rapid dissemination of research output.

Indexed by SCOPUS, DBLP, WTI Frankfurt eG, zbMATH, SCImago.

All books published in the series are submitted for consideration in Web of Science.

More information about this series at http://www.springer.com/series/13304

Fevzi Belli · Ferdinand Quella

A Holistic View of Software and Hardware Reuse

Dependable Reuse of Components and Systems

 Springer

Fevzi Belli
University of Paderborn
Paderborn, Germany

Ferdinand Quella
Siemens AG
Munich, Germany

ISSN 2198-4182   ISSN 2198-4190 (electronic)
Studies in Systems, Decision and Control
ISBN 978-3-030-72260-9   ISBN 978-3-030-72261-6 (eBook)
https://doi.org/10.1007/978-3-030-72261-6

This Springer imprint is published by the registered company Springer Nature Switzerland AG
The registered company address is: Gewerbestrasse 11, 6330 Cham, Switzerland

To my wife, my best friend and advisor,

Bettina Belli

Foreword

The book covers all considerable areas of REUSE, such as software, hardware, materials technology, industrial and legal standards. It provides an exhaustive description of the technologies employed by REUSE. The book also offers a constructive assessment related to the environmental impact of REUSE in contemporary society.

The book signifies the essence of the roles that REUSE plays in consumer products. The descriptive information collected by the authors' comprehensive investigation on the misuse of REUSE provides leeway to reusability-driven software development, REUSE engineering, applications of REUSE in organizations, and REUSE testing to enable effective management and application of REUSE.

The authors have tremendous experiences in both academia and industry. I believe this book can help researchers and practitioners gain the most relevant, applicable, and state-of-the-art knowledge on REUSE.

Editor-in-Chief of *IEEE Transactions on Reliability*, https://personal.utdallas.edu/~ewong/

Prof. Eric Wong
Department of Computer Science
University of Texas at Dallas
Dallas, USA

Preface

It is a good tradition that authors apologize for increasing the huge number of books by adding yet another one and then justify its writing. We, of course, were convinced there are compelling reasons for a book to be written that asks the following questions and seeks to contribute to the finding of constructive answers:

- Software reuse can considerably reduce development costs. Some authors claim: About € 30.00 returned for every € 1.00 invested (especially when "harvested" from Internet without validation to be reused in a composed software). What about the *dependability* of those software? Furthermore, moving to cloud computing can reduce energy consumption and carbon emissions by 30% or more. Software is a major actor in energy consumption, since it also controls and affects cloud computing. What about the greenability of software?
- Electrical and electronic (E&E) waste is growing at a very fast rate; in 2019 it reached 53.6 million tons and is projected to grow even faster. What about the *environment*?

Dependable, environment friendly reuse of E&E components, especially in new products, is an appropriate counter-measure to meliorate this situation. Efficient technologies are available to develop, validate, and re-validate reusable hardware and software. There is an urgent need to integrate and unify the existing and emerging knowledge on reuse in different disciplines, considering also recommendations and requirements of materials technology and industrial & legal standards. The authors believe the present book will help in offering a holistic view.

The authors also believe that not enough is being undertaken to promote reuse in the E&E industry. A considerable amount of social, environmental, and political engagement is vital for the merits of available technologies to be transferred to real-life.

Reuse has already been established for many years in many sectors, for example, medical, aircraft, telecommunication equipment, and military, because it is well-understood in these sectors that reuse can greatly help in decreasing production costs and time to market and thus increasing profit. Here, again the questions: What about the dependability of the composed products and the protection of nature's precious goods? And what about avoiding hazardous material that is proliferating, while allegedly un-reusable goods are incinerated?

Unfortunately, the role of reuse for consumer products is still not completely understood. Therefore, these products will usually not be treated carefully in a reuse context during their production, usage and take-back phases, which makes a reuse often less profitable. This is an unbearable situation that needs to be immediately taken care of.

Almost all E&E products contain software. This is another factor why reuse is not yet profit-able in some sectors, since very often this software is desperately obsolete and cannot to be updated or upgraded at a reasonable cost when it is time to reuse it or its components. This situation needs to change. Legal rules and their enforcement are necessary that require software embedded in long-lasting products be upgradeable and revisable to not only improve its performance but also its environmental impact, for example, its energy consumption.

Reuse is actually already an object of European legislation. It is odd, however, that software is not included, barring some vague references. Rules of reuse and revalidation of software in E&E products must be explicitly included in these laws.

The authors hope this book will help not only students, researchers, practitioners, and anybody who is interested in reuse and environment-friendly development of E&E products. They also hope politicians will not ignore it.

Paderborn, Germany	Fevzi Belli
Munich, Germany	Ferdinand Quella
January 2021	

Prologue—Introductory Remarks

Objectives and Topics of this Book

The objective of *reuse* is to create new systems using existing ones. Thus, reuse helps avoiding building new systems from scratch by applying existing material and knowledge to contribute to the manufacturing of new systems. This fact might explain the economic importance of reuse from industrial and financial aspects, but there is also the factor of its ecologic impact on environmental protection, the avoidance of unnecessary production, and thus the wasting of precious natural resources.

Compared with other industrial sectors, electrical and electronic (E&E) waste has the fastest growth rate in the world; in 2019 it reached 53.6 million tons and is, according to the United Nations Global E-waste Monitor 2020 (www.globalewa ste.org), considered to be growing even faster. On a global level, only 20% of E&E-waste is treated properly and very little data is available to explain what happens to the rest [1]. One of the major concepts of the circular economy is to keep products, their components, and materials in use as long as possible and, therefore, to prevent premature obsolescence. The best way to realize this concept is to *reuse* those systems and, if not entirely, then as many components of them as possible (See Sect. 4.3 for a precise definition of *component*).

This book is about reuse of technical and industrial systems, specifically E&E products, preferably computing systems. Difficile, multi-facetted interactions of these systems with their environment, including human beings, require a *holistic* view. This view requires considering not only their *hardware*, the utilization of which is provided by *software*, but, moreover, features of *materials* included in those systems that might largely impact their reusability. Last but not least, legal and industrial norms and regulations are also to be considered. This holistic view covers *dependability* aspects as an integrating factor, especially to compensate for the dominance of commercial goals of reuse, since undependable reuse can lead to disastrous situations. For example, the reuse of existing software in a new Roentgen system that caused the appalling death of nine people [2], or the crash of the ESA satellite system that caused damages in excess of a billion euro [3], likewise due to improper reuse of software.

Any system's dependability centers on its ability to provide *services* that can be depended on, that is, defensibly be trusted within a pre-defined period of time [4]. Economic and environmental constraints necessitate, even force, extending the scope of dependability by considering further critical features; for example, energy consumption, inclusion of hazardous material, recyclability, and electromagnetic compatibility. The present book follows this extended, *big picture* view of dependability.

The service, delivered by an E&E system, is nowadays largely realized by software. Moreover, moving to cloud comprehensively reduces energy consumption and carbon emissions; software controls and affects also the cloud [5, 6]. These are the reasons this book starts with software and its dependability aspects.

An interesting fact: Reuse of software was one of the driving forces behind the creation of the relatively young discipline of *Software Engineering* and it follows that one of its major objectives is *reuse*, which was clearly expressed from the very beginning [7]. The average saving potential of software reuse is estimated to be up to 40% of the total development costs; a fact that encouraged many companies to embark on serious reuse initiatives and install appropriate organizations and technologies (see, for example, [8–12, 13]).

The goal of this book is to introduce this holistic view and explain the techniques and tools to realize it.

Why is a Holistic View of Reuse Necessary?

From the dependability side, this book discusses how high development standards and a comprehensive quality assurance procedure can ensure the development of more durable and longer lasting, robust products by lowering information asymmetries between manufacturers and consumers. These dependability procedures should take the main issues of product failures and obsolescence into account, such as product reliability and durability.

E&E systems and components are commonly complex; their reuse is even more complex, given that several facts and factors are to be considered at the same time.

First, one has to take into account that hardware needs software, and vice versa. Moreover, they both are embedded in systems that consist of different kinds of components and materials.

Second, the role of software is often underestimated in hardware reuse; for example, by not sufficiently testing its energy consumption that is a major environmental impact factor. For the purposes of ecodesign, which is discussed in many books or standards [14–16, 21], the role of hardware and software together and their interactions need to be comprehensively discussed to increase the ratios of products reused and to make them environmentally compatible.

Third, hardware may necessitate the reuse of a fairly aged product as a component; for example, the reuse of a mechanical device to maintain the functioning of energy transfer in a plant. However, reuse can be practiced also in a new product, where components will be exchanged and reused, such as peripheral components of a printer.

At any rate, all components, hardware, software, and materials involved must fit together. For stand-alone hardware, this might not be too difficult. But for software, old and new components must cooperate without failures, in mutual dependency. Therefore, in this book, both sides, that is, reuse of hardware and reuse of software till recycling, is covered. Thus, the techniques reviewed and discussed for software must be considered to create a functioning hardware with reused components. Additionally, facts of materials technology also need to be considered that impact the reuse of any system, such as their likely hazardousness under certain conditions, as well as environmental factors, such as electro-magnetic compatibility.

Another aspect to be considered is that of industrial and legal standards. This is actually a tremendously difficult, nevertheless worthwhile endeavor. While industrial standards exploit optimal earnings given by technical possibilities, legal standards set clear limits to the utilization of those possibilities, especially concerning environmental aspects. For example, the European WEEE law (Annex I) requires a collection scheme of between 55% and 80% of E&E-waste products to be prepared for reuse and recycling. Gas and discharge lamps need only to be recycled.

To sum up, a holistic view is definitely necessary, under consideration of the following disciplines.

- Hardware and software
- Material technology
- Environmental constraints
- Industrial and legal standards

This book explains and reviews elementary notions and methods of these four disciplines, provides hints for technical and financial opportunities and, more importantly, takes into account interferences and consequent constraints imposed by environmental protection.

Structure of this Book

This book consists of two parts, each concerning an important asset: Software aspects (Part I) and hardware aspects (Part II), which includes consideration of aspects of materials engineering and legal conditions, as well. Industrial standards will be mentioned in both Part I and II, wherever appropriate.

The objective, *dependable, environmental-compatible reuse of assets*, runs like a thread right through all chapters of the book and integrates them. In this context, an *asset* can be material, such as an electronic device, or immaterial, such as an idea, concept, or algorithm (see Chap. 4).

Figure 1 roughly reflects the concept of the book, illustrating its structure and revealing the interactions between Part I and Part II.

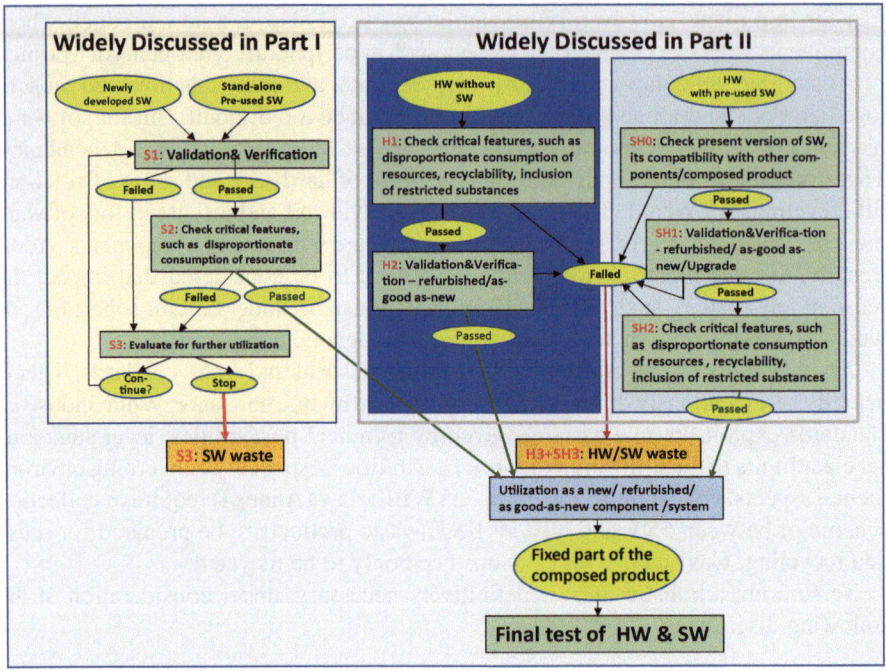

Fig. 1 Reuse of software only, hardware only, and hardware-software-together

Each of the three blocks in Fig. 1 consists of ovals and rectangles (boxes). *Ovals* represent conditions and requirements as states (situations) and inputs, as well as post-conditions and results as outputs. *Boxes* represent actions and activities as transitions. The adept reader might have recognized this representation style as a rudimentary Petri net (see Appendix B).

Part I, figuratively represented by the left block, reviews and discusses aspects of software, focusing on verification and validation (V&V) of two critical features: Firstly, the dependability (the box "S1"). Secondly, other critical features, such as disproportionate consumption of resources; for example, energy, but also memory, run time (the box "S2").

A software *component under consideration* (CUC) for reuse, no matter whether yet to be developed as a new component or already existing, that is, one formerly used, has to pass both S1 and S2. If this CUC fails one or both of the checks S1 and S2, the decision is to be made whether to improve this CUC or to waste it (the box "S3").

In the "good, old times," that is, many decades ago, hardware to be reused included no software, or only to an insignificant, lesser extent, so that software aspects could be entirely neglected. This situation is now rare, nevertheless sketched in the central block of Fig. 1. It includes the activities for verification and validation of dependability and critical features symbolized by the boxes "H1" and "H2", in analogy to the boxes S1 and S2 of the left, software-related block. If the hardware CUC fails one or both of H1 and H2, the decision is to waste it (leading to the box "H3+SH3" that corresponds to "S3" of the left block. There is no need to check whether to improve this hardware CUC or to waste it, because software is "soft", that is, flexible, thus commonly easy and inexpensive to rework and improve, while hardware is "hard," expensive or infeasible to be modified for correctional measures. If CUC passes both H1 and H2, it is suitable for reuse. In this case, the hardware CUC can be a brand-new one, a refurbished one, or, as a special case, a "qualified as-good-as-new (QAGAN)" one. The latter CUC is called, in accordance with IEC 62309, a "QAGAN" component. This aspect will be discussed in detail in Part II.

The right block of Fig. 1 discusses the most interesting case. Hardware-CUC, no matter whether new or refurbished or QAGAN, includes pre-used software to an extent that cannot be neglected. An upstream check, represented by the box SH0, determines the version and state of this pre-used software CUC to clarify the following question: Is this software compatible with other components (software and hardware), and with the targeted composed system, that is, the final product? If not, this component will be wasted (box H3+SH3). Otherwise, the CUC undergoes tests for verification and validation of dependability and critical features summarized in the boxes SH1 and SH2, in analogy to the boxes H1 and H2. If the CUC fails one or both of SH1 and SH2, the decision is to waste it (again, the box "H3+SH3"). If CUC passes all of SH0, SH1 and SH2, it is suitable for reuse.

After passing the above-described comprehensive checks, nothing stands against reusing components in a new, QAGAN, or refurbished product. Eventually, a final test must confirm that obvious failures have not been overseen; that is, the developed asset, either a composite one containing reused components or a refurbished one, needs to be checked thoroughly before being put up for sale.

The V&V activities symbolized in the boxes S1 and S2 will be reviewed in Part I; the activities H1, H2, SH0, SH1, and SH2 will be reviewed in Part II, together with the aspects of ecodesign. It goes without saying that Part II relies on the quality and qualification procedures explained in Part I whenever software is included and thus interferes with the hardware. The old (pre-used) software might have to cooperate with the latest, cutting-edge software, combined with a new component to be added. For example, an old printer is supposed to work with a new computer.

To sum up, a plan is required for potential tests, dependability, or environmental risks. It could also be distinguished between the target that the refurbished product works sufficiently and the objective that the product is equivalent-to-new.

The overall aim for sustainability is to meet the needs of the present without compromising the ability of future generations to meet their needs [17, 18]. This book is supposed to modestly contribute to this goal with a brief, uniform introduction into reuse, its terminology, and techniques. The authors suggest what needs to be considered to develop high quality products and to enable better reuse of software and hardware components and systems.

How to Read and Work this Book

The reader of this book is not required to have in-depth knowledge of software, hardware, material technologies, or environmental protection. Part I refreshes and deepens elementary topics of software engineering that is necessary to understand the dependability requirements for software to be developed and, more interestingly, to be reused. Moreover, in Part I, procedures are discussed to qualify this software. These requirements are also valid for reuse of software as a constituent of hardware, whether as a stand-alone system or as a part of a heterogeneous cyber-physical system.

Accordingly, the reader with a solid background in software engineering can immediately start with the Part II and understand and apply the procedures for the reuse of existing hardware and software to compose complex systems. On the other hand, a reader who is interested only in software reuse, can focus on Part I, thoroughly ignoring Part II.

Topics and Aspects Discussed

In accordance with the goal set, the book covers and discusses following the topics and aspects.

- Terminology of reuse, including dependability and software aspects.
- Software reuse technology.
- Hardware aspects, especially ecodesign of E&E systems in combination with software, such as compatibility of old and new hardware with old and new software.
- Green IT and further concepts entailing environmental impacts; for example, excessive high energy consumption, electromagnetic (EM) burdens and damages.
- Materials engineering, considering also physical, chemical, and logic notions and processes, as well as recycling aspects.
- Industrial and legal, multi-national standards and aspects of their usage.

Note that Part I of this book firstly discusses and reviews features and techniques to develop, verify, and validate software in general, as needed to test software that comes with hardware to be used or reused, as explained in Part II. Secondly, Part I discusses and reviews techniques of software reuse and its V&V, which is a specific area of software engineering, and not in the focus of Part II.

Too Many Figures, Too Many Tables?

The first author has made extensive use of schematic representations by figures and bulleted lists in Part I to explain complex topics, sometimes indented to several levels, in the hope that this will aid the reader in intuitively understanding the complicated interactions and interrelations, leading to a bigger picture of the topic. The author, certainly no friend of wordy explanations, acknowledges Ikea's success in communication based on a picture being worth a thousand words and can only hope for a modicum of the same success.

Last but Not Least: Targeted Readers

This book is supposed to be useful for scientists and practitioners in computer engineering, environmental and sustainability engineering, and industrial engineering and management. Nevertheless, the book can be considered also as a textbook for undergraduate and graduate students of these disciplines. The exercises for each chapter are supposed to support the students and instructors. Depending on the discipline and prior knowledge, the subject can be taught in one or two terms.

The authors assume that no prior knowledge is necessary for understanding the techniques reviewed and discussed. Thus, anybody interested in the subject can read and get benefits from this book. References at the end of each chapter are for those wanting to deepen their knowledge.

To sum up, depending on interests, objectives, and state of knowledge, the reader can choose an individual way to use this book, or any part of it. The authors welcome any reader, including (but not limiting to) the following ones.

- Students and instructors
- Researchers & developers

  - Software specialists, hardware specialists, quality and environmental engineers,
  - Legal practitioners, lawyers,
  - Newcomers interested in any or all of the subjects listed in the section "Topics and Aspects Discussed" above.

Additional Remarks

About the Authors
As an electrical engineer with a doctorate in computer science, Fevzi Belli spent about ten years in industrial development of large communication systems. Following this

period of valuable insights and experiences, he was appointed as a professor of computer science and information technology. He continues his F&D activities in close cooperation with the industry.

Ferdinand Quella has a doctorate in polymer chemistry and, for many years at Siemens AG, headed corporate groups for material science and testing and introduced total quality management also EFQM and later headed the corporate function for product-related environmental protection including eco-design.

Based on their academic, scientific and practical activities and experiences in dependability and reuse, the authors founded and headed the workings groups that developed the industry standards on dependable reuse of hardware and software, IEC 62309 and IEC/PAS 62814, respectively (see [19, 20]). The authors have made use of these industrial standards while preparing this book.

The artificial word "QAGAN" (qualified-as-good-as-new") was coined in IEC 62309, in a serious attempt to standardize, and hopefully help to legalize, the utilization of previously used components in new products.

Answers to Exercises and Support after Purchasing the Book

Answers to the exercises are not included in the book since they can easily be found within the text of the corresponding chapters.

The reader might find additional information, references, and updates on the authors' website: www.ivknet.de.

References and Literature: Mix of Old and Recent Ones

Somebody who only reads newspapers and at best books of contemporary authors looks to me like an extremely near-sighted person who scorns eyeglasses. He is completely dependent on the prejudices and fashions of his times, since he never gets to see or hear anything else. And what a person thinks on his own without being stimulated by the thoughts and experiences of other people is even in the best case rather paltry and monotonous.

Albert Einstein (1879–1955)

The discerning reader will notice that references are a mixture of old - even ancient - and current literature. Thus, their age is of no relevance. The older ones are from primary literature, written by pioneers of their field. For example, one cannot speak about the General Net Theory (also called Petri Nets) without mentioning its founder, to wit, the immortal and eternal Carl Adam Petri and his timeless Ph.D. thesis (see Appendix B). Others are rather recent and are meant to help the reader catch up on the state-of-the-art.

Acknowledgments

The authors are grateful to Viktor Klippenstein (Diebold Nixdorf) for his valuable help with the formatting of the manuscript.

David Andersen, (my old, dear friend in Berlin), and Philip John Newcombe (Maksimum Ingilizce, Ankara), deserve our thanks for proofreading, editorial suggestions, and styling the text for Part I and Part II, respectively.

A good portion of the first part of this book was refined and written during the COVID-19 period, during which the first author was stuck in Germany. Unfortunately, the preponderance of his comprehensive, hand-written material was in his office in Turkey. Erhan Ayrilmaz (Ege University, Izmir) patiently searched for and forwarded these documents. Sincere thanks to him for this great help. A very special thanks to Bettina Belli for perfecting the figures in Part I.

Fevzi Belli
Ferdinand Quella

References

1. Forti, V., Baldé, C.P., Kuehr, R., Bel, G.: Global e-waste monitor 2020: quantities, flows and the circular economy potential, United Nations University (UNU)/United Nations Institute for Training and Research (UNITAR)—co-hosted SCYCLE Programme, International Telecommunication Union (ITU) & International Solid Waste Association (ISWA), Bonn/Geneva/Rotterdam, ISBN Print: 978-92-808-9115-7 (2020) [See also www.globalewaste.org (Global E-waste Monitor of the United Nations University)]
2. Leveson, N.G., Turner, C.S.: An investigation of the therac-25 accidents. IEEE Comput. **26**(7), 18–41 (July, 1993)
3. Nuseibeh, B.: Ariane 5: Who dunnit? IEEE Softw. **14**(3) 15–16 (1997)
4. Avizienis, A., Laprie, J.-C., Randell, B., Landwehr, C.: Basic concepts and taxonomy of dependable and secure computing. IEEE Trans. Depend. Secur. Comput. **1**(1) (January–March, 2004), available at http://ieeexplore.ieee.org/xpls/abs_all.jsp?arnumber=1335465
5. Acar, H., Benfenatki, H., Gelas, J.-P., da Silva, C.F., Alptekin, G., Benharkat, A.-N., Parisa Ghodous, P.: Software greenability: a case study of cloud-based business applications provisioning. Proceedings of the IEEE 11th International Conference on Cloud Computing (CLOUD), pp. 875–878, (2018). https://doi.org/10.1109/CLOUD.2018.00125. hal-01887065
6. Microsoft, Accenture and WSP Environment & Energy Study; https://news.microsoft.com/2010/11/04/microsoft-accenture-and-wsp-environment-energy-study-shows-significant-energy-and-carbon-emissions-reduction-potential-from-cloud-computing/
7. Naur, P., Randell, B. (eds.): Software Engineering, Report on a conference sponsored by the NATO Science Committee, Garmisch, Germany, 7th to 11th October 1968, available at http://homepages.cs.ncl.ac.uk/brian.randell/NATO/nato1968.PDF
8. Ezran, M., Morisio, M., Tully, C.: Practical software reuse. Springer Practitioner Series (2002)
9. Hallsteinsen, S., Paci, M.: Experiences in software evolution and reuse—twelve real world projects. Springer Science & Business Media (1997)
10. van der Linden, F.J., Schmid, K., Rommes, E.: Software product lines in action. Springer (2007)
11. Poulin, J.S., Caruso, J.M., Hancock, D.R.: Business case for software reuse. IBM Syst. J. **32**(4), 567–594 (1993)
12. Frakes, W.B., Isoda, S.: Success factors of systematic reuse. IEEE Softw. **11**(5), 14–19 (1994), available at http://ieeexplore.ieee.org/stamp/stamp.jsp?tp=&arnumber=311045
13. Mili, A., Chmiel, S.F., Gottumukkala, R., Zhang, L.: Managing software reuse economics: an integrated ROI-based model. Ann. Softw. Eng. (Kluwer Academic Publishers), **11**, 175–218 (2001)
14. Belli, F., Quella, F.: Utilization of Used Components, VDE Verlag: Berlin-Offenbach (2015), ISBN 978-3-8007-3898-4

15. Wimmer, W., Lee, K.M., Quella, F., Polak, J.: Ecodesign—the competitive advantage. Springer Dordrecht et al. Niederlande (2010), ISBN 978-90-481-9126-0

16. ISO/TR 14062:2002: Environmental Management—Integrating Environmental Aspects into Product Design and Development

17. Kienzle, J., et al.: Toward model-driven sustainability evaluation, Commun. Assoc. Comput. Assoc. **63**, 80–91 (2020)

18. Gomes, C., et al.: Computational sustainability: computing for a better world and a sustainable future. Commun. Assoc. Comput. Assoc. **62**, 56–65 (2019)

19. IEC 62309: Dependability of Products Containing Reused Parts—Requirements for Functionality and Test (2004)

20. IEC/PAS 62814: Dependability of Software Products Containing Reusable Components—Guidance for Functionality and Tests (2012)

21. [IEC_19] IEC 62430:2019: Environmentally Conscious Design (ECD)—Principles, Requirements and Guidance

Contents

Abbreviations

ADL	Architecture Definition Languages
AE	Application Engineering
ANDL	As-New-Designed Life
ANSI	American National Standards Institute
bi	Boundary-Interior
BS	British Standard
CB	Component-Based
CBSE	Component-Based Software Engineering
CES	Complete Event Sequence
CESG	Complete Event Sequence Graph
CFC	Control Flow Chart
CFG	Control Flow Graph
CM	Configuration Management
CMM	Capability Maturity Model
CORBA	Common Object Request Broker Architecture
COTS	Commercial-off-the-Shelf (COTS)
CSR	Corporate Social Responsibility
CUC	Component under Consideration,
CVL	Common Variability Language
DA	Domain Analysis
DE	Domain Engineering
DfR	Design for Recycling
DO-178B/C Level A (Aerospace)	Software Considerations in Airborne Systems and Equipment Certification
E&E	Electro (Electrical) and/or Electronic
ELV Directive	End of life Vehicles
EMC	Electro-Magnetic Compatibility
EN	Europe Norm
EP	Event Pair
ES	Event Sequence
ESG	Event Sequence Graph
ETN	Equivalent-to-New

FAST	Family-Oriented Abstraction Specification and Translation
FC	Formal Concept
FCA	Formal Concept Analysis
FCES	Faulty Complete Event Sequence
FDA	Food and Drug Administration (USA)
FEP	Faulty Event Pairs
FES	Faulty Event Sequences
FSA	Finite-State Automaton
HW	Hardware
IDL	Interface Definition Languages
IEC	International Electro-Technical Commission
ISO	International Standardization Organization
LCA	Life Cycle Assessment
LCSAJ	Linear Code Sequence and Jump
MBC	Multiple Branch Coverage
MIL	Module Interface Languages
NDL	New Designed Life
OOP	Object-Oriented Programming
OVM	Orthogonal Variability Model
PAS	Publicly Available Specification
PL	Product Line
PLE	Product Line Development
PUT	Program Under Test
QAGAN	Qualified-As-Good-As-New
RIC	Remanufacturing Industries Council
RoHS Directive	Restriction of Hazardous Substances Directive
RoI	Return on Investment
SM	State Machine
SO	Service-Oriented
SPICE	Software Process Improvement and Capability Determination
SPL	Software Product Line
SPLE	Software Product-Line Engineering
STD	State-Transition Diagram
SUC	System/Software Under Consideration
SUT	System/Software Under Test
svo	Subject-Verb-Object
SW	Software
TER	Test Effectiveness Ratio
TSD	Test Suite Designer
TSP	Traveling Salesman Problem
V&V	Validation and Verification
V&V	Verification and Validation
VDI	Verein Deutscher Ingenieure

WEEE	Waste of Electro- and Electronic Equipment
WIMP	Windows, Icons, Menus, and Pointers
ZVEI	Zentralverband der Elektrotechnik und Elektroindustrie

Part I
Dependable Reuse of Software

Software reuse has happened so quickly that we do not yet understand the best practices for choosing and using dependencies effectively, or even deciding when they are appropriate and when not.
Russ Cox [1]

Software realizes sophisticated, complex applications, which explains why it is one of the key factors for the reuse of E&E systems. The first part of this book reviews the features of software and techniques for its development, discusses dependability aspects, and introduces a uniform terminology and specific techniques of software reuse and its validation.

Reuse can help to dramatically save development and production costs, but it is a well-understood fact that a software component, which functions in an application, can completely fail, or worse, work incorrectly, sometimes with fatal impacts. Chapter 1 of the Part I, with five sections, discusses reusability in general, and takes the dependability aspects into account. Existing verification and validation techniques, especially test methods, are reviewed, while also covering fault tolerance aspects. The general aspects of software and its dependability, as will be discussed in the Sects. 1.2 and 1.3, are also valid for reuse. This means that they need to be understood to comprehend verification and validation of reuse. Nevertheless, specific discussions and reviews are included in later sections wherever reuse-specific adaptation is necessary, for example, testing of reuse (Sects. 5.1 and 5.2).

Terms and definitions of software reuse and its ingredients conclude the first chapter.

In spite of the partly detailed review of existing techniques, the first chapter is only a refresher introduction to software and dependability. As such, it is not intended to replace a comprehensive text book on the topic. In case the reader feels uncomfortable, the references at the end of the chapter might help to acquire further, necessary insight.

Chapter 2, consisting of three sections, focuses on application aspects and organization of reuse. Different types of reuse-oriented architectures will be introduced, methods to assure dependability, and legal aspects of reuse are discussed.

Chapters 3 and 4, totaling nine sections, review techniques of reusability-driven software development and best practices for software reuse; for example, component-based software engineering, commercial-off-the-shelf software, and product-line technique. While methods of Domain Engineering help to systematically develop reusable components, techniques of Application Engineering, as its counter-part, enable the selection of the right component for the application under consideration. Organizational aspects conclude Chap. 4.

The last chapter of Part I critically reviews the methods for verification and validation discussed in Sect. 1.3, and selects the ones that are suitable to test software reuse, both product families in general and specific products in particular. Economical aspects of reuse and its validation conclude Part I.

To rephrase the words of R. Cox, this part is meant to help in understanding, choosing and using best practices of software reuse, and deciding when they are appropriate and when not [1].

Reference

1. Cox, R.: Surviving Software Dependencies, Commun. of the Assoc. of Computing Assoc., **62**, 36–43 (2019)

Chapter 1
Reusability, Dependability, and Software

Reuse is the objective that must lead to a dependable target system composed of reused and new components. Systems nowadays consist of different types of E&E components that will be integrated by software. Therefore, the authors suggest to start with software aspects, without neglecting its interaction with hardware.

1.1 Reuse and Reusability of Software and Electronic and Electrical Systems

Remember that *reuse* is the process of creating new systems from existing ones, rather than building them from scratch [64]. Reuse is thus not limited to the utilization of particular components; it has, moreover, to consider all of the information that is related to the product generating processes, including also documents from requirements definition, analysis, design, implementation to test cases, and test procedures.

Reusability is, generally speaking, the ability of an asset to be reused. This ability will usually be interpreted qualitatively. However, reusability can also be interpreted quantitatively, especially for a comparison of several assets concerning the degree of their capability of reuse. This second interpretation leads to a metric. Kim and Stohr [63] have defined software reusability as "a measure for the ease with which the resource can be reused in a new situation." This view will be discussed in Sect. 5.3.2.

Long period market analyses encourage reuse that tends to have a very high return on investment (RoI); for example, in the case of software engineering, a return of about € 30.00 for every € 1.00 invested [44, 81]. In the case of hardware, about 25% of the existing electrical and electronic (E&E) components are suitable for reuse. This means savings of billions of euros every year, apart from saving precious environmental resources. In Europe, E&E reuse is considered to be the best waste

© The Author(s), under exclusive license to Springer Nature Switzerland AG 2021   3
F. Belli and F. Quella, *A Holistic View of Software and Hardware Reuse*,
Studies in Systems, Decision and Control 315,
https://doi.org/10.1007/978-3-030-72261-6_1

handling procedure. This explains why reuse of software and hardware components and systems has a long tradition in the industry.

Nevertheless, whereas a component may be perfectly suited to one application, it may prove to cause severe faults in reuse applications. Therefore, it is fundamental to provide the certainty that a reused component, or system, is dependable in its new application and does not need to be re-designed from scratch. Consequently, an adequate validation process considering the changed purpose and the different application configuration in combination with new, reused, or further used components is needed.

Systematic approaches to reuse are available for both software and E&E. Industrial standards for reuse, that is, IEC 62309 (*Dependability of products containing reused parts—Requirements for functionality and tests*) [50] and IEC PAS 62814 (*Dependability of software products containing reusable components—Guidance for functionality and tests* [53]) support the reuse process. The authors have been chairing and co-chairing these international standards. This book is intended to critically review these approaches; it introduces a uniform terminology, summarizes and improves well-known techniques, and provides suggestions and recommendation for the practice.

1.1.1 Reuse of Software and Reuse of Hardware

It is a well-known fact that the hardware of long-lasting E&E systems becomes obsolete more quickly than the software. Wirth's law says "Software is getting slower more rapidly than hardware becomes faster" [90]. This phenomenon can be explained by the maintenance efforts and modifications that enable the durability, longevity, and especially the reuse of systems and components in larger, more complex applications, whereby software needs to be adapted to changing user requirements more often and faster than hardware does. This explains why software costs increase faster while hardware costs stagnate (Fig. 1.1).

1.1.2 Reusability as an Overall Requirement for Systems Design and Architecture

The vision of software reuse is as old as software itself—it was first introduced back in 1968, the same year the term "software engineering" was coined during the constitutional NATO conference in Germany [74]. Nevertheless, not every "copy and paste" action programmers do daily while constructing their programs qualifies as a software reuse this book has in mind. Also calling an internal or external function and even a remote-procedure call do not necessarily constitute a reuse this book would accept [46, 86].

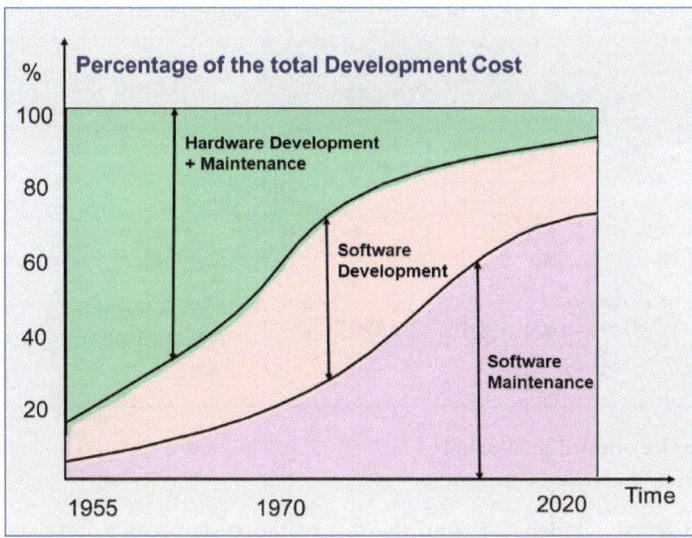

Fig. 1.1   Tendency: increasing maintenance costs

All the examples above suggest that the *context* and *domain* of the called software do not change. Therefore, there is no need for them to perform *pre-store* and *pre-use* activities that will be described in this book (Sect. 2.1).

1.1.3   When Is It Reuse? An Example

Contrary to the examples above, using a service in a s*ervice-oriented* (SO) landscape or in "*Common Object Request Broker Architecture* (CORBA)" is of more interest here because the *context* and *domain* of the software that delivers a service might change. The typical constellation of SO architecture consists of services for constructing, offering, selecting, and validating. More specifically, a service has to be registered and "published" before it will be offered. Infrastructural services are offered to realize a broker, etc. [36] (Fig. 1.2).

A variant of the service-oriented architecture (SOA) structural style is defined by *microservices* leading to a software development technique that arranges an application as a collection of loosely coupled services [60, 76, 78].

Microservice-based applications consist of multiple services that can evolve independently. When services are modified, they have to be tested before being deployed. As usual, the test suites that are executed are usually designed without exact knowledge of how the services will be reused. Gazzola, et al., introduced an approach to analyzing the execution of deployed services at runtime in the field, in order to generate test cases for future versions of the same services. This approach exploits cloud technologies, containers in particular, and counterfeits an environment to

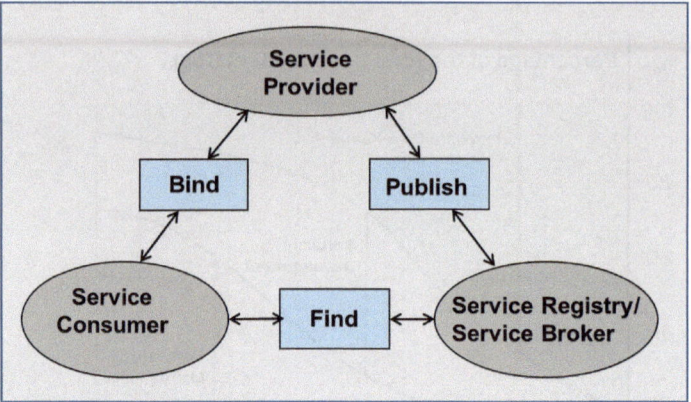

Fig. 1.2 Service-oriented architecture

isolate the service under test from the rest of the system. As a next step, service interactions are reproduced as previously analyzed, enabling the testing of the new version of the service against usage scenarios that capture the field usages of its earlier versions [39].

1.1.4   What Is Really Software Reuse?

At the risk of trying the reader's patience by being repetitive: Software reuse is the process of creating software systems from existing software rather than building software systems from scratch. Thus, software reuse is not limited to the source or object code; it has, moreover, to consider all of the information that is related to the product generating processes, including, as well, requirements, analysis, design, documents, and test cases apart from the code.

1.2   Dependability—Factors and Quantification

Before starting with quality aspects of reuse, basic elements of dependability and quality will be discussed in this section.

1.2.1   What Is Critical?

The safeguarding of basic dependability requirements to protect human health and life and valuable goods, relative to the risks involved and the *degree of safety* expected from the product in question, should be carefully taken into account when considering reuse of existing software [52]. It is important that the control is executed by monitoring the product's behavior in the field after sales.

Transferring the safety requirement into reuse: Before reusing a software component, the context and domain it was built for should be carefully compared with the context and domain it is intended to be built in, including the hardware and physical and organizational aspects.

1.2.2 Quality and Its Non-functional Characteristics

The long period quality of a product or a service depends to a great extent on the objectives of the users, on how they intend to deploy it. It is, of course, also crucial that the constraints given by usage and available technology be considered (Fig. 1.3, see also Capability Maturity Model of SEI, [80]). See also ISO/IEC 15504 (Sect. 1.3.3.14). A brief discussion in Sect. 5.3.1 associates CMM with Reuse.

The most important feature of a product (or a service) is, first, its *operation*; in other words, that it works at all, fulfilling all the requirements specified. Second, in the case of long-life products, it is important that the product can be adapted to changing requirements (*product revision*). Finally, a long-life product is supposed to be operated in different environments and also on future platforms (*product transition*). Figure 1.4 sketches these orthogonal characteristics and their elements with brief explanations.

The "root" element of quality is "fault." Faults are defined as deviations between specified requirements and realized implementations. A fault can, however, be *latent*; that is, unperceived if/when its occurrence does not cause an error, which manifests the fault. An error can cause a *malfunction* of a component that may, or may not, endanger the fulfillment of the system's mission. Finally, a failure is an event, for

Fig. 1.3 What is quality?

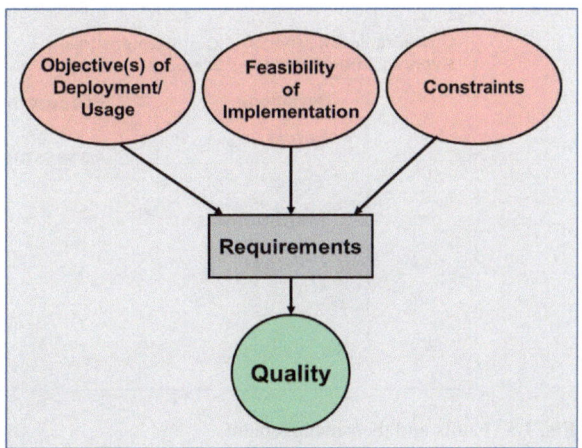

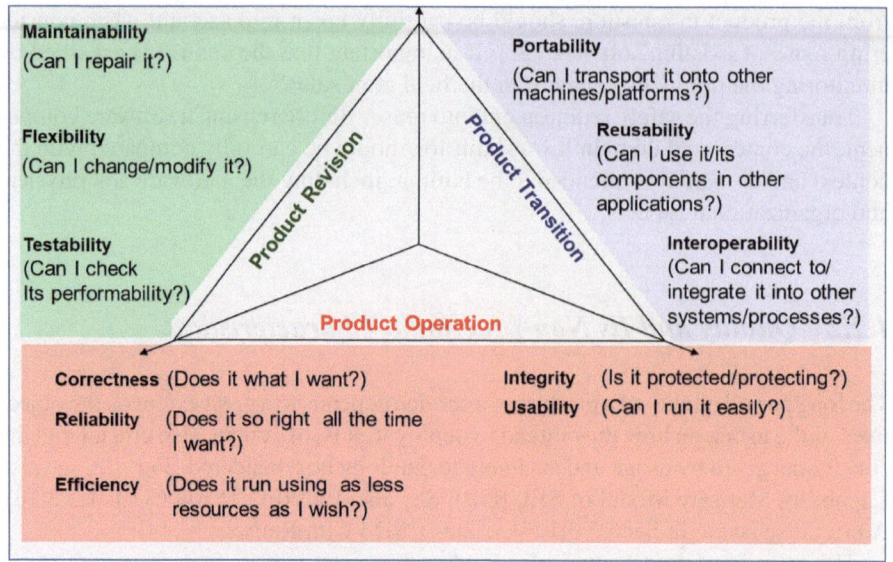

Fig. 1.4 Non-functional characteristics as quality factors

example, a system crash, which cannot be hidden and will consequently be perceived [55].

Figure 1.5 illustrates the chain of *Deviation-Fault-Error-Malfunction-Failure*. *Reliability Engineering*, a young technical-scientific branch, offers countermeasures to cope with faults and their impacts. These are firstly techniques for construction of reliable systems, that is, means for procurement. When a system is to be sold over the

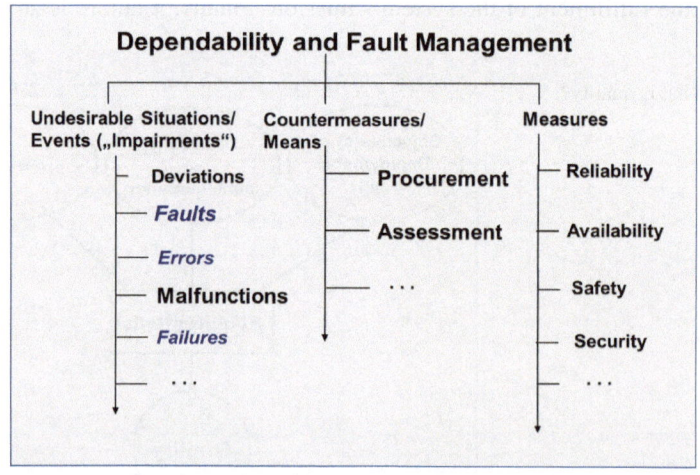

Fig. 1.5 Faults and their management

counter, its quality cannot be assured using construction techniques that have already been completed. Moreover, the quality can be assessed using validation techniques.

Finally, it is necessary to quantify the "goodness," that is, to measure the extent to which the constructive and assessment techniques were successful. These measures are based on probability theory and, its specific branch, reliability theory. *Reliability* is, informally defined, the probability that a system will survive a mission under given conditions and a given period of time [55].

Availability measures the probability that the system is still "alive," that is, correctly and properly operating at a given point in time. Safety and security are similar features concerning harms the system can cause to its environment (*safety*) and it can experience from outside (*security*).

Figure 1.6 expands Fig. 1.5 and refines activities of fault management. The best way to deal with faults is to avoid or prevent them. As this is not always affordable, one has to remove them; that is, correct the product so that the faults do not appear in the operational phase. In practice, however, this is very difficult (and expensive) to accomplish. Therefore, it is more realistic to construct systems in such a way that they can deliver acceptable services even though faults appear; in other words, they tolerate faults. Forecasting techniques are supposed to validate these critical features; that is, that faults are removed or will be tolerated.

A step further, Fig. 1.7 introduces the most important and thus comprehensive feature: Dependability as the property, which combines and adjusts all the critical features, is outlined in the Figs. 1.5 and 1.6 [11, 65].

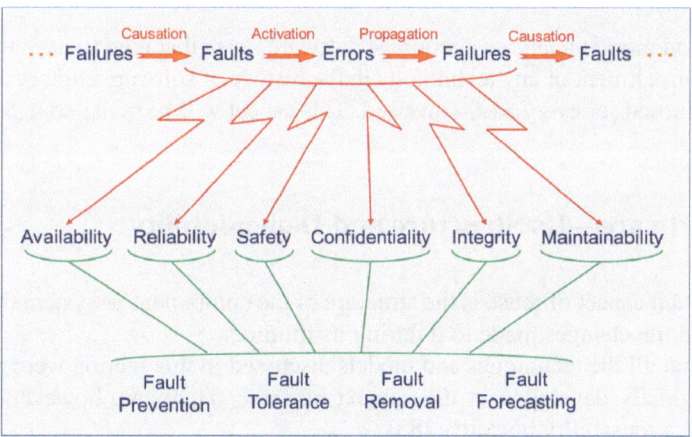

Fig. 1.6 Fault-related critical features (based on [65])

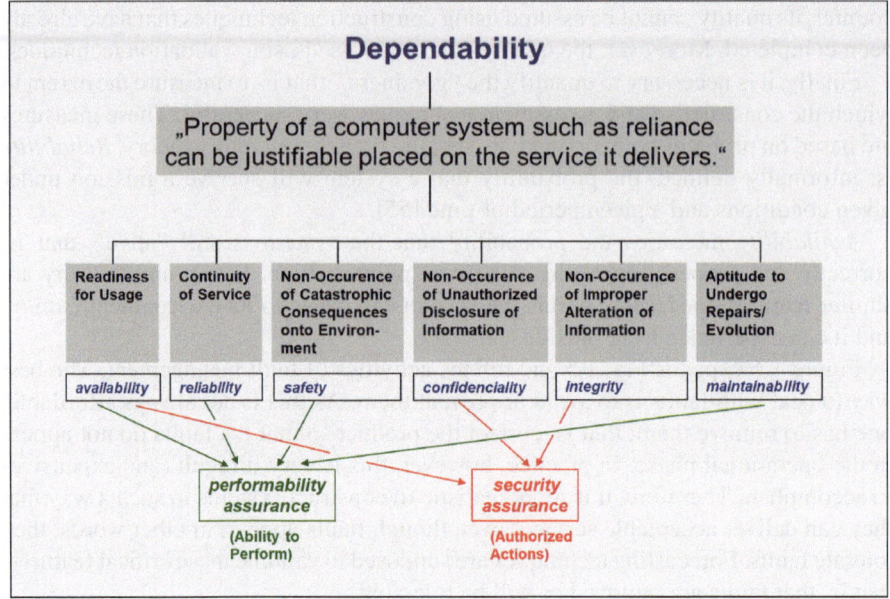

Fig. 1.7 Dependability (based on [65] and [11])

1.2.3 Last-But-Not-Least—Return on Investment

Long period market analyses encourage software reuse that tend to have the highest return on investment of any technology in the history of software engineering: about $30.00 returned for every $1.00 invested. This aspect will be revisited in Sect. 5.3.1.

1.3 Software—Its Structure and Dependability

An important aspect of reuse is the structure of the component or system that will be reused and the changes made to it during its lifetime.

Note that all the techniques and models discussed in this section were not necessarily originally developed in the context of reuse. They can, however, be easily adapted to a reusability objective [85].

1.3.1 Configuration Management

Expensive products are durable and often survive for generations. However, they require appropriate adjustments to adapt to the changing user and/or environmental requirements. These changes and adjustments are to be systematically documented to enable a precise tracking of the modifications of the structure and configuration of components over the course of their life. This aspect is also essential for reusable systems and components (see Sect. 3.1).

Configuration is the constellation of the integral elements of a component or a system and their structural relationships at a given point in time during its development and operation. *Baseline* is the "frozen" system configuration at a well-defined point in time ("milestone"), until the release of the next baseline.

Configuration management (*CM*) identifies and controls the system configuration at a desired state during the development and use of the system. CM has the following functions to support project and quality management.

- *Identification* of system components and baselines through meaningful identifiers, with reference to hierarchy, version and serial membership, etc.
- *Change Control* from definition until delivery—and even later, until the scrapping of the system and/or its components (managed by the *Change Board* as a part of project and quality management).
- *Change/Release Management* for accounting changes by versions and cooperating with the change board.

CM usually has its own tools that are integrated into the production environments and supports project/product management and quality management. Customer participation is included to a great extent, so that all stakeholders can be informed at any point in time of the product development and use [4, 10].

1.3.1.1 Component Identification

Large systems, both hardware and software, are built modular, that is, not in one single step and by only one developer but in many steps by many developers, mostly working simultaneously. Therefore, a well-defined system has many components that cooperate and are integrated into a well-defined structure, mostly hierarchically. Other structures can be meshed, centralized, decentralized, etc.

Figure 1.8 depicts a hierarchical system structure. Horizontal lines group elements of the same hierarchical level. The vertical lines should be read as "uses the services of"; for example, a subroutine (or subsystem) utilizes functions that are implemented by a section.

Well-defined functions will be implemented as *modules* (segments) that consist of a set of statements. The bottommost elements are *program statements* and can be considered as the leaves of a tree. In accordance with its size, the system can be broken down into several hierarchical levels.

An important issue is the naming of the system elements, so they can be identified unambiguously. A naming strategy has to fulfill the following requirements:

- Identification of the element itself.
- Determination of its position within the structure, for example, hierarchy.

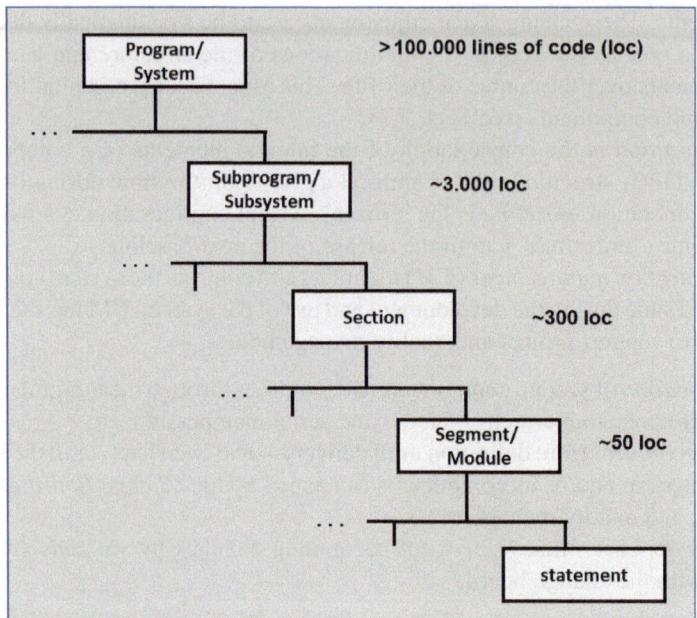

Fig. 1.8 Typical program structure

- Also, determination of its predecessors and successors.
- Indicating its version, which shows the number of changes made to this element.

  Figure 1.9 exemplifies a naming system that fulfills these requirements.

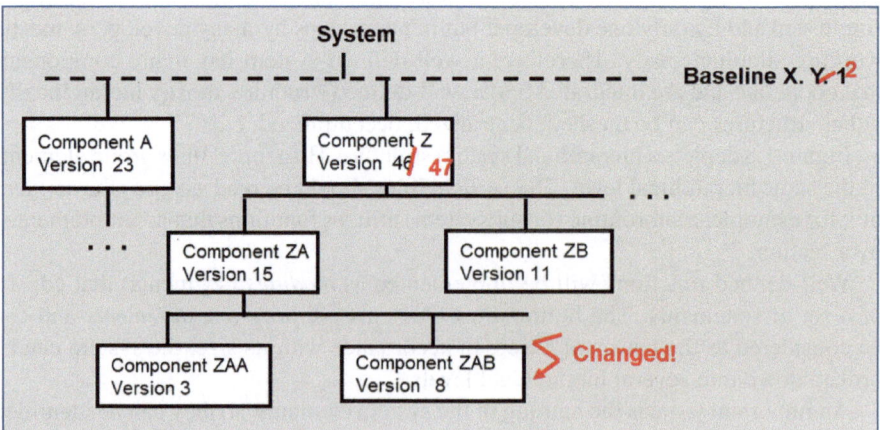

Fig. 1.9 Identification, baseline, change

1.3.1.2 Change Management

Bookkeeping of the component changes, and both the reasons and impacts of these changes, is one of the important tasks of Configuration Management (CM) that will be carried out by its sub-component, *Change Management*. A change to a component propagates to connected components all the way up to the top level. This chain of changes leads to a new version of the current baseline. Figure 1.9 explains this change propagation (see also Sect. 1.3.1.3).

A *Change Control Board*, which includes representatives of project management, developers, and customer, is necessary for the coordination and control of the changes.

1.3.1.3 Types of Changes

Changes are inevitable for long-life products. Figure 1.10 summarizes changes of a component as follows:

- *Regular development* includes refinement, definition of additional and/or complementary components.
- Different operational environments and user profiles require *alternative* components; for example, for different languages, different specifications of markets and countries.
- *Corrections* and *revisions* lead to modified versions of a components.

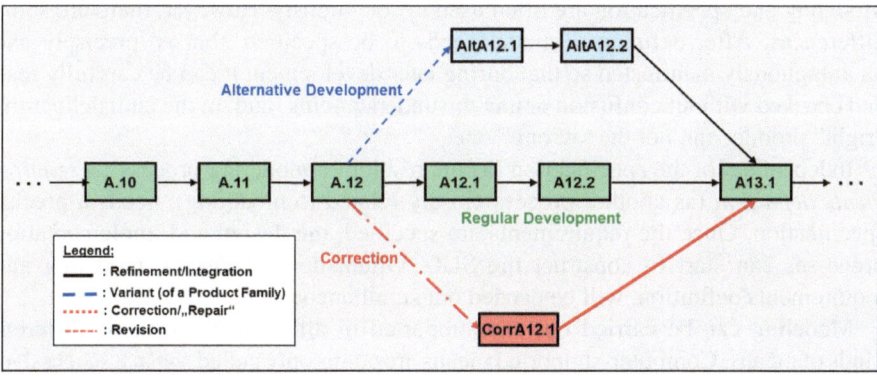

Fig. 1.10 Version group, development history

1.3.2 Modeling and Development Process

The development of large software systems, like developing other long-life products, is a costly and risky process. Moreover, misunderstandings of user requirements are likely and usually happen unexpectedly.

Modeling enables to focus on usage-relevant features of the system under consideration (SUC), and thus can be helpful in validating the requirements in an early stage of the development (see also Sect. 5.2). This helps with reducing risks and likely costs. Also defining milestones, an important function of project and configuration management, and dividing the development process into meaningful phases enables the further reduction of the risks.

Note: The acronym SUC used in this book denotes a program, software, or system "under *consideration*" that likewise includes a program/software/system "under *test*" (mostly abbreviated as "SUT"), since "consideration" subsumes "testing," but not vice versa. An SUC can also be a model, as considered in model-based testing, wherein a model can only be statically analyzed but not executed and thus not (dynamically) tested in the sense defined in this book (see Sects. 1.3.3.2 and 1.3.3.3).

Figure 1.10 summarizes various kinds of changes, from alternative development of the same product—for example, to be marketed in different countries—to changes caused by the correction of detected faults.

1.3.2.1 Modeling and Specification—Methods, Tools

Modeling and specification are often used synonymously. However, there are some differences. After defining, a model needs to be specified, that is, precisely and unambiguously manifested so that during later development it can be carefully read and checked without confusion or any misunderstanding, and, in the end, deliver the "right" product and not the "wrong" one.

Independent of the specification (as the result of a modeling process), a *requirements definition* (as another process closely related to modeling), needs a precise specification. Once the requirements are specified, the design and implementation processes can start to construct the SUC. Often, both processes, modeling and requirements definition, will be carried out simultaneously.

Modeling can be carried out and supported in different ways, using different kinds of means. Computer-supported means are commonly called *tools*; a successive, coordinated set of them a *tool chain*. Means and tools can be reviewed in a hierarchy, with increasing power from bottom to top.

- *Notational* means can be

  - *descriptive*, that is, to explain and depict the whereabouts of a process, for example a *use-case diagram*;
  - *prescriptive*, that is, to explain and stipulate how it is to be realized and implemented, for example a *structured check-list*.

A model can be both descriptive and prescriptive, for example, *decision tables.*

- *Rationalizing (amplifying)* means can be

  - *generating*, that is, to produce items of different kinds, for example, *test cases, product types;*
  - *recognizing*, that is, to accept or refuse a product based on its correctness, for example, *finite-state automata.*

- *Cognitive* means can combine information and knowledge to extend the present knowledge, for example, a *knowledge base* that processes and replies queries, implemented by a *Prolog program.*

It almost goes without saying that the most powerful ones, cognitive means, are commonly extremely complex, and thus, for uneducated, inexperienced novices, not always easy to understand, learn, and use. Therefore, it is recommended to look for the simplest means that is satisfactory for the purpose.

Modeling can be carried out and supported in different ways, using different kinds of means. Computer-supported means are commonly called *tools*; a successive, coordinated set of them a *tool chain.* Means and tools can be reviewed in a hierarchy, with increasing power from bottom to top.

- *Notational* means can be

  - *descriptive*, that is, to explain and depict the whereabouts of a process, for example a *use-case diagram*;
  - *prescriptive*, that is, to explain and stipulate how it is to be realized and implemented, for example a *structured check-list.*

A model can be both descriptive and prescriptive, for example, *decision tables.*

- *Rationalizing (amplifying)* means can be

  - *generating*, that is, to produce items of different kinds, for example, *test cases, product types;*
  - *recognizing*, that is, to accept or refuse a product based on its correctness, for example, *finite-state automata.*

- *Cognitive* means can combine information and knowledge to extend the present knowledge, for example, a *knowledge base* that processes and replies queries, implemented by a Prolog program.

It almost goes without saying that the most powerful ones, cognitive means, are commonly extremely complex, and thus, for uneducated, inexperienced novices, not always easy to understand, learn, and use. Therefore, it is recommended to look for the simplest means that is satisfactory for the purpose.

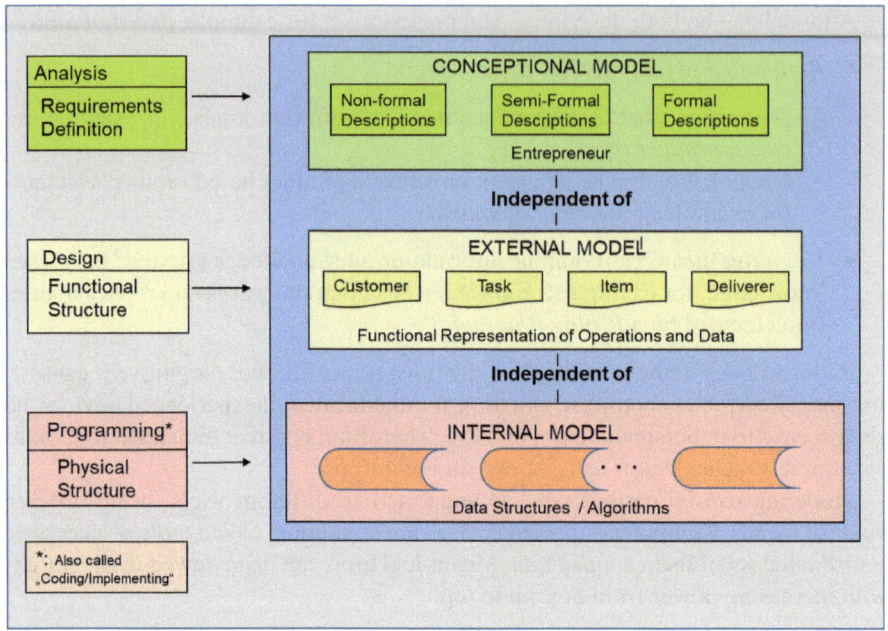

Fig. 1.11 Three layers of modeling

1.3.2.2   Three Layers of Software Modeling

Modeling usually starts with an *analysis* that delivers a *concept*, the core document that represents the idea for usage and functionality of the planned product.
Depending on the purpose and qualification of the development team, this concept
can be noted using formal, mathematical means or less formal ones. *Design* is the
construction of the *external* model that includes customer-oriented usage aspects.
Finally, *programming* entails an internal model delivering developer-oriented aspects
for implementing the system under consideration (Fig. 1.11).

1.3.2.3   Modeling of Software Development Process

It has already been mentioned that large software systems are not developed in a
single step, but in a series of well-coordinated, consecutive steps called *development
phases*. Since the nineteen-seventies, several models for development processing
have been suggested.

The first and primarily accepted process model is the Waterfall model, as depicted
in Fig. 1.12 [84]. The characteristic feature of this model is that the work items are
carried out in phases that are symbolized by boxes. These items, like water drops,
"trickle" from the top box to the bottom one. Further, before one phase starts, the

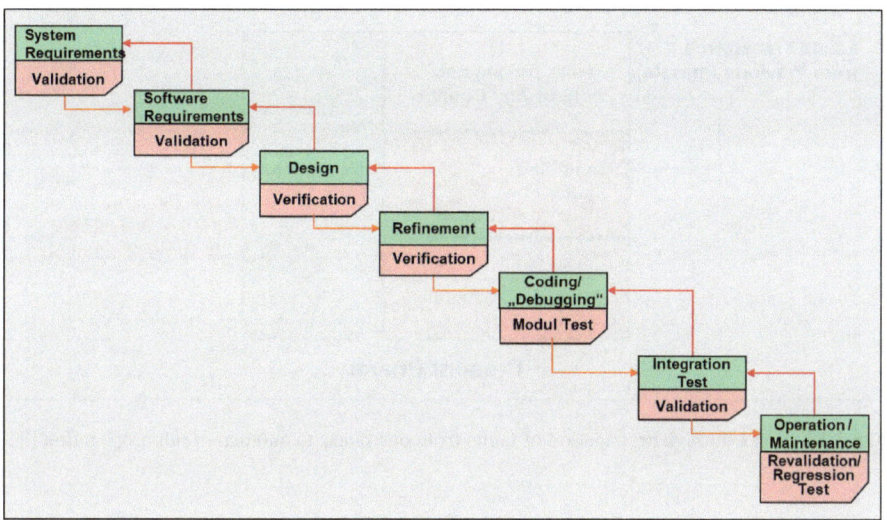

Fig. 1.12 Waterfall model of software development process

phase before must be concluded (its corresponding box "filled"), and deliver the necessary inputs of the current phase.

Another decisive characteristic is that the next phase cannot be launched before the results of the present phase are checked. In the course of the phases, this checking process has different names, depending on the stage of the development. *Validation* includes all of the analyses and checks that directly correspond to the external model, in other words, those that consider usage and customer aspects (see Fig. 1.12). *Verification*, on the other hand, focuses on the assessment of the results achieved by the internal model, that is, consider the aspects of the analysts and programmers. Consequently, validation checks whether or not the *right* (correct) system will be constructed. Verification, on the other hand, checks whether or not the development steps will be carried out *right* (correctly). To sum up, verification valuates the intermediate products to find defects, while validation tests the end product to find defects. Verification and validation will often be abbreviated as V&V.

Characteristic for the waterfall model is that the neighbored phases are checked against each other. Next, we focus on and analyze a single phase of this model to find out likely fault sources and what we can do to render them harmless.

1.3.2.4 Fault Propagation

Generally speaking, a phase under consideration, or the *present* phase, handles and processes the input(s) delivered from the last phase. In the end, this present phase produces its own output(s), in other words, its service(s) (Fig. 1.13).

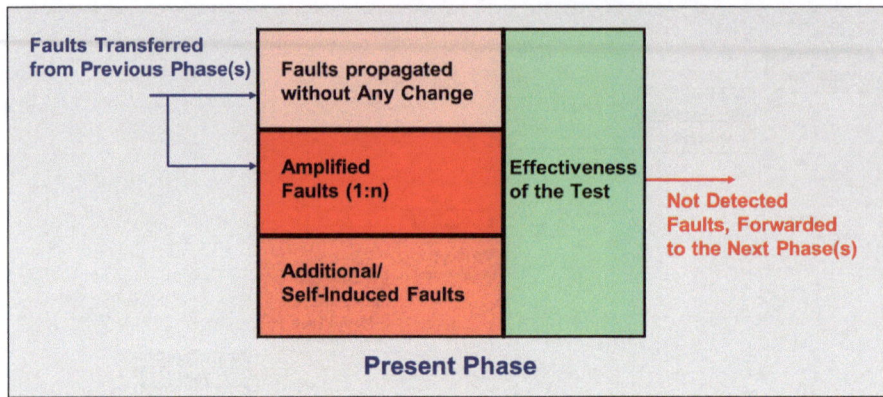

Fig. 1.13 Risks through propagation of faults from one phase to another—fault propagation [82]

The cautious assumption is that the input is likely to contain not only correct results of the phase before, but also the faulty ones. Depending on the effectiveness of the test, the present phase can pass these faulty results on to the next phase without any change. It is also possible that the mechanisms of the present phase amplify these faults because, among other things, they will be spread to many other components that then process the inputted faulty results to produce their own results—in this case also faulty ones (see Sect. 1.3.1.2).

Last but not least, the present phase can produce its own faults that will be transferred to the next phase.

It is evident that the validation mechanism of each phase, especially those of the early phases, are crucial for reducing the rate of faults passed from one phase to the next and, in the end, to the final product.

Another interesting finding is that the latent, propagated faults can form a flood (or avalanche) that form a ticking time bomb, ready to cause severe damages (see Fig. 1.14; [41]).

1.3.2.5   V-Model as Extension of the Waterfall Model

The waterfall model requires the neighbored phases be checked against each other. However, it is possible that early, undetected faults propagate and cause latent faults that can be revealed in later phases. The Waterfall model cannot help much in this situation. Therefore, not only direct neighbors but phases far from each other should be checked against each other, which will be realized in V-model (Fig. 1.15; [58, 59]).

The left leg of Fig. 1.15 represents a typical waterfall model up to the implementation phase. The right leg represents the follow-on activities, from component testing to the operational phase. So, the reader can bend down the right leg so that a bigger waterfall model can appear, covering all phases in a descending order. Shaping this

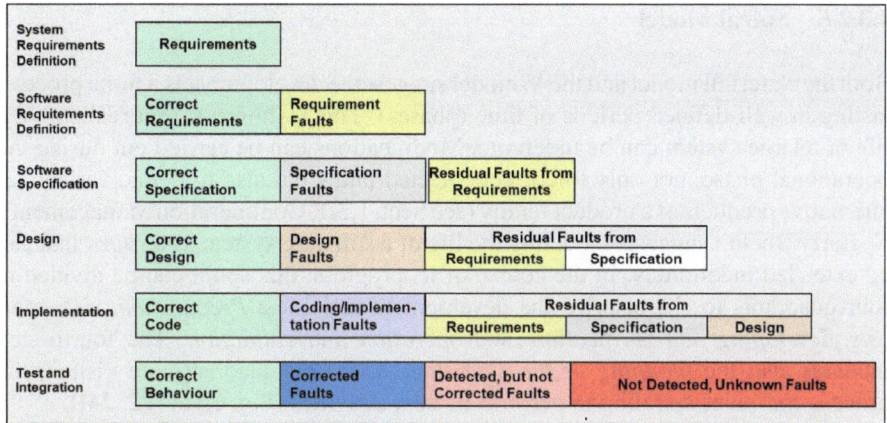

Fig. 1.14 Flood (or avalanche) of faults (based on [41])

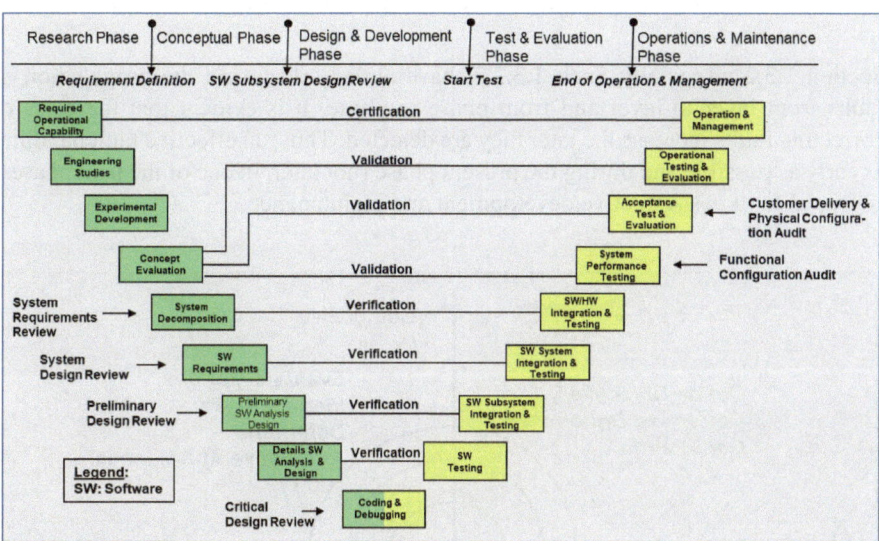

Fig. 1.15 V-model

bigger waterfall model into a V-form enables the identification of correlating phases that will be checked against each other.

1.3.2.6   Spiral Model

Both the waterfall model and the V-model suggest the development is a finite process, lasting in well-defined periods of time (phases). This is, however, not realistic; the life of a large system can be indefinite. Modifications can be carried out during the operational phase, not only to correct detected faults but also to define variants as alternative products of a product family (see Sect. 1.3.1, Configuration Management).

Barry Boehm suggests modeling the life of a software system as a *spiral* that can be extended indefinitely. In the course of its progress, this spiral can be divided in four quadrants to characterize the development activities: *Preparation, risk analysis, developing and verification,* and *operation and validation.* The fourth step includes also the *planning of the next stage.* Once all quadrants are visited, the process continues spiraling to perform its next activities (Fig. 1.16; [22, 24]).

1.3.3   Verification and Validation (V&V) and Fault Tolerance

Section 1.3.2, especially Sect. 1.3.2.3, have already discussed the propagation of faults from layer to layer and from phase to phase. It is evident that the costs of correcting faults increase the later they are detected. Thus, an effective fault handling as early as possible and during the present phase (not later, in one of the next phases) is crucial for a cost-effective development and maintenance.

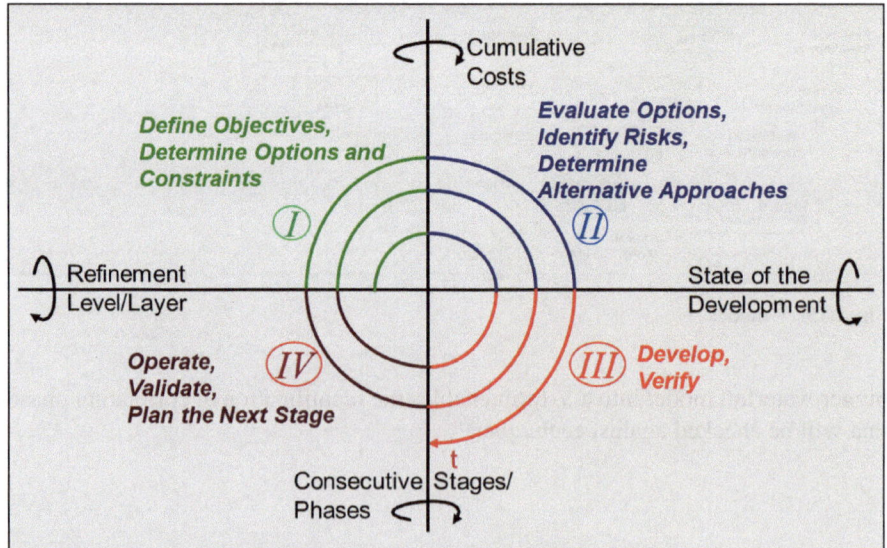

Fig. 1.16   Spiral model

Modeling the faults before refining and specializing fault handling techniques will help to understand how and when and which kind of faults will be produced (caused) and when and in which form they arise.

1.3.3.1 Fault Modeling

One of the most important steps in testing is detecting the fault and determining its kind. Since the programs are formed out of symbols (signs/signals/icons), semiotics, a traditional area of linguistics, can help to categorize faults (Fig. 1.17). *Semiotics* is the study of the meaning of symbols and communication with them. So, semiotics focuses on *symbols* that can be acoustic, optic, or electronic (for example, digitalized).

Symbols are put together to build *meta*-symbols. For example, the letters of an alphabet used in a natural language can be viewed as individual symbols. Words, on the other hand, are meta-symbols composed of these letters, with *syntactic* rules controlling the construction of correct words out of letters (symbols). For example, "end" is a valid word in English but not in German, which requires a capital "E" at the beginning and an additional "e" at the end: "Ende." Any deviation of these syntactic rules leads to a *syntactic* fault.

Meta-symbols—in our example, words—are supposed to possess a shared meaning, that is, to make sense to both the producer (sender) and consumer (receiver). This leads to the *semantics* branch of semiotics. If a word is not used correctly, we have a *semantic* fault. An example of this would be instead of saying the sentence "it is five o'clock," saying the sentence, "it is five watches" (the English words "clock"

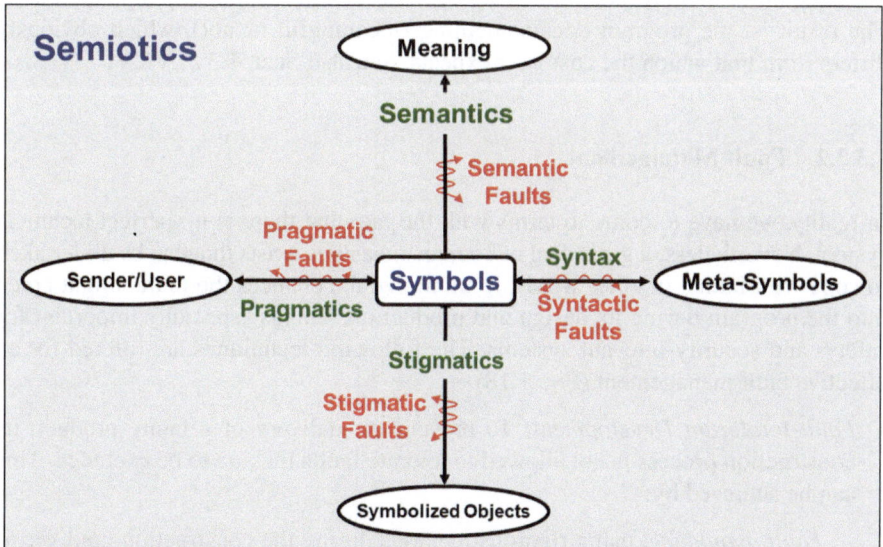

Fig. 1.17 Faults revisited—a semiotic approach

and "watch" have only one counterpart in German, namely "Uhr"). Furthermore, words are supposed to form sentences in accordance with precise syntactic rules. For example, the subject-verb-object (SVO) sentence structure in English.

Syntactic faults in programming are nowadays not really critical since they can mostly be detected automatically by modern compilers. Also, some kinds of semantic faults can be detected based on information included in the context of a faulty construct.

A symbol, or a meta-symbol, such as a sentence constructed and spoken by an individual, usually has a meaning that this person clearly has in mind and intended. Unfortunately, this meaning does not always conform to the meaning the audience understands. In this case, a *pragmatic* fault arises since the sender's intended meaning differs from how the receiver interprets (understands) the meaning.

Last but not least, symbols have *stigmatic* relations to the signs that symbolize them. For example, the letter "o" and the form of a person's mouth when this letter is pronounced are similar, namely the form of a circle.

Semiotics has far more areas, such as *enigmatics*, which studies the relationships between symbols and their semantics and pragmatics, the rules of which are intentionally kept secret and ideally known only by the sender and legal receiver. The rules may or may not be known—or only partly known—by other, perhaps illegal receivers. This area is important, not only for puzzle and game construction, but also for cryptology; for example, coding/ciphering versus decoding/deciphering.

It is evident that the most critical faults during software development are pragmatic faults since they are caused by misunderstandings between sender/creator and receiver/consumer of the information. For example, a programmer can misunderstand a user (customer) and deliver a syntactically and semantically correct program, but which is, unfortunately, not able to satisfy the user's needs due to pragmatic faults. The result is, the program does something (meaningful or not), which obviously differs from that which the customer expects. See also Sect. 1.3.3, V&V.

1.3.3.2   Fault Management

In reality, we have to come to terms with the fact that there is no perfect technical system. Nevertheless, a great deal of counter-measures exists that can be undertaken to cope with faults; in other words, to minimize the chances the faults might creep into the program during its design and production. This is especially important for safety- and security-relevant systems. The following techniques are offered for an effective fault management (Fig. 1.18).

- *Fault-Intolerant Development*: To prevent the delivery of a faulty product, its construction process is not allowed to tolerate faults that are to be excluded. This can be achieved by:

  - *Fault avoidance*: using rigorous methods during the construction (and verification) to avoid faults, that is, not to make them, and

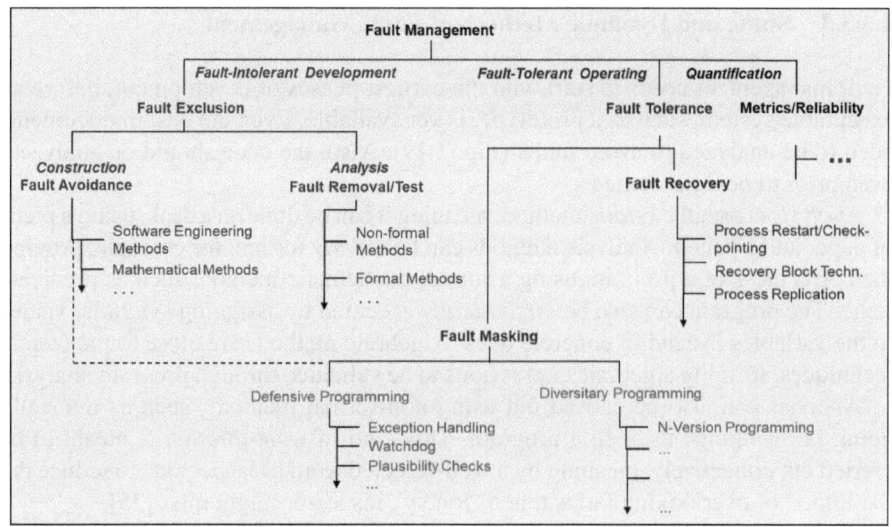

Fig. 1.18 Techniques of fault management

- *Fault removal*: analyzing and testing the (interim) product to detect and then eliminate faults, assuming that some faults could have sneaked into the development, despite whatever measures were taken to avoid them.

- *Fault-Tolerant Operating*: Even though everything feasible has been done to avoid and remove faults, it is always safer to assume that faults have nevertheless been made and can still arise during the operation. In such cases, the product is expected to tolerate the emerging of faults, that is, it has to continue operating and delivering services, even if with less but still acceptable efficiency (see Sect. 1.3.3.16). This can be achieved by:
  - *Fault masking*: isolating the defect component of the system in order to prevent it from harming the functionality of the entire system;
  - *Fault recovery*: developing more than one version of components (stored in sort of a spare parts storage); whenever a defect component is detected, replacing the defect one with a "healthy" one out of the storage.

- Last but not least, we have to measure the "goodness" of the product (as a non-functional feature, for example, reliability, availability) to determine to what extent we can trust it (in the sense of validation), or to what extent the developer did a good job (in the sense of verification).

Fault tolerance is of increasing interest, not only for developing dependable systems but also to enable user-friendliness. Even though it is expensive to be realized, fault tolerance is relevant for reusability. Therefore, we will revisit aspects and techniques of fault tolerance in Sect. 1.3.3.16.

1.3.3.3   Static and Dynamic Methods of Fault Management

Fault management needs to start with the earliest phases of development, before an executable system, such as a prototype, is yet available. Even the design documents need to be analyzed to avoid faults (Fig. 1.19). Also, the code should be analyzed, even prior to being executed.

Analysis is usually a *static* method, meaning it can be done on a desk, using a piece of paper and a pencil. Analysis methods can be *strictly* formal, for example, proving the correctness of a program using a sound, mathematic method, such as predicate logic. The program can also be *symbolically* executed by assigning symbolic values to the variables instead of concrete ones. Algebraic methods are close to the formal techniques, forming algebraic expressions to be validated through program analysis.

Analysis can also be carried out using non-formal methods, such as manually going or "walking" through a program. This kind of *walk-through* is meant to be carried out collectively, meaning by a well-selected team of inspectors, to reduce the likelihood of overlooking faults that a "lonely" inspector might miss [35].

Testing, on the other hand, is a *dynamic* process that necessitates a tool, most suitably a computer to run a program under consideration.

Both static and dynamic techniques have a unique objective: detecting faults. That is the subject of the next section.

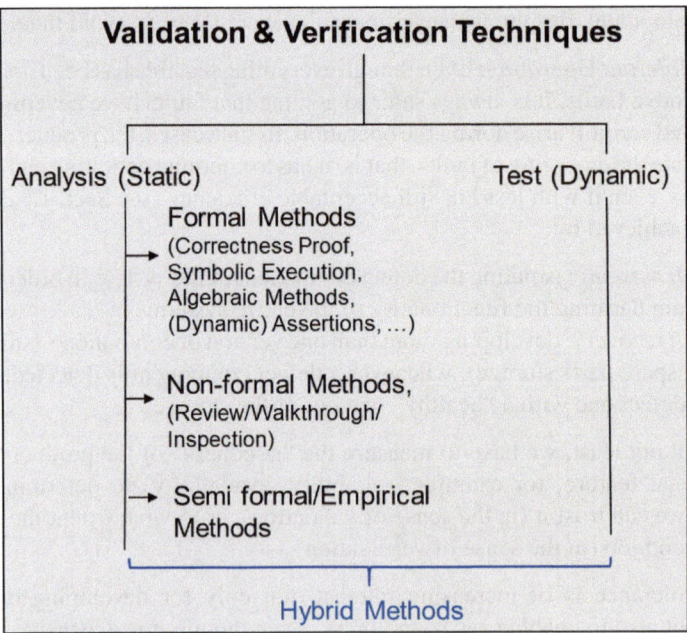

Fig. 1.19   Techniques of fault detection

1.3.3.4   Fault Detection

The previous Sects. 1.3.3.1 and 1.3.3.2, attempted to point out the importance of fault detection by testing, which is also expected to entail localizing the detected fault, with the goal to correct it.

Figure 1.20 briefly presents the process of detecting and correcting a fault, assuming that the SUC, here a program with the function p, has the input area (domain) I, and output area (range) O.

The domain and range of a program can be infinite or indefinite. While constructing a program, the programmer usually has a small, finite domain and range in mind, often represented by a set of exemplary inputs and outputs. Figure 1.20 notes this "programmer-minded" *regular* domain as I_R. Any regular input i_r will be mapped to a regular output o_r of the regular output range O_R of the program p.

Any good programmer is aware of the fact that unexpected inputs are also likely to appear; that is, the user inputs something that does not belong to I_R, for example, an *exceptional input* i_e out of the *exceptional domain* I_E. In this case, a good designed program outputs a warning o_e out of the *exceptional range* O_E, activating the *exception handling* mechanism of the program p (see Sect. 1.3.3.16.2, Sect. A.7 of Appendix A, [43]).

As long as the user does not leave the regular domain I_R and exceptional domain I_E the programmer has had in mind, that is, uses the "right" inputs, the program p behaves well. And it functions "right."

However, when/if the user changes his/her habits of using the program p, or a new user joins, it is possible that p will be given a "false" input i_f outside of the regular and exceptional domain, that is, out of $(I/(I_R + I_E))$, which was not anticipated by the programmer. In this case, instead of delivering corrupted, faulty outputs as results, the program will usually do something unexpected, in most cases (hopefully!) leading

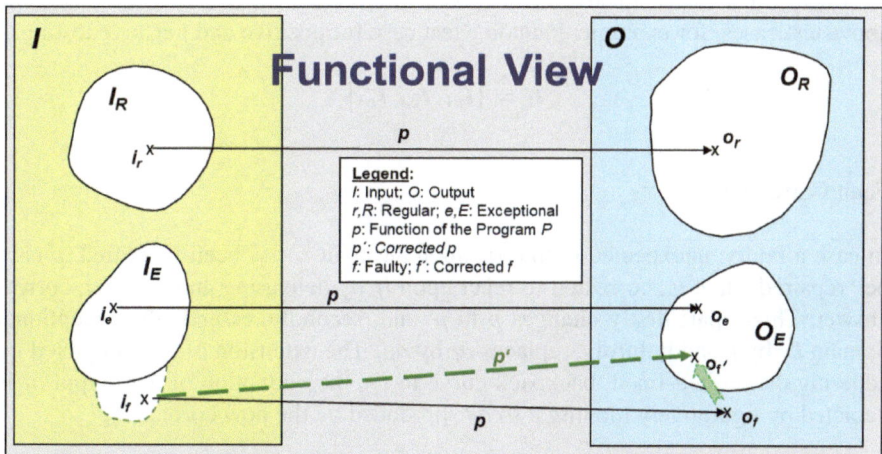

Fig. 1.20 Functional view of fault modeling

to a crash that clearly alarms the user that something is definitely wrong. Referring to Fig. 1.17 of Sect. 1.3.3.1, we likely have here a *pragmatic* fault.

1.3.3.5   Testing—First Encounter: Positive Testing, Negative Testing, Regression Testing

We formalize the cases we observed in the last section by constructing pairs of input and expected output upon this input and call this pair a *test case*. That is:

$$\text{test case} = (\text{input}, expected \text{ output}), \text{ for example } t_{c1} = (i_r, o_r)$$

This sort of testing forms a *positive test* of regular cases, for example, checking whether or not the SUC complies with its specification, meaning whether or not it functions correctly, in the sense of validation (Fig. 1.20).

On the other hand, we can use the *exceptional* input i_e, that leads to corresponding output exceptional output o_e.

$$t_{c2} = (i_e, o_e)$$

Finally, we can use the *false* input i_f that lead to corresponding *false* output o_f.

$$t_{c3} = (i_f, o_f)$$

The last two cases with t_{c2} and t_{c3} form *negative tests* for handling to determine how the SUC behaves in exceptional and unexpected situations. A positive test, together with the complementary negative test, form a *holistic* view (see also Appendix A, Sect. A.5).

A set of test cases constructed for a specific purpose is called a *test suite*. The above test cases, for example, include a test case for positive and negative testing.

$$t_s = \{t_{c1}, t_{c2,} t_{c3}\}$$

Fault Correction

In case a faulty, unexpected output o_f upon an input i_f has been identified, p is to be "repaired," that is, corrected to react upon i_f by delivering an expected, correct answer. This repair firstly changes p to p' and, secondly, extends the exceptional domain I_E by i_f, and, thirdly, replaces o_f by o_f'. The extension of I_E is depicted by adhering the dashed-lined, backpack curve to I_E; the correction of the output o_f is depicted by figuratively moving it to o_f', produced by the now corrected p'.

Regression Testing

Note that once we have repaired p, it is now a different program, having been transferred to p', and the overall behavior of p' can unexpectedly differ from p. In other words, the repair can interfere with other functionalities of p, so that p' behaves correctly in the case of i_f, but no longer for the cases i_r and i_e, to use our test suite that was constructed in Sect. 1.3.3.5. Thus, we have to check whether or not p' functions correctly, not only for i_f but also in all cases where p has functioned correctly, for example, for i_r and i_e. This way, a *regression testing* will be carried out.

Summary

To sum up, the following types of inputs are possible.

- Expected inputs can be
  - regular inputs,
  - exceptional inputs.
- Unexpected inputs that cause fault(s) to occur.

1.3.3.6 Historical Remarks, Testing-in-the-Small Versus Testing-in-the-Large

Program testing can be used to show the presence of bugs, but never to show their absence.

W. Dijkstra [31].

We prove a fundamental theorem showing that properly structured tests are capable of demonstrating the absence of errors in a program.

J. Goodenough, S.L. Gerhart [42].

... it is possible to use testing to formally prove the correctness of programs ...

W.E Howden [47].

Testing is a much discussed and hotly debated discipline. Below are quotes from some pioneers of computer science to provide an idea of the situation.

To introduce a little bit of calm to this fervently (unfortunately verging on religiously) heated discussion on testing, let us consider alternative methods of testing.

Generally speaking, testing involves running the software under consideration (SUC) in order to:

- detect, locate, and correct faults (testing in the *small*, also called *debugging*): This might be what Dijkstra had had in mind in 1970. This is not the primary subject of interest in this book [29].
- determine the level of non-functional features (reliability, security, etc.) of SUC (testing in the *large*): This is the subject of interest in this book.

1.3.3.7   Towards a Formal Definition

The construction of a program p and its V&V require an exact description of the functionality of p, which also includes the problem and its solution s as an algorithm. Thus, p realizes a function f_s, such as

$$f_s : I \to O$$

as defined in last section (Fig. 1.20).

An execution e of an (assumingly correct) program p is expected to fulfill the requirements of the specification of this solution s, such as

$$f_p : I \to O \text{ with } f_p(x) = f_s(x), \quad \text{for all } x \in I.$$

We have a *formal verification* if this equivalence can be proofed formally. Remember that I can be infinite.

A *test* is formed in case a proof cannot be delivered for all $x \in I$, but for some $t \in T$, where T is a finite set with $T < I$.

The "art" of testing is for the tester to construct a test in such a way that $f_p = f_s$ can be deducted from $f_p(t) = s_a(t)$ for those $t \in T$. Thus, the program p delivers an output $f_s(t) = o \in I$ for any $t \in T$. This pair of (input, *expected* output), specifically $(t, f_s(t))$, describes a special case of the execution of p, and will be called a *test case*.

Id.: _____

Objective: _____

Requirements/Pre-Conditions: _____

Input(s): _____

Expected Outcomes: _____

Post-Conditions: _____

Involved/Invoked Other Tests: _____

Date   Result/Score   Software Under Test (SUT)/Version   Performed/Run by Tester

_____   _____   _____   _____

Fig. 1.21 How to specify a test case

1.3.3.8   Test Case Specification

In practice, testing is an *experimental* validation method: An input is given to run the SUC, and the expectation, that is, the output of the execution of the SUC, is determined beforehand. Given this expectation is fulfilled, the test was *successful*, even if the expectation was that the SUC fails, since one of the objectives in running this test case was to reveal a fault (in the sense of what Dijkstra says about testing, see the beginning of Sect. 1.3.3.6).

An experiment is to be precisely prepared and carried out. First, apart from precisely defining inputs and expected outputs to these inputs, the whereabouts of the test environment is also to be specified.

Figure 1.21 specifies very simply a test case as often used in the practice.

1.3.3.9   Principles and Folklores

As stressed in the previous sections, testing is an experimental way of program validation, no more, no less. Nevertheless, there are very different kinds of expectations. The following are some of those expectations that vary between principles and folklores.

1. *Testing needs hypotheses to show the presence of bugs.*
2. *Exhaustive testing is not always feasible.*
3. *Early testing exposes expensive late faults.*
4. *Learn from users: form operational profiles, relate them to clusters of defects.*
5. *Where one fault is, there is also another.*
6. *Minor faults signal major faults.*

7.   *The more test case sets (test suites) are executed, the faster they become obsolete/ineffective (pesticide paradox).*
8.   *Consider the context when you test.*
9.   *Correct programs need not be valid (absence-of-fault fallacy).*

We leave the validation of these claims to the reader's own experiences.

1.3.3.10   Testing Is Difficult, Because …

As mentioned in the last section, testing requires proper preparation that also includes the determination of the expected results, the good ones (*success*) and the bad ones (*fail*).

The example below is the control flow chart of a simple program that includes two selections (*if… then …* clause). The figure to the right of the control flow *chart* depicts the control flow *graph* that symbolizes the paths (events, or statements), along with the alternative decisions *sel1* and *sel2* (Fig. 1.22). Note that the jump from *s2* to *s1* also represents an event, namely an *empty* event that will be symbolized as *c*.

A popular test technique relies on executing each path of the program at least once (C1 test [71, 72]; see Sect. 1.3.3.12 for explanation in more detail). In this case, we have

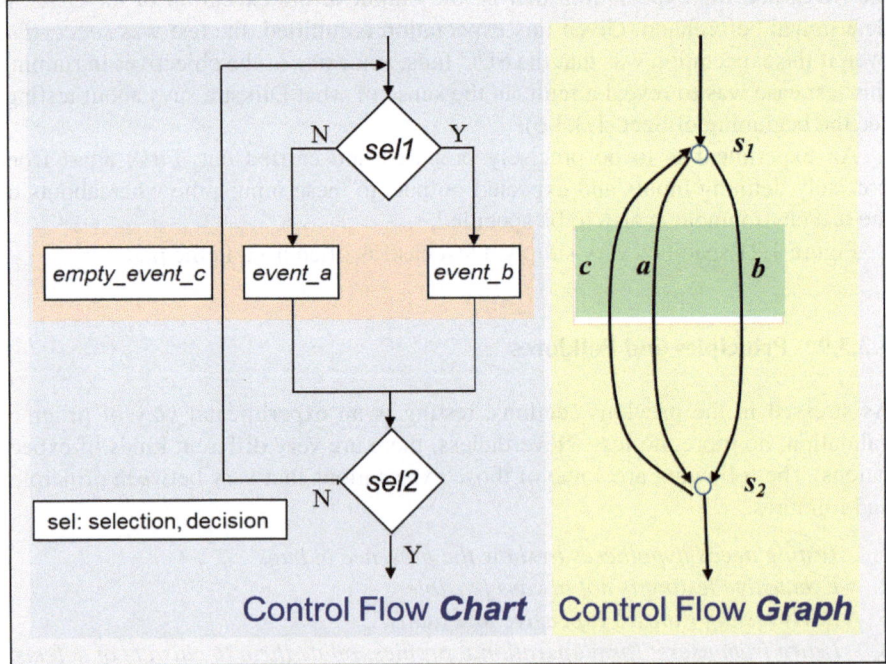

Fig. 1.22   A program with two sections and its control flow graph

- 2 selections, executed (iterated) once: $\{s_1bs_2, s_1c\,s_2\} \rightarrow 2^1 = 2$ **paths**

However, it is a widely accepted belief that the more path combinations executed, the better the chances are to detect latent faults. Given that we decide to execute the paths multiple times to produce different combinations of path execution sequences, we have following situation:

- 2 selections, executed twice: $\{s_1bs_2, s_1cs_2, s_1bdbs_2, s_1bdcs_2, s_1cdbs_2, s_1cdc\,s_2\} \rightarrow 2^2 + 2^1 = 6$ **paths**

- 2 selections, executed three times: $\{\ldots\} \rightarrow 2^3 + 2^2 + 2^1 = 14$ **paths**

General: $$p := \sum_{i=1}^{n} d^i$$

p: number of paths; d: number of selections; n: number of executions.
When three selections are iterated 25 times:

- $3^{25} + 3^{24} + \cdots + 3^2 + 3^1 = 1.2 * 10^{12} \approx 1$ million paths!

Obviously, this kind of exhaustive testing is an infeasible, brute-force method to gain confidence in the reliability of testing. It is better to look for methods that have good ideas behind them.

1.3.3.11 Test Methods—A Brief, Rigorous Review

Testing is, especially in industrial practice, a widely used validation method. Nevertheless, it has severe weaknesses concerning the singularity of the test results since a test is valid only for the executed use case. Therefore, many techniques are proposed to extend the validity and thereby increase the trustworthiness of testing. This section summarizes and briefly reviews existing popular testing techniques. For a comprehensive study of the reviewed techniques, see references at the end of this chapter, for example [20, 23, 68].

Figure 1.23 attempts to categorize testing techniques in three aspects as dimensions that are *orthogonal* in the sense that the methods and activities covered in one dimension are independent of the ones covered by either of the other dimension. For example, the "Black-Box" method in the dimension "Hypothesis/Method; ...," can be used in "Module/Unit Test" of the dimension "Development Level/Phase," or in "Top-Down" Test Direction. These orthogonal dimensions are:

- *Hypotheses*: This aspect concerns the method to be selected for the generation and selection of tests (Note that terms "tests" and "test cases" are used synonymously).

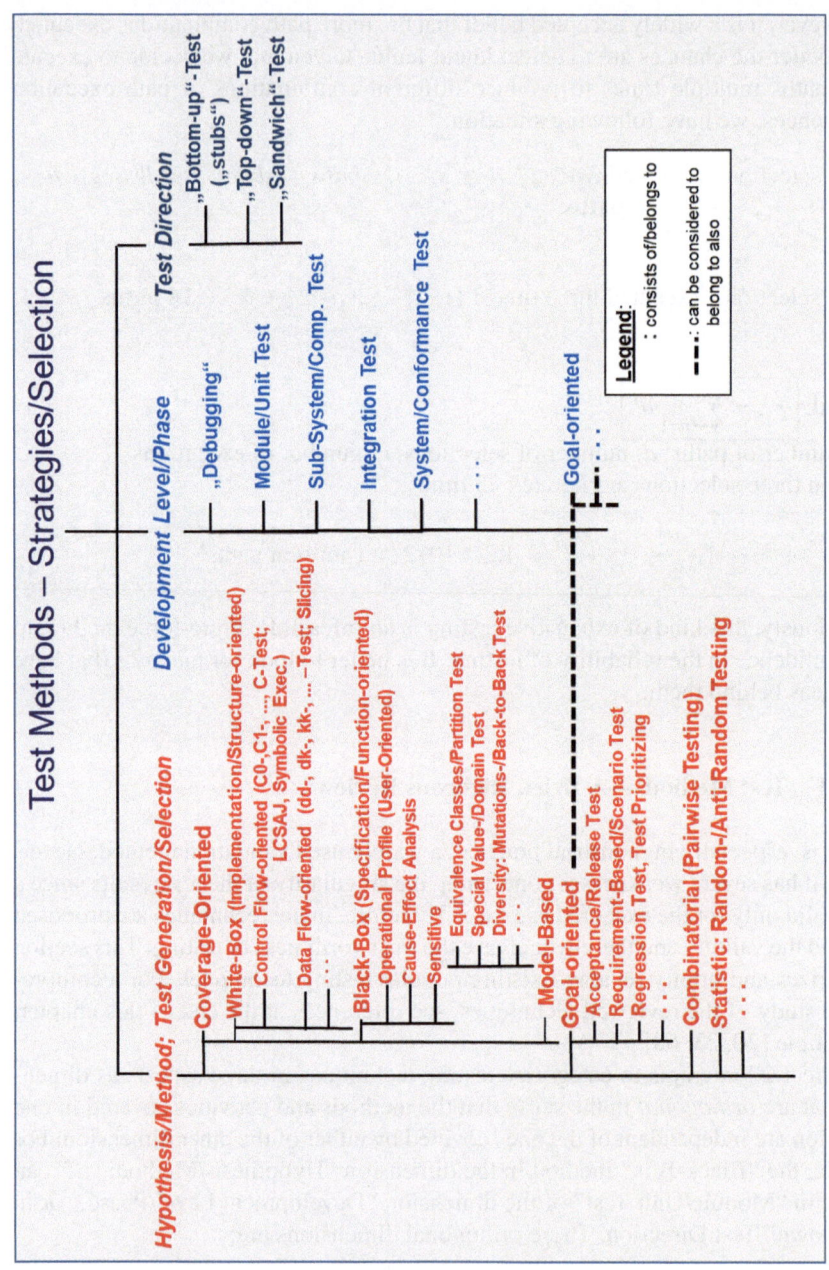

Fig. 1.23 An orthogonal view of test methods

- *Development level/phase*: This aspect concerns the stage of the development and testing activities.
- *Test direction*: Testing can start at the bottom, module level, and work upwards, or start at the top, system level, and work downwards.

The methods grouped in the above explained dimensions will be discussed in the following.

Hypotheses to Generate and Select Tests

The idea is: The more a test engineer knows about the SUC and the programmer who developed it, the better hypotheses he/she can set up to select appropriate method(s) to construct tests. Following alternating views are existing.

- *Coverage-oriented testing*. Tests will be constructed to cover the SUC, that is, to execute as many elements of the SUC using as few test cases as possible. The elements can be chosen from the *structure* of the code (*white-box*) or from the usage-oriented requirement *specification* (*black-box*) of the SUC, whichever is available. Well-known techniques of coverage-oriented white-box and black-box testing will be reviewed in the following sections.
- *Model-Based testing*. The object under consideration here is not the program itself, but its model that represents its function, behavior, structure, etc. This view will be discussed in Sect. 5.2 in detail [15].
- *Goal-oriented* testing. In this view, the test objectives determine the test method to be selected, which can be developer-oriented or customer/user-oriented. A user/customer-oriented *acceptance test*, for example, is performed executing as many of the application cases as possible to assure that all of the customer require-ments will be fulfilled so that the delivery of the product can take place. Those application cases can be generated using model-based testing, whereby the models are to be constructed by or with the user; for example, based on use-case *scenarios* derived from the requirements definition of the SUC [14].

  Release test, on the other hand, is to be performed on the developer's side to prepare the acceptance test.

  Regression tests are to be performed upon a modification of the SUC to re-validate its correct performance (Sect. 1.3.3.5.2). A *prioritization* scheme of the test activ-ities can be necessary if running a complete set of tests (component test that will be followed by an integration test that, in turn, will be followed by a system test, etc.; see below in Sect. 1.3.3.11.2).

  Note that the goal-orientation can also be viewed under the aspect of development phase. Well-known techniques of goal-oriented testing are reviewed in the next sections.

  - *Combinatorial Testing*. Combinations of test cases that differ slightly from each other are built to "touch" different neuralgic parts of the SUC. A special *combinatorial test technique*, *pairwise testing*, is of special interest for reuse (see Sect. 5.1.3 [28]).

- *Statistic Testing*. Contrary to the other test methods, *statistic testing* is not concerned with any criteria, rather it tests the SUC by selecting and/or constructing test cases *randomly* or *anti-randomly* [90].

Development Level/Phase and Test Direction

In the following, the other two aspects for classifying test methods will be discussed.

- *Development level/phase*: Test activities can be classified considering the life cycles of the software development, for example, in accordance with the waterfall model (Sect. 1.3.2.3). Programmer efforts for fault detection and correction during coding, commonly known as *debugging*, usually do not form a testing activity in the sense of the definition of testing-in-the-large, but does in testing-in-the-small, which is not being addressed in this book (Sect. 1.3.3.6, see testing-in-the-small). So, *module testing*, or *unit testing*, is the first level of systematic testing, followed by *sub-system* testing (also called *component testing*). Testing all of the components after putting them together leads to *integration testing* that takes place in subsequent steps of the test direction, primarily to validate the interactions of the components. Finally, *system test* validates the *conformance* of the functionality of the SUC with respect to the requirements specification and the environment of the software under test, such as user, hardware, network, inter-operation with other systems, etc.
  The methods discussed in the category *goal-oriented* testing in the left column of Fig. 1.23, for example acceptance testing, can also be seen as a test stage. Therefore, this category appears also in the middle column of Fig. 1.23.
- *Test direction*: One can start integration testing at the bottom (module) level and work upwards (*bottom-up* testing), or in the other direction, from the top down (*top-down*), depending on which kind of experience is available.

  - Bottom-up testing is recommended if the test team is not familiar with the application and its development/testing environment. By starting at an easily manageable level, the tester can learn the features of the SUC and become familiar with its environment.
  - Top-down direction is recommended if the test team is familiar with the application and its development/testing environment. The top level of the software will then be tested each time a module is tested. In the case the development is not yet finished and only a subset of the modules is ready for testing, the modules under construction will be replaced by "stubs" that simulate the input/output behavior of the SUC, in accordance with the test case(s) for this module.
  - It is also possible to start anywhere and to work a step upwards, another step downwards, etc. (*sandwich testing*).

Test direction is important for test automation, not only because of the requirement to realize the stubs (in the case of top-down testing), but also to get adjusted to the policy and strategy of the project management and product structure.

White-Box Testing

If the SUC code is available, the tester can use its structure twofold to generate test cases, either focusing on the control flow or data flow. The idea is to "touch," that is, to execute the "neuralgic" parts of the program code that might be suspicious instances, where the control flow or data flow will be disrupted upon critical test inputs and not deliver the expected outputs.

Control-flow-oriented testing is done by identifying the instances in which control-flow is subject to manipulation, for example, jumps caused by selecting an *if …* *then … else*, or a *case*-statement. External interrupts and sub-routine calls manipulate the control flow, as well. The assumption (test hypothesis) is: Chances are that programmers will occasionally choose the wrong outcomes for a selection, leading to wrong destinations within the SUC. Or, they forget to consider potential interrupts. Therefore, those parts of the SUC where decisions are made are to be carefully tested.

The best-known control-flow-oriented test methods are based on test case selection criteria. Exemplary: *C0 testing* that requires each statement of the SUC be executed by at least one test case or by a set of test cases. Similarly, *C1 testing* requires each branch of the SUC be executed at least once (Sect. 1.3.3.12).

The *LCSAJ (Linear Code Sequence and Jump)* is a test technique that is a very good fit for the control flow strategy. LCSAJ generates test cases to run the statements of the SUC to determine where the control really "jumps" to and to check whether the destination is correctly determined and selected (Sect. 1.3.3.12) [45].

Another well-known white-box testing technique is *symbolic execution* that replaces real test input/output values by symbols and algebraically determines their values [62]. More specifically, this technique executes SUC on symbolic inputs, which represent the concrete, real inputs. The program behavior will be represented using mathematical constraints over these symbolic inputs. Solving these constraints with off-the-shelf solvers yields inputs that exercise different paths of the SUC. Typical applications of symbolic execution are test input generation and fault detection.

Data flow-oriented testing focuses on the instances where the SUC's data are manipulated; for example, when its value is modified by an assignment. An interesting classification scheme leads to a unification of control flow and data flow orientation. This view generalizes "use" of the present program structure that can be predicate-oriented or calculation-oriented [37, 83]. We will discuss this technique in Sect. 1.3.3.13.

White-box testing is implementation-oriented, that is, the implementation aspects are in the foreground during development, as is the case [68].

Black-Box Testing

In case the SUC code is not available—or the programming knowledge and experience of the tester is limited—the requirement specification is in the foreground that describes the functionality of the SUC. The SUC's *operational profile* can be determined by observing the behavior of typical users that indicate which functionalities are to be tested more intensively. In case some applications are rarely used, but have higher priorities or high follow-on fault costs, the weighing of those critical functionalities can help.

Cause-effect analysis is popular for generating useful test cases [73]. This method uses cause-effect graphs to reduce the number of test cases based on Boolean operations and has been formalized and generalized [13].

Another group of test methods, forming the *sensitive testing* group, attempts to determine *equivalence classes* of test cases that stimulate similar or critical behavior of the SUC, which form *partitions* in the domain and corresponding rank of the test cases. Representatives of those classes can then be chosen to build appropriate test suites.

Special cases and values, such as division by zero, can be used to generate test cases as *special values* [20]. The SUC's data flow is manipulated by assignments, that is, when the value of a variable is changed. Again, the hypothesis is that the programmer will make the mistake of taking the wrong value.

Slicing, or *program slicing*, forms a typical data-flow-oriented testing that takes a group of program statements (a slice) of the SUC that can be separately compiled, executed, and thus tested. The objective is to affect the value of a particular element or a specific outcome of the SUC as a point of interest, such as the value of a variable.

Mutation testing technique belongs to the group of diversity-oriented sensitive testing tests. The assumption is, programs constructed by experienced programmers are close to being correct; they might contain only minor faults. Thus, it is likely that they form *mutants* of the correct program. Accordingly, the SUC will be slightly and systematically modified and tested together with the mutants. If the SUC and a mutant produce different outputs in a test case, the modification has been discerned, that is, the mutant has been "killed." If the mutant reacts the same way as the original SUC, meaning their outputs do not differ, a suspicious situation has emerged that must be analyzed. Most likely, the test case is not sensitive enough to detect the change and thus recognize (kill) the mutant. Note that mutation testing can be used for determining the effectiveness and efficiency of a test suite by the ratio of the number of dead (killed) mutants to the total number of generated mutants (*mutation score*).

Originally, mutation analysis and testing technique was suggested for implementation-oriented testing [30]. However, it has been extended to also be applied to model- and specification-oriented testing (Sect. 5.2.6; [25, 88]).

Another special member of sensitive testing group, *back-to-back testing*, uses redundancy given by the set of recent versions of a component. Similar parts of these versions are executed with corresponding, similar inputs. The results are compared in order to detect impairments. Different outputs produced by the corresponding parts upon the same input are a signal for likely faults. Those parts are then to be analyzed.

Black-box testing is frequently carried out using model-based testing techniques. Some of the well-known test methods of this kind will be reviewed in the next sections and in Sect. 5.2 [23].

Gray-Box Testing

Gray-box (also spelled as *Grey-box*) Testing combines white-box testing and black-box testing. Thus, both the internal structure of the SUC, at least partially, and its specification are available. A gray-box tester partially knows the internal structure, which includes access to the documentation of internal data structures as well as the algorithms used. [23]. So, a gray-box tester needs both high-level (application-oriented) and low-level (implementation-oriented) documents in order to construct test suites[20, 23, 61].

1.3.3.12 Some Broadly Accepted Sets of White-Box Coverage Criteria

Some of the broadly accepted sets of white-box-oriented coverage criteria are listed below and briefly explained. For more information see [23, 71, 72].

- *Statement Coverage* (C_0) necessitates execution of all of the statements of the SUC at least once.
- *Branch Coverage* (C_1) necessitates execution of the branches (decisions/predicates) of the SUC at least once.
- *Path Coverage* (C_2) necessitates execution of all of the paths of the SUC at least once.
- *Boundary-interior Path Coverage* (C_2-bi) necessitates execution of all of the loops (iterations) of the SUC at most twice.
- *Structured Path Coverage* (C_2-str) necessitates execution of all of the loops (iterations) of the SUC at most n-times.
- *Decision Coverage* (C_3) necessitates execution of both outcomes (TRUE and FALSE) of all of the branches.
- *Multiple Branch Coverage* (*MC*) necessitates execution of the nested branches of the SUC at least once.
- The coverage criterion, *LCSAJ* (*Linear Code Sequence and Jump*), based on a concept similar to C_i-Criteria ($i \geq 0$), is popular. It defines *Test Effectiveness Ratios* (*TER*) that strongly correspond to *C0, C1, MC*, etc., leading to TER_1 (comparable with C1), TER_2 (comparable with C2), and TER_3. LCSAJ begins at the start of the program or branch and ends at the end of the program or branch, focusing on locations where the linearity of the code is disrupted, for example, by a jump command [45].

Table 1.1 Use-oriented classification of testing

Operations along the paths		
d: defined, created, initialized, etc. k: killed, undefined, released u: used for something	c: used in a calculation (data flow) p: used in a predicate (control flow)	
dd: suspicious[1] dk: ?bug?[2] du: normal	kd: normal kk: ?buggy?[3] ku: bug[3]	ud: bug? normal? uk: normal uu: normal
$*k$: ?anomalous? $*d$: normal $*u$:?anomalous?	$k*$: normal $d*$:?anomalous? $u*$ normal	

Legend: 1: Something likely wrong; 2: something wrong; 3: something definitely wrong; *: something (nothing of interest), don't care; ?: check; ??: check again

A relation " > " ("stronger/more powerful") of the coverage criteria is defined as follows.

$C_i > C_j$ (i, j: natural numbers) means that all the test cases that can be generated by C_j are included in the set of test cases generated by C_i. More precisely:

$x > y$ (x Stronger y): <=> {$a|a$: test case produced by x} → {$b|b$: test case produced by y}.

EXAMPLE: $C_2 > C_1 > C_0$

1.3.3.13 Toward a Unified View of Unit Testing

A general classification differs between using a variable in a calculation (data flow) and a predicate (control flow). Variables, such as pointers, can be defined, used, and released ("killed") [37, 83].

Combinations of these operations *defined/used/killed* can refer to regular (normal) cases and anomalies in programs. For example, while *du* (*define* a variable and then *use* it) is normal, *dk* is obviously wrong (*define* a variable and then *kill* it before using) (Table 1.1).

Now, the strength of these test criteria can be analyzed and compared with each other using the definition of Stronger relation (Fig. 1.24, see also Sect. 1.3.3.12).

1.3.3.14 Some Relevant Standards for Validation

Standards are essential for any engineering discipline, including for software reuse and software testing. Many organizations set up company-internal technical standards, often supported by global industrial standards, mainly in order to limit the

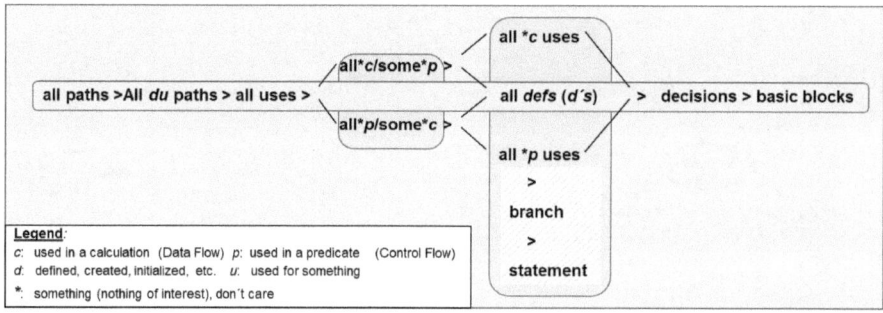

Fig. 1.24 Power hierarchy of test criteria

choices their software engineers can make while selecting components to integrate in systems [69]. For example, the developer is given the option to select between database-1 or database-2, and Web-browser1 or Web-browser2, but has to use specific, identified middleware or spreadsheet products whenever either is necessary. This limitation is supposed to reduce costs that can be incurred by a variety of programming alternatives, which must be learned and carefully applied to avoid faults.

Technical and industrial standards define constraints to increase interoperability and to decrease costs of the selection, adaptation, integration, and maintenance of commercial components [26, 53, 69].

Back to testing: the test selection criteria discussed in the last sections are widely accepted, resulting in some of them having found a place in industrial standards. Figure 1.25 figuratively positions some of these standards in correspondence with the coverage criteria of different power (Sect. 1.3.3.12).

- DO-178C (ED-12C) (Aerospace)—*Software Considerations in Airborne Systems and Equipment Certification* [33]: This document provides guidelines for the production of software for airborne systems and equipment that performs its intended function with a level of confidence in safety that complies with airworthiness requirements. The levels are defined in terms of the potential consequence of an undetected error in the software certified at this level. Here are such consequences for each defined level:

  - Level A: *Catastrophic*: prevents continued safe flight or landing, many fatal injuries
  - Level B: *Hazardous/Severe*: potentially fatal injuries to a small number of occupants
  - Level C: *Major*: impairs crew efficiency, discomfort or possible injuries to occupants
  - Level D: *Minor*: reduced aircraft safety margins, but well within crew capabilities
  - Level E: *No Effect*: does not affect the safety of the aircraft at all

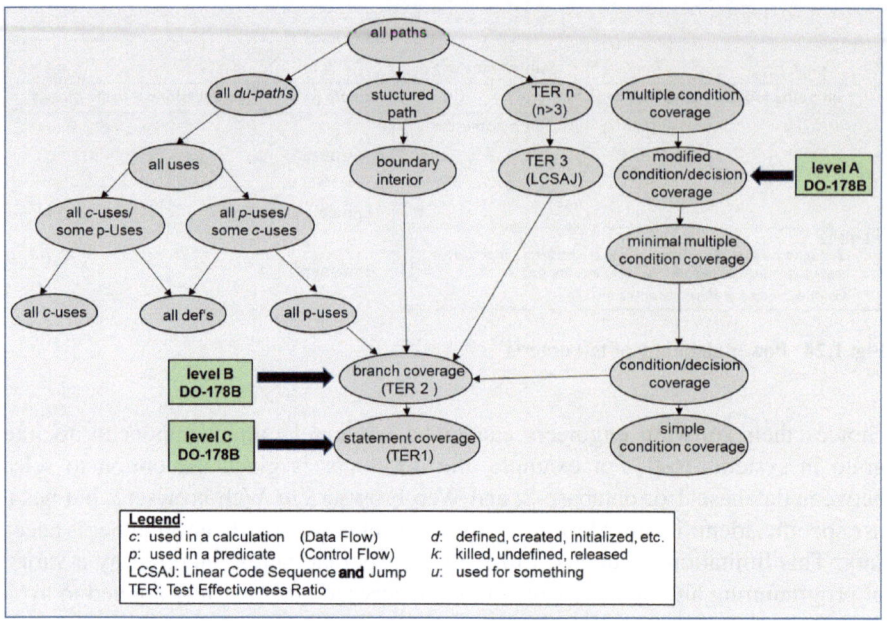

Fig. 1.25 Comparison of the strength of use-oriented test techniques and industrial standards

- IEC 61508 (Industrial)—Functional safety of electrical/electronic/programmable electronic safety-related systems—Part 2: Requirements for electrical/electronic/programmable electronic safety-related systems [52]: This document sets out the requirements for ensuring that systems are designed, implemented, operated, and maintained to provide the required safety integrity level (SIL). Four SILs are defined according to the risks involved in the system application, with SIL4 being used to protect against the highest risks.
- ISO 26262 (Automotive)—*road vehicles—functional safety* [57]: This document provides a general design framework for safety-critical systems. It is more or less an adaptation of the IEC 61508 that addresses possible hazards in automotive systems caused by malfunctioning behavior of electrical and/or electronic (E&E) safety-related systems, including interaction of these systems. The ISO 26262 standard consists of ten parts. Part 1 provides the vocabulary; Part 2 introduces management of functional safety; Part 3 to Part 7 (main parts) describe the activities of different phases of the development process using the classic V-model (Sect. 1.3.2.5). Finally, Part 8 to Part 10 explain how to use the standard.
  Luo et al. use ISO 26262 to propose a model-based approach for assuring compliance with safety standards to facilitate reuse in the assessment, qualification, and certification processes [66].
- EN 50128 (Rail Transportation)—*Railway applications—Communication, signaling and processing systems—Software for railway control and protection systems* [34]: This document specifies procedures and technical requirements for

the development of programmable electronic systems, which are used in railway control and protection applications (SIL 3/4). The international version of this standard is IEC 62279. EN 50128 requires software testing to contain information about test coverage and test completeness and more. There is a choice of tools for verification and validation, including tools for static analysis and test coverage.

- IEC 62304 (Medical Devices)—*Medical device software—Software life cycle processes* [51]: This document defines the life cycle requirements for medical device software. The set of processes, activities, and tasks described in this standard establishes a common framework for medical device software life cycle processes.
- The ISO 9000 family of standards focuses on quality systems and can be used for external quality assurance purposes. ISO 9001:2015—*Quality management systems—Requirements*: specifies requirements to introduce, install, and continually improve a documented management system.
- *Capability Maturity Model for Software* (CMM) was developed by the Software Engineering Institute (SEI) with the sponsorship of the U.S. Department of Defense in the early nineties of the last century [4, 80]. CMM is not an industrial standard. However, it has gained importance in helping software organizations improve the maturity of their software processes. It introduced an evolutionary path in five maturity levels with following characterizations:

1. *Initial* level is characterized by an ad hoc and occasionally even chaotic software process. Only few processes are defined, and success depends on individual rather than the organized effort of the developers.
2. *Repeatable* level is characterized by basic project management processes to track cost, schedule, and functionality based on successes with similar projects.
3. *Defined* level is characterized by a documented, standardized software process that is integrated into a standard software process defined for the organization.
4. *Managed* level is characterized by detailed measures, which enable a quantitatively understanding and control of the software process and product quality.
5. *Optimizing* level is characterized by quantitative feedback from the process, steadily taking innovative ideas and technologies also into account to enable a continuous process improvement.

- ISO/IEC 15504—*Information technology—Process assessment*, also termed *Software Process Improvement and Capability Determination (SPICE)*, was initially derived from process lifecycle standard ISO/IEC 12207 ([56], Sect. 2.3) and from maturity models like Bootstrap, Trillium, and the Capability Maturity Model (CMM).

1.3.3.15   Test-Driven Development and Test-Driven Reuse

Test-driven development (*TTD*), as an agile software practice [16], suggests constructing test cases that realize a well-defined functionality [17] prior to implementing a component. This is a reasonable proposition in the sense of a *holistic testing* approach as suggested by Belli [21]: Consider not only what the SUC is supposed to and required to perform, but also, complementarily, what it is *not* supposed to do. In other words, validate the SUC by both positive and negative tests (see also Sect. 1.3.3.5 and Appendix A, Sect. A.5). This is also the reason why this section began with fault modeling (Sect. 1.3.3.1).

In analogy to TDD, test-driven reuse practice recommends leveraging the activities for locating and reusing source code by test cases that are meaningfully constructed to describe the functionality as desired by the developer who intends to reuse this component [49, 77].

1.3.3.16   Fault Tolerance

> *After a complicated repair of his car, Michael Ellis DeBakey, the renown cardiac surgeon, scientist, and medical educator, was asked by his mechanic, "Roughly speaking, we have been doing the same job. For example, I open the hood to check the engine, and immediately see the valves to be cleaned. I can even exchange the engine. Tell me, how come you charge tens of thousands of bucks, and I charge you only fifty?" DeBakey replied, "Try to do the job with the engine running".*

As discussed in Sect. 3.3.2, faults that occur in the operation phase are to be tolerated since the system continues delivering services, perhaps possibly of a lower quality, but still acceptable. This feature is called *fault tolerance* [11, 65] and will be realized by additional structural and/or functional elements that are not primarily necessary for an appropriate operation but will be activated and used in case a fault occurs. These additional elements are called *redundancy,* which are usual in natural systems to ensure their survivability and behavioral autonomy. Human beings have two kidneys, but need only one; two eyes, even though one would also do, etc. Fault-tolerant technical systems are constructed, copying the concepts and features we find in biological systems [12, 18, 75, 84].

Elements of Fault Tolerance

Figure 1.26 sketches how redundancy is used to realize fault tolerance. As always in fault handling, the fault first has to be diagnosed, that is, it must be clearly identifiable, detectable, and uniquely correctable. Once integrated into the system, this

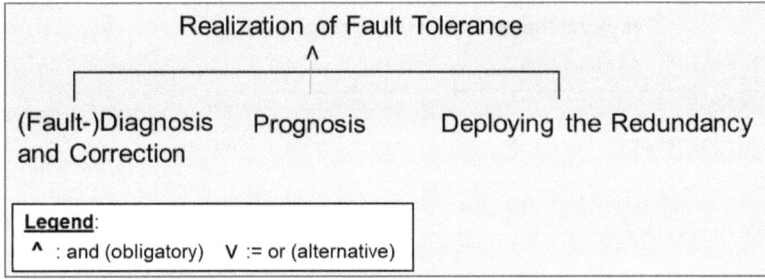

Fig. 1.26 Elements of fault tolerance

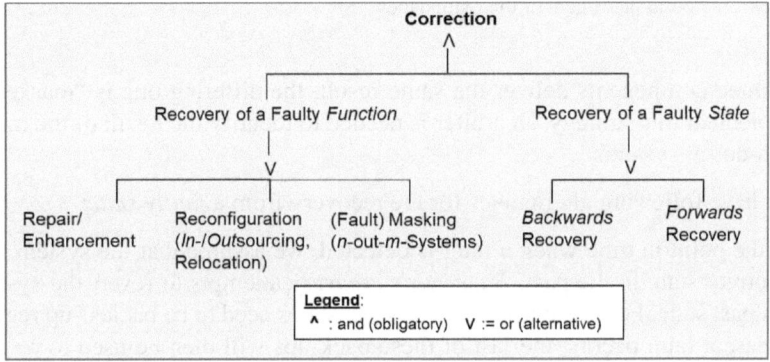

Fig. 1.27 Function recovery and state recovery

additional information is used during its operation to prognose and self-correct the corresponding fault, if and when anything goes wrong.

Recovery from Faults

To put the above-mentioned steps into effect, that is, to enable a self-correction, the system must be recovered from a faulty function or from a faulty state. For the recovery from a *faulty function*, we have following alternatives (Fig. 1.27):

- The system itself repairs the faulty defect component.
- The defect component causing the misfunction is replaced by an intact one, which necessitates outsourcing the defect one and insourcing the intact one. In case an intact component is not available, and the defect component is removed, a possible solution is resorting to reconfiguring the system for a redistribution/relocation of the load.
- Once the defect component has been identified, it is then "masked," that is, isolated. In a 2-out-3 systems example of n-out-m, three versions of a function are run simultaneously with the results being constantly compared. When two out

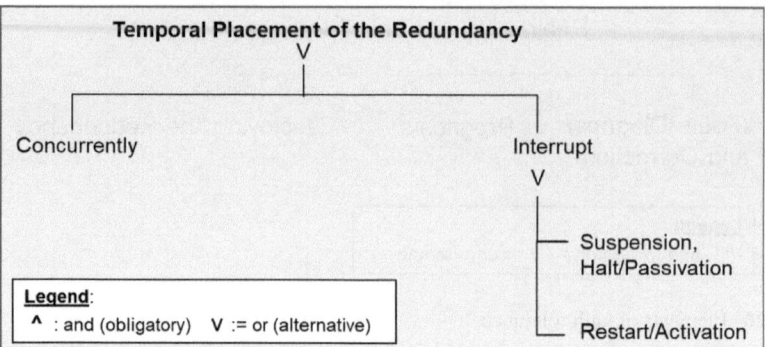

Fig. 1.28 Temporal placement of the redundancy

of three components deliver the same result, the differing one is "masked." To implement this strategy, an arbiter is needed to identify the result of the majority of n-out-m systems.

We have following alternatives for the recovery from a *faulty state*:

- At the point in time when a fault is detected, we assume that the system was in a correct state in the past. *Backwards recovery* attempts to revert the system to this past state. For this purpose, the correct states need to be backed-up regularly. In case a fault occurs, the last of these back-ups will then be used to revert the system to a secure, correct state.
- In case a back-upping is too expensive, or even not feasible, *forwards recovery* can attempt to transfer the system into a correct state in the future. *Exception handling* is a good example that comprises special responding processing to the occurrence of anomalous or unexpected conditions during operation (Sect. 1.3.3.4).

Insertion of Redundancy

Placing, that is insertion and activation of the redundancy to realize fault tolerance at the right time, is one of the key processes of fault tolerance. One strategy is to run the redundant units simultaneously, together with the functional units. This requires additional resources and, more importantly, additional processing time, which might be critical. The other strategy is to interrupt the process and insert the redundancy. For this purpose, the redundancy is kept idle until it is to be activated. After this is accomplished, the system can be restarted (Fig. 1.28).

Forms and Activation of Redundancy

One can compare the concurrent strategy with the dual tires of a truck that are running in sync. If one of these tires goes flat, the other one can still keep the vehicle running.

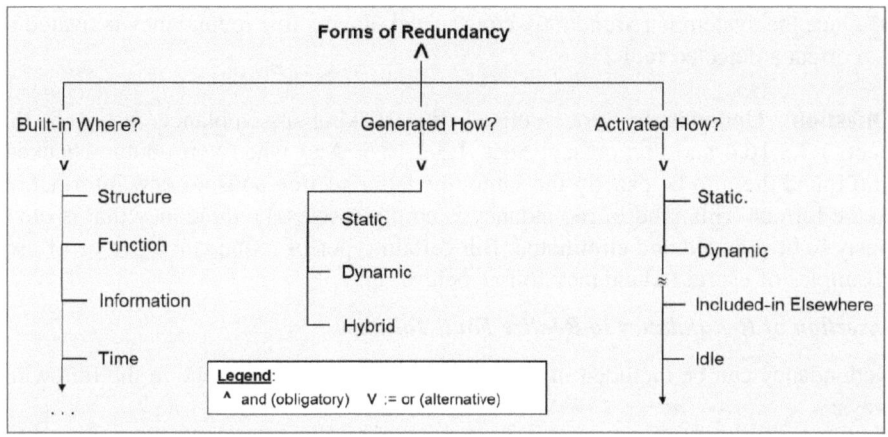

Fig. 1.29 Forms and activation of redundancy

Using the spare tire in the trunk, on the other hand, necessitates stopping the vehicle, taking out the spare tire and replacing the flat one with it, and then, finally, restarting the vehicle.

Last but not least, it is important to use the correct form of the redundancy. While constructing the redundancy, the following decisions need to be made beforehand (Fig. 1.29).

- Which element is to be formed redundant? That is, where is the redundancy to be inserted?
- How is the redundancy generated?
- Once the system is redundantly constructed, how is this redundancy activated to correct a detected fault?

More About Redundancy

One can compare the concurrent strategy with the dual tires of a truck that are running in sync. If one of these tires goes flat, the other one can still keep the vehicle running. Using the spare tire in the trunk, on the other hand, necessitates stopping the vehicle, taking out the spare tire and replacing the flat one with it, and then, finally, restarting the vehicle.

Last but not least, it is important to use the correct form of the redundancy. While constructing the redundancy, the following decisions need to be made beforehand (Fig. 1.29).

- Which element is to be formed redundant? That is, where is the redundancy to be inserted?
- How is the redundancy generated?

- Once the system is redundantly constructed, how is this redundancy activated to correct a detected fault?

Question: Undoubtedly you noticed the striking resemblance between the Sect. 1.3.3.16.4 and this present Sect. 1.3.3.16.5. And when you double-checked, you found them to be exactly the same; the latter section adds no new information to the former. This kind of redundancy is empty (useless) redundancy that is obviously to be avoided and eliminated. But certain types of redundancy can be of use. Examples of useful redundancy follow below.

Insertion of Redundancy to Realize Fault Tolerance

Redundancy can be included in a system to realize fault tolerance in the following ways:

- *Structural Redundancy* is added to the system by additional components that are to be activated if a corresponding fault is detected. Thus, this works similar to a spare parts storage.
- *Functional Redundancy* makes use of functional similarities of components so that if one fails, another intact one can replace it.
- *Information Redundancy* is the additional information to enable and, wherever possible, realize a self-correction. For example, if I tell my friend, "We'll meet on Christmas Eve, the 25th of December," my friend will know that I made a mistake because Christmas Eve is on December 24. He/she can then automatically correct the wrongly specified date. The theory of self-correcting codes is a good example for using information redundancy for constructing self-detecting/-correcting codes for data transmission.
- *Time Redundancy* is the extra time between the latest expected point in time and the realized point in time for the delivery of the service by the system under consideration. This extra, spare time is necessary to validate the result and, if a fault has been detected, to correct it.
  Time redundancy is the most important kind of redundancy since no self-correction can be made if there is not sufficient time to detect a fault and determine and activate the redundancy to correct said fault.

Generation of Redundancy

Different kinds of redundancy can be constructed in the following ways:

- *Static Redundancy* is running continuously. It is used by the service of interest, regardless whether faults are present or not. Compare this with the dual tires of a truck (see above, Temporal Placement of the Redundancy). If one goes flat, the other one *masks* this fault.
- *Dynamic Redundancy* is activated on demand when a corresponding fault, which can be compensated by this redundancy, is detected. Remember the example with the spare tire in trunk.

Table 1.2 Static versus dynamic redundancy

Resource utilization/activation	
Static redundancy	*Resource-intensive*: Resources are always in use, regardless whether faults are present or not. Faults are tolerated by fault masking
Dynamic redundancy	*Resource-friendly*: Resources are used on demand in presence of faults by reconfiguration
Timing behavior	
Static redundancy	Requires no fail-over time, that is, the time consumption is low
Dynamic redundancy	Additional fail-over time required
Reliability	
Static redundancy	Provides high short-term reliability
Dynamic redundancy	Provides high long-term reliability

- *Hybrid Redundancy* is a mixture of static and dynamic redundancy to overcome their drawbacks; EXAMPLE: fault masking until a certain threshold, followed by reconfiguration.

Activation of Redundancy

- Redundancy can be activated the way it was generated: *static, dynamic.*
- *Redundancy Included Elsewhere*, that is, in an element performing another function, can be used to compensate a detected fault. For example, if one of our eyes (or ears) fails, the other one can continue to supply the service of seeing (or hearing), even if no longer three dimensional (or stereo).
- **Idle Redundancy** is unused and, therefore, useless since there is no way for fault compensation to use it. In the case of software, a copy of a module detects only idle redundancy because when a fault is detected in the running module, its copy will have the same fault, leading to no fault-tolerant behavior.

Next, different types of redundancy are compared to enable the developer to select the right one (Table 1.2).

Examples for Fault-Tolerant Systems

The following two examples are meant to explain the realization of fault-tolerant software.

Example (Fig. 1.30): *Recovery Block Scheme* utilizes dynamic redundancy. The example below has two components; each uses different methods to determine the greater of two given integers. The process is: The first component delivers an output and a third component checks this output. If the result is correct, the main program continues. If the result of the first component is not correct, the second, redundant

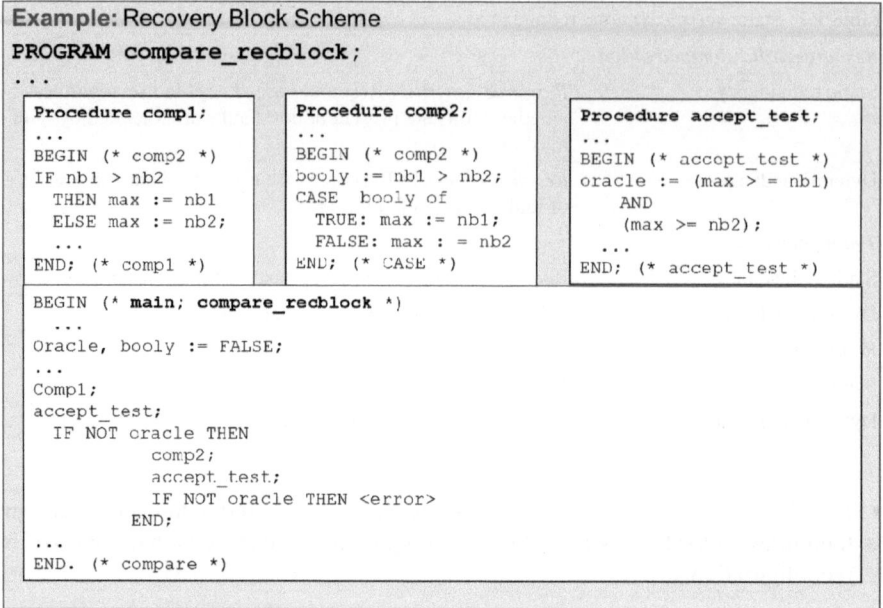

Fig. 1.30 Dynamic redundancy

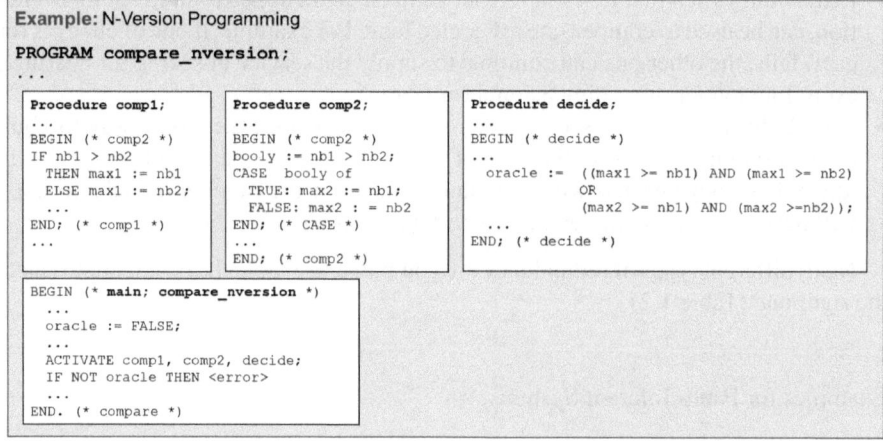

Fig. 1.31 Static redundancy

component will be activated. If it also cannot deliver a correct output, an error message will be output [19].

Example (Fig. 1.31): *N-Version Programming* utilizes static redundancy. The two components run simultaneously and a voter compares the results. If they are the same, the main program continues. If not, an error message will be output.

Summary of the Examples

- **Recovery Block Technique**

    - Fault Diagnosis: *Acceptance* test (*absolute* test) in the sense that the correctness will be validated.
    - Fault Correction: Undo (reset) and activation of an alternative component (if available).
    - Activation of (dynamic) Redundancy: upon fault detection (dynamic). Generally: subsequent execution of the alternatives.

- **N-Version-Programming**

    - Fault Diagnosis: *Majority decision* (*relative* test) in the sense that voting determines a fault when simultaneously running alternative components delivers different results.
    - Fault Correction: Majority result (fault masking)
    - Activation of Redundancy: always running.
    - Simultaneous execution of components.

1.3.4   Verification and Validation (V&V) and Fault Tolerance for Reuse

The aspects of software and its dependability, as discussed in Sects. 1.2 and 1.3, are also valid for reuse. This means that they need to be understood to understand V&V of reuse. However, this book includes specific discussions and reviews wherever an adaptation is necessary, for example, system architecture and reuse (Sect. 2.1), dependability and legal aspects of reuse (Sects. 2.2 and 2.3), or testing and reuse (Sects. 5.1 and 5.2).

1.4   Terms, and Definitions of Reuse

The previous sections reviewed areas of software engineering relevant to software reuse, such as development, configuration management, and quality of software. Reuse-related terminology and a general taxonomy of software reuse will be discussed and summarized in the next subsections.

The following features will be used to introduce and classify the reuse terminology used in this book (Table 1.3, for explanation of the terms used, see Sect. 1.4.3; [38, 53, 70, 79]):

- *Reuse assets* and *entities* can be product-oriented and, thus, material and concrete, such as components. They can, however, also be ideal, that is, not or only indirectly product-oriented and, therefore, non-material, such as concepts, ideas, algorithms, etc.
- *Domain scope* refers to application area (Sect. 3.1.1).
- *Development scope* refers to platform and origin of the component.
- *Modification* refers to additional work required prior to reuse.
- *Managerial aspect* refers to whether and which kind of organizational and technical work is to be done in performing reuse.

A reuse approach is:

- *compositional* if existing components are reused (such as the Unix shell);
- *generative* if application or code generators are required (such as refine and meta tools);

  - *direct reuse* approach requires no "glue code," that is, an additional software that intermediates between the reusable component and the receiving system (Sect. 1.5.3);
  - *indirect reuse* necessitates an intermediate entity.

1.4.1   Terms Related to Software Engineering

To avoid likely misunderstanding, the terminology often used in this book will be defined and explained (based on [53]).

- *Software unit/Software module*: Software element in programming codes that can be *separately* specified, compiled, documented, and tested to perform a task or activity to achieve a desired outcome of a software function.

  - Note: The terms "unit" and "module" are often used interchangeably or defined to be sub-elements of one another in different ways, depending upon the context. The relationship of these terms is not yet standardized.

- *Software (configuration) item*: Software item that has been configured and treated as a single item in the configuration management process (see Sect. 1.3).

  - Note: A software configuration item can consist of one or more software units to perform a software function.

- *Software function/(software) function block*: Elementary operation performed by the software module or unit as specified or defined as per stated requirements to fulfill a well-defined user or system function or a part of it.

Table 1.3 Classification of reuse terminology (based on [53])

Reuse asset	Reuse entity	Domain scope	Development scope	Modification	Management	Approach
Ideas, concepts	Architectures	Vertical	Internal	Adaptive	Accidental	Compositional
Artefacts, components	Requirements	Horizontal	External	Black box	Systematic	Generative
Procedures, skills	Designs			White box		Indirect
	Specifications					Direct
	Source code					
	Object code					
	Test cases					

- EXAMPLE: Calculation of the sinus function of a given angle is a function block of a unit to calculate trigonometric functions; providing the address to buy a ticket is a function block of a web portal.
- Note 1: Software units consist of function blocks.
- Note 2: A function block contains input variables, output variables, through variables, internal variables, and an internal behavior description of the function block.

- *Software system*: Defined set of software items that, when integrated, behave collectively to satisfy a requirement.

  - EXAMPLES: Application software for accounting and information management, system software for linking library functions, application-oriented system software for text processing, performance analysis, programming tools.

- *Product line*: Collection of systems potentially derivable from a single architecture.

1.4.2   Terms Related to Software Dependability

Software dependability is the ability of the software to perform as and when required when integrated in the system operation. Based on this definition, following features can be defined:

- *Reuse dependability*: Ability of a composite system containing reusable components to perform as and when required to meet users' service needs.
- *Qualification*: Process of validation and verification (V&V) used to demonstrate that the product is capable of meeting its specification for all the required conditions and environments.
- *Quality Target*: Specified level of quality as a goal; quantified, wherever possible, using a software metric.

  - EXAMPLE: The overall reliability of the composite system in terms of the required MTBF (mean time between failures), or the requirement that the cyclomatic complexity of a software unit be kept below a specific value.

1.4.3   Terms Related to Software Reuse

This is the core area of the terminology used in this book (based on [53]).

- *Software reuse*: Using a software asset, that is, software or software knowledge, in the solution of a different problem in order to construct new software.

- Note 1: This notion is very broad and covers both "heritage" and "legacy," and it is refined into categories: "black-box," "white-box," "adaptive," "systematic," and "accidental" reuse software (definitions of these terms follow in this section, near the end).
- Note 2: The opposite of "software *re*use" is "software *one*-use," which requires being developed from scratch and is not necessarily intended for reuse.
- EXAMPLE: Some dedicated software routines, such as security codes, are not designed for reuse; they are one-use components.

- *Software (reuse) asset*: Software configuration item that has been designed for use in multiple contexts and domains.

  - EXAMPLE 1: Design, specification, source code, documentation, test suites, manual procedures, etc.
  - EXAMPLE 2: Availability of the information that a specific navigation function uses an algorithm based on Kalman filter.
  - Note: A software-based or software-oriented knowledge is also an asset.

- *Context*: Software environment tied to mission and software requirements.
- *Domain*: Problem space or application area.
- *Software reusability*: Degree to which a (reuse) asset can be used in more than one software system or in building other assets.

  - Note 1: In a (reuse) repository software, reusability represents the characteristics of an asset that makes it easy to reuse vertically or horizontally.
  - Note 2: Usability is a measure of software unit's or system's functionality, ease of use, and efficiency.

- *(Software) Component*: Constituent of a software system with specified interfaces and explicit context and domain dependencies.

  - Note 1: A software component can consist of one or more software units to perform a software function.
  - Note 2: A software component can be utilized independently and is subject to composition, also by third parties.
  - EXAMPLE: An individual component is a software package, a web service, or a module that encapsulates a set of related functions (or data).

- *(Software) Commercial-off-the-shelf (COTS)*: Commercially available components.

  - Note 1: A COTS software can consist of one or more software units to perform a software function.
  - Note 2: Components in governmental use are called "government off-the-shelf (GOTS)."
  - Note 3: COTS and GOTS software usually represent components; they can also be stand-alone applications.
  - Note 4: COTS and GOTS typically realize reuse incorporation or integration.

- Note 5: COTS and GOTS are designed to be implemented easily into existing systems without the need for customization ("glue code," "wrappers", Sect. 1.5.3).
- EXAMPLE 1: Microsoft Office is a stand-alone COTS application that is a packaged software solution for businesses. An operating system, a word processor, a compiler, etc. are further examples of stand-alone COTS.
- EXAMPLE 2: Also, libraries that need linkage to an application code, such as graphic engines, Windows DLLs, etc., are COTS components.
- EXAMPLE 3: Software that is used to create software, but is not part of a composite software system, is not COTS software; it is a development tool. However, development environments with runtime modules are COTS (for example, Visual Basic™, Sybase™), or information retrieval applications (for example, hypertext and data mining tools), or operating system utilities (for example, for file operations and memory management).

- *(Reuse) Repository*: Storage of a collection of reusable components

  - Note 1: The repository is the place where reusable software assets are stored, along with the catalogue of assets. Every employee should be aware that it contains important company know-how and should be able to access for easy use.
  - Note 2: In a narrower sense, "software libraries" have the same function as software repositories; for example, building sets of reusable software units, such as trigonometric functions.

- *Accidental reuse*: Reuse without strategy, typically reusing software components not designed for reuse.

  - Note: Also known as "ad hoc" or "opportunistic" reuse.

- *Systematic reuse*: Developing software components intended for reuse and/or building new applications from those reusable components, following a formal plan of product line, also known as "planned reuse."
- *Adaptive reuse*: Using previously developed software that is modified only for portability; for example, a new application on a different operating system.
- *Black-box reuse*: Reuse of unmodified software components, incorporating existing software components into a new application without modification.
- *White-box reuse*: Modifying and integrating software (function) blocks into new applications.
- *Vertical reuse*: Reuse in the same domain.
- *Horizontal reuse*: Reuse in different domains.
- *Internal reuse (in-house reuse)*: Reuse of a company or government software unit developed within that company or government.
- *External reuse*: Reuse of a software unit from another company or government.

  - Note: A "third party software" is usually written by another company as a legal entity. It incorporates external reuse if it is to be used in a context or

domain other than that for which it has been designed and developed. It can be, however, also a dedicated, one-use software component.
- EXAMPLE 1: Open-source software (OSS), mostly for external reuse.
- EXAMPLE 2: A service-oriented architecture (SOA) can operate on components in internal or external reuse.

- *Heritage software*: Inherited software reused from a previous mission that is currently in usage.
- *Legacy software*: Software reused from a previous mission that is currently out of usage, or in restricted usage.
- *Component to be reused*: Software component that is intended to be reused in a different context or domain other than its original development context or domain, for any kind of software reuse.
- *Reusable component*: Software component to be reused after being qualified for reuse.
- *Receiving system*: Software system into which reusable component(s) will be integrated.
- *Integration of reusable components*: Process of installation/assembly of reusable components into the receiving system, including integration validation to ensure the proper functionality of the final system.
- *Composite system*: Final system resulting from the integration of reusable components.
- *Qualified composite system*: Composite system after qualification.

1.5 Software Reuse—General

1.5.1 Ingredients of Software Reuse

Software reuse, that is, the use of existing software components or knowledge to build a new software system, is meant to realize benefits, such as improved productivity, but also not negatively impact the dependability.

Reuse process that consists of identifying an appropriate reusable component, its analysis, and integration into the receiving system, in accordance with the definition requirements (Fig. 1.32).

1.5.2 Dependability-Related Objectives of Software Reuse

Reuse dependability is the ability of a *composite* system containing reusable components to perform as and when required to meet users' service needs. Thus, for fulfilling the requirements of assessment of reuse dependability, the identification

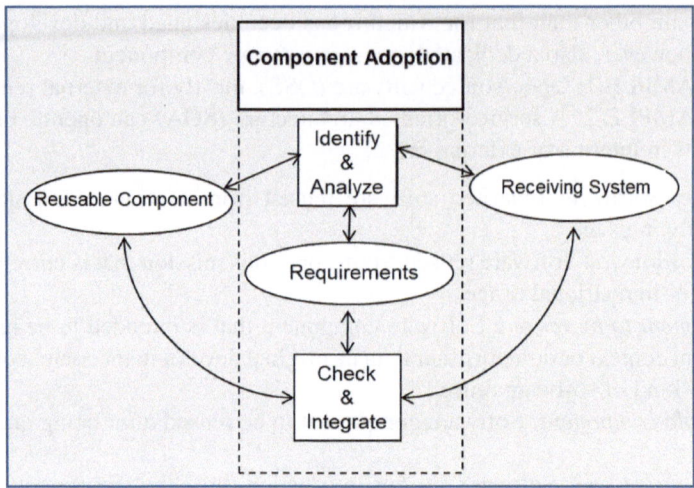

Fig. 1.32 Ingredients of software reuse (based on [53])

and deployment of the relevant methods and techniques of reuse dependability are indispensable.

Several activities influence the dependability of reuse, including the following ones:

- development, operation, and maintenance of reusable components as reuse assets and composite systems with reuse assets, and
- management of practice and assets of reuse.

The dependability of the composite systems with reuse assets is influenced by the following factors:

- software functions to serve and satisfy user needs, and
- dependability of these services.

We focus on requirements on functionality and tests of software products containing reusable components.

1.5.3   Integration of the Reusable Component into a Receiving System to Produce a Composite System

This process requires that both parts be "prepared," for example, by connectors and glue code, which means additional work and, consequently, additional dependability risks (Fig. 1.33). A traditional approach is to develop "adapters", as recommended by the "Gang of Four" [40]; see also [48, 87].

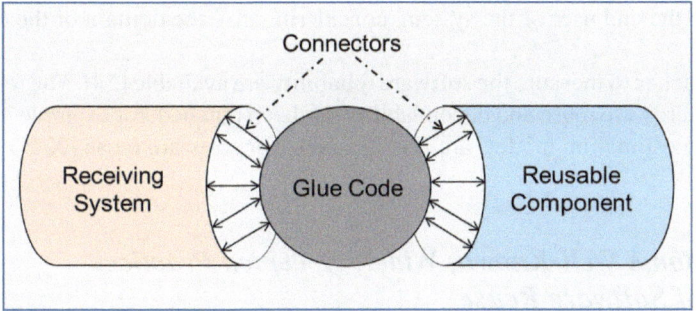

Fig. 1.33 Connectors and glue code (based on [53])

- *Glue code*: Software that intermediates between the reusable component and the receiving system.
- *Connector*: Interface elements of composite system to receive reusable components.
- *Wrappers*: Additional software to complete the functional and interface requirements, if they are not priorly fulfilled (see also Sect. 2.2.3.1).

1.5.4 Hypotheses for Reuse Dependability

The desirable, albeit naïve, demand about reuse can be stated as follows:

Increased software reuse can significantly improve the dependability of a software system.

This desire necessitates fulfilling the following preconditions to succeed on the market:

- D1: Reusable components have to be more dependable than their one-use equivalent.
- D2: Composite systems built by reusable components have to be more dependable than their equivalent built by one-use components.
- D3: Generating a system from a high-level, user-oriented, behavior-based specification realized by reusable components has to be more dependable than one built by hand.

To make these requirements operable, D1 and D2 can be studied by means of reliability measures. They can be determined using software reliability models that are available for industrial practice [9, 54].

From the view point of reliability engineering, it is evident that developing and validating the glue software and determining and realizing its relation with connectors of the receiving system are to be considered as a part of the reuse process. Thus, D2 is influenced by the reliability of the glue software to an extent that is comparable to the reliability of the reusable component. D3 concerns the domain, which can be best judged by the end user of the system, considering also the domain of the composite system.

Approaches to measure the software reliability are available [54]. The relationship between hardware reuse and dependability has been pursued, for example in [50, 53]. Reliability estimation models are also available for software reuse [32].

1.5.5 Some Well-Known, Widely Accepted Practices of Software Reuse

As reminded at the beginning of this chapter, software reuse is as old as software engineering. Therefore, many techniques for realizing reuse have been developed, mostly independent of each other. One of them, service-oriented architecture, has already been mentioned in Sect. 1.1.2, Fig. 1.2. Some other techniques are briefly described in the following (Fig. 1.34) [8, 67, 82, 85].

- *Object-oriented programming*: Implementing applications using "objects" that consist of data structures, methods (algorithms), and their interactions with other software.
- *Aspect-oriented software development*: Weaving shared components into an application at different places, if separation of concerns is feasible.
- *Program generators*: To support the automatic production of required and specified programs.

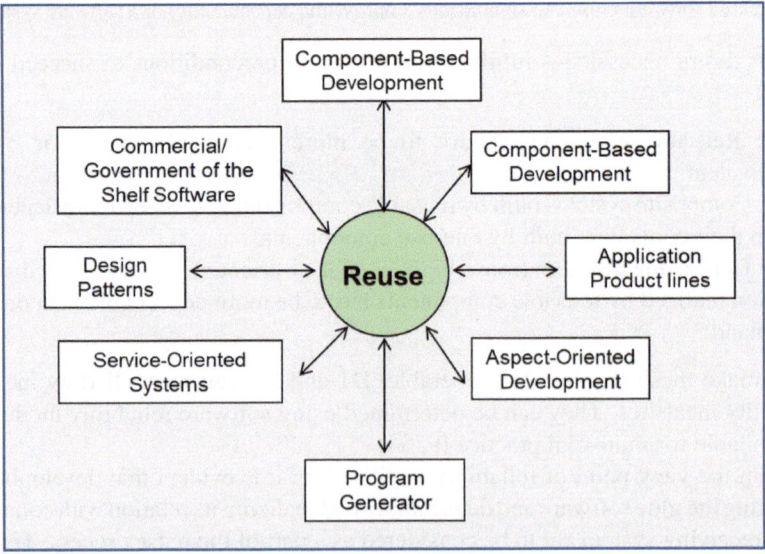

Fig. 1.34 Some well-known practices of software reuse (based on [53])

- *Design Pattern*: A reusable solution to a commonly occurring problem within a given context in software design.
- *Component-based development*: A reuse-based approach to defining, implementing and composing loosely coupled, independent components into systems.
- *Commercial/Governmental-off-the-Shelf Software*: Ready-made software, alternative to custom software or one-off developments.
- *Application Product Lines*: Creating a collection of similar software systems from a shared set of software assets using common means of production.

The first four techniques are rather programming-oriented; also called "reuse in-the-small." The following three (Component-based Development Software, Commercial/Governmental-off-the-Shelf, and Design Pattern) define strategies and practices for reuse and will be called "reuse in-the-large." They will be reviewed and discussed in Sects. 3.2–4.3).

Key Points, Exercises, Recommended Further Reading, References

Key Points

- A software system or a software component used successfully in one application can cause severe problems in another one.
- A successful reuse that fulfills the user requirements under defined conditions and for a defined period of time is considered dependable. Dependability is more than reliability, availability, etc. It is a well-tempered concert of those critical features.
- The software to be reused most often needs to be modified. Moreover, additional software can be necessary to "glue" it to the receiving system with the help of "connectors."
- Any modification, even a minor one, is a development that is to be carried out properly.
- Software development is a process that must be modeled and implemented.
- There are many techniques for modeling software and its development. The waterfall model was the first one introduced; V-model and spiral model are the more recent ones.
- Software configuration management defines the structure and controls and disciplines modifications of software. Therefore, it is a good idea to also utilize techniques developed there for software reuse.
- There are many kinds of software reuse, from accidental to systematic. It is important to consider dependability aspects from scratch, that is, during the construction if the software under consideration is supposed to be reused.
- Verification and Validation (V&V) are more than testing; verification checks whether the developers are working correctly and are in compliance with the

specification of the current phase; validation checks whether the system developed works properly and is in compliance with the user requirements.

- The first step of V&V is a precise definition of faults that are the root events that occur whenever the realized item differs from the required and/or specified one.
- An error is a manifestation of a fault, that is, the impact of a fault that could not be tolerated.
- Finally, a failure occurs whenever a part of the system under consideration is active that contains an intolerable fault. If this part is idle, the fault is a latent one.
- Faults can be classified based on linguistic or functional aspects. Accordingly, countermeasure against the consequences of a fault are to be activated.
- The best way to cope with faults is not to make them, that is, to exclude them during development.
- Nevertheless, there are still chances that faults can creep in. Therefore, countermeasures are necessary to assure that the system can still deliver reduced, yet acceptable services, even though faults occur within the system or its environment.
- Structured and systematic testing is one of the widely accepted V&V methods in industrial practice.

Exercises

1. Why is dependability important for reuse?
2. Is a hardware reuse possible without considering software? Why?
3. Is a software reuse possible without considering hardware? Why?
4. Name the documents you refer to learn more about the terms and techniques of reuse for hardware and software.
5. Give three examples of software reuse in your environment.
6. Why is configuration management (CM) important for reuse?
7. Give a naming strategy to identify the departments and members of your company, school, club, etc., which considers the reusability of items they work on and/or with.
8. What is the most important issue of dependability?
9. Give the key features of reuse dependability.
10. Why is the waterfall model not sufficient to develop dependable software?
11. Why is the spiral model necessary?
12. How can fault tolerance can be utilized for reuse?
13. Give examples of systematic reuse.
14. Name and discuss the elements of software reuse.
15. Give three hypotheses for reuse dependability.
16. Explain reuse-in-the-small versus reuse-in-the-large.
17. Which additional efforts are necessary to realize reuse?
18. Is reuse a very novel approach? How did it started?
19. Compare well-known Practices of Software Reuse. Which one is most efficient?

Recommended Further Reading

1. Ammann, P., Offutt, J.: Introduction to Software Testing. Cambridge University Press, UK (2008)
2. Beizer, B.: Software Testing Techniques, 2nd edn. International Thomson Computer Press (1990) and Dreamtech (2003)
3. Binder, R.V.: Testing Object-Oriented Systems: Models, Patterns, and Tools. Addison-Wesley, Boston (2006)
4. CMMI® for Services, Version 1.2, CMMI-SVC, V1.2 CMMI Product Team: Improving processes for better services. Technical Report CMU/SEI-2009-TR-001 ESC-TR-2009-00 (2009). Available at https://resources.sei.cmu.edu/asset_files/TechnicalReport/2009_005_001_15092.pdf. See also https://www.sei.cmu.edu/search.cfm#stq=cmm&stp=1
5. Leach, R.J.: Software Reuse: Methods, Models, Costs, 2nd edn. (2012)
6. Many relevant papers and books. See http://ivknet.de/index.php/en/publications
7. Mathur, A.P.: Foundations of Software Testing, 2nd edn. Pearson, London (2013)
8. Sommerville, I.: Software Engineering, 10th edn. Pearson, London (2015)

References

9. AIAA R-013-1992: Recommended practice: software. Reliability (1992)
10. Aiello, R.: Configuration Management Best Practices: Practical Methods that Work in the Real World. Addison-Wesley, Boston (2010)
11. Avizienis, A., Laprie, J.-C., Randell, B., Landwehr, C.: Basic concepts and taxonomy of dependable and secure computing. IEEE Trans Dependable Secure Comput. 1(1) (2004). http://ieeexplore.ieee.org/xpls/abs_all.jsp?arnumber=1335465
12. Anderson, T., Randell, B.: Computing Systems Reliability—An Advanced Course. University of Newcastle upon Tyne, Cambridge University Press, Cambridge (1979)
13. Ayav, T., Belli, F.: Boolean differentiation for formalizing Myers' cause-effect graph testing technique. In: 2015 IEEE International Conference on Software Quality, Reliability and Security—Companion, pp. 138–143 (2015)
14. Bertolino, A., Fantechi, A., Gnesi, S., Lami, G.: Product line use cases: scenario based specification and testing of requirements. In: Käkölä, T., Duenas, J.C. (eds.) Software Product Lines Research Issues in Engineering and Management, Chap. 11, pp. 425–445. Springer, Berlin (2006)
15. Belli, F., Budnik, Ch.J., Hollmann, A., Tuglular, T., Wong, W.E.: Model-based mutation testing—approach and case studies. Sci. Comput. Program. **120**, 25–48 (2016)
16. Beck, K., Andres, C.: Extreme Programming Explained, 2nd edn. Addison-Wesley Professional, Boston (2004)
17. Beck, K.: Test Driven Development: By Example. Addison-Wesley Professional, Boston (2002)
18. Belli, F., Echtle, K., Görke, W.: Methoden und Modelle der Fehlertoleranz. In: Informatik Spektrum, vol. 9, no. 2, pp. 68–81. Springer, Berlin (1986)
19. Belli, F., Jedrzejowicz, P.: Fault-tolerant programs and their reliability. IEEE Trans. Reliab. **39**(2), 184–192 (1990)
20. Beizer, B.: Software Testing Techniques, 2nd edn. Van Nostrand Reinhold (1990) and Dreamtech (2003)
21. Belli, F.: Finite state testing and analysis of graphical user interfaces. In: Proceedings 12th International Symposium on Software Reliability Engineering, pp. 34–43 (2001)
22. Benington, H.D.: Production of large computer programs. IEEE Ann Hist. Comput. **5**(4), 350–361 (1983). http://doi.org/10.1109/MAHC.10102. Retrieved 2011-03-21
23. Binder, R.V.: Testing Object-Oriented Systems: Models, Patterns, and Tools. Addison-Wesley, Boston (2006)

24. Boehm, B., Lane, J. A., Koolmanojwong, S.: The Incremental Commitment—Spiral Model: Principles and Practices for Successful Systems and Software. Addison Wesley, Boston (2014)
25. Belli, F., Budnik, Ch.J., White, L.: Event-based modelling, analysis and testing of user interactions: approach and case study. In: Software Testing, Verification & Reliability, pp. 3–32 (2006)
26. Clements, P., Northrop, L.: Software Product Lines—Practices and Patterns. SEI Series in Software Engineering. Addison-Wesley, Boston (2002)
27. Cox, C.: Surviving software dependencies. Commun. Assoc. Comput. Assoc. **62**, 36–43 (2019)
28. Cohen, D.M., Dalal, S.R., Fredman, M.L., Patton, G.C.: The AETG system: an approach to testing based on combinatorial design. IEEE Trans. Softw. Eng. **23**(7), 437–444 (1997)
29. DeRemer, F., Kron, H.: Programming-in-the large versus programming-in-the-small. In: Proceeding of ACM International Conference on Reliable Software, pp. 114–121 (1975). http://doi.org/10.1145/800027.808431
30. DeMillo, R.A., Lipton, R.J., Sayward, F.G.: Hints on test data selection: help for the practicing programmer. IEEE Comput. **11**(4), 34–41 (1978)
31. Dijkstra, E.W.: Notes on structured programming. In: On the Reliability of Mechanisms, Sect. 3. T.H.-Report 70-WSK-03. Technology University Eindhoven (1970)
32. Dimov, A., Punnekkat, A.: On the estimation of software reliability of component-based dependable distributed systems. In: Proceedings QoSA-SOQUA 2005, LNCS 3712, pp. 171–187. Springer, Berlin (2005)
33. DO-178C ED-12C:2011: Software considerations in airborne systems and equipment certification (available 2012)
34. EN 50128:2012: Railway applications—communication, signalling and processing systems—software for railway control and protection systems
35. Fagan, M.E.: Design and code inspections to reduce errors in program development. IBM Syst. J. **15**(3), 182–211 (1976). https://doi.org/10.1147/sj.153.0182
36. Frakes, W.B., Isoda, S.: Success factors of systematic reuse. IEEE Softw. **11**(5), 14–19 (1994). Available at http://ieeexplore.ieee.org/stamp/stamp.jsp?tp=&arnumber=311045
37. Frankl, P.G., Weyuker, E.J.: An applicable family of data flow testingcriteria. IEEE Trans. Softw. Eng. **14**(10), 1483–1498 (1988)
38. Frakes, W.B., Kang, K.: Software reuse research: status and future. IEEE Trans. Softw. Eng. **31**(7), 529–536 (2005)
39. Gazzola, L., Goldstein, M., Mariani, L., Segall, I., Ussi, L.: Automatic ex-vivo regression testing of microservices. In: Proceedings of 1st IEEE/ACM International Conference on Automation of Software Test (2020)
40. Gamma, E., Helm, R., Johnson, R., Vlissides, J.: Design patterns. In: Elements of Reusable Object-Oriented Software. Addison-Wesley, Boston (1994)
41. German Assoc. for Quality: Deutsche Gesellschaft für Qualitätssicherung: Software Qualitätssicherung, DGQ-NTG-Schrift Nr. 12-51. Frankfurt (1986)
42. Goodenough, J., Gerhart, S.L.: Toward a theory of test data selection. ACM SIGPLAN Not. **1**(6), 156–173 (1975). https://doi.org/10.1145/390016.808473
43. Goodenough, J.B.: Exception handling—issues and a proposed notation. Comm. ACM **18**(12), 683–696 (1975)
44. Hallsteinsen, S., Paci, M.: Experiences in Software Evolution and Reuse—Twelve Real World Projects. Springer Science & Business Media, Berlin (1997)
45. Hennell, M.A., Woodward, M.R., Hedley, D.: On program analysis. Inf. Process. Lett. **5**(5), 136–140 (1976)
46. Hossain, Sh.: Rework and reuse effects in software economy. Glob. J. Comput. Sci. Technol. C Softw. Data Eng. **1**(4) (2018)
47. Howden, W.E.: Theoretical and empirical studies of program testing. In: Proceedings 3rd International Conference on Software Engineering—ICSE '78, pp. 305–311 (1978)
48. Hummel, O., Atkinson, C.: The managed adapter pattern: facilitating glue code generation for component reuse. In: Proceedings 11th International Conference on Software Reuse (ICSR), LNCS 5791, pp. 211–224. Springer, Berlin (2009)

49. Hummel, O., Janjic, W.: Test-driven reuse: key to improving precision of search engines for software reuse. In: Finding Source Code on the Web for Remix and Reuse, pp. 65–80 (2013)
50. IEC 62309:2004: Dependability of products containing reused parts—requirements for functionality and tests
51. IEC 62304:2006 (amended 2015): Medical device software—software life cycle processes
52. IEC 61508:2010: Functional safety of electrical/electronic/programmable electronic safety-related systems: Part 1: General requirements. Part 2: Requirements for electrical/electronic/programmable electronic safety-related systems. Part 3: Software requirements. Part 3-1: Software requirements—reuse of pre-existing software elements to implement all or part of a safety function (2016). Part 4: Definitions and abbreviations
53. IEC/PAS 62814-2012: Dependability of software products containing reusable components—guidance for functionality and tests (withdrawn, however available)
54. IEEE 1633:2016: Recommended practice for software reliability
55. IEEE Std 610.12-1990: IEEE standard glossary of software engineering terminology (Reaffirmed 2002)
56. ISO/IEC/IEEE 12207: 2017, Systems and software engineering—software life cycle processes
57. ISO 26262 (1 to 12): 2018, Road vehicles—functional safety
58. Jensen, R.W., Tonies, Ch.: Software Engineering. Prentice-Hall, Hoboken (1979)
59. Jensen, R.W.: Improving Software Quality. Prentice-Hall, Hoboken (2014)
60. Jamshidi, P., Pahl, C., Mendonça, N.C., Lewis, J., Tilkov, S.: Microservices: the journey so far and challenges ahead. IEEE Softw. **35**(3), 24–35 (2018). https://doi.org/10.1109/MS.2018.214 1039.ISSN0740-7459
61. Kaner, C., Falk, J., Nguyen, H.Q.: Testing Computer Software, 2nd edn., Wiley, Hoboken (1999)
62. King, J.C.: Symbolic execution and program testing. Commun. ACM 385–394 (1976)
63. Kim, Y., Stohr: Software reuse: survey and research directions. J. Manage. Inf. Syst. **14**, 113–147 (1998)
64. Krueger, C.W.: Software reuse. ACM Comp. Surv. **24**(2), 131–183 (1992)
65. Laprie, J.-C. (ed.): Dependability: Basic Concepts and Terminology in English, French, German, Italian and Japanese. Springer, Berlin (1992)
66. Luo, Y., Brand, M., Engelen, L., Favaro, J., Klabbers, M., Sartori, G.: Extracting models from ISO 26262 for reusable safety assurance. In: International Conference on Software Reuse, pp. 192–207 (2013)
67. Lim, W.C.: Managing Software Reuse. Prentice Hall PTR, Hoboken (1998)
68. Mathur, A.P.: Foundations of Software Testing, 2nd edn. Pearson, London (2013)
69. McClure, C.: Software Reuse: A Standards-Based Guide. Wiley, Hoboken (2001)
70. Mili, A., Chmiel, S.F., Gottumukkala, R., Zhang, L.: Managing software reuse economics: an integrated ROI-based model. In: Annals of Software Engineering, vol. 11, pp. 175–218. Kluwer Academic Publishers, Dordrecht (2001)
71. Miller, E.F., Howden, W.E.: Tutorial: Software Testing and Validation Technique. IEEE Computer Society Press, IEEE Catalog No. EHO 180-0 (1981)
72. Miller, E.F.: Structurally based automatic program testing. In: Proceedings EASCON-74 (1974)
73. Myers, G.: The Art of Software Testing, 3rd edn. Wiley, Hoboken (2012)
74. Naur, P., Randell, B. (eds.): Software engineering. Report on a conference sponsored by the NATO Science Committee, Garmisch, Germany, 7–11 Oct 1968 available at http://homepages. cs.ncl.ac.uk/brian.randell/NATO/nato1968.PDF
75. Nelson, V.P.: Fault-tolerant computing: fundamental concepts. IEEE Comput. **23**(7) (1990)
76. Newman, S.: Building Microservices: Designing Fine-Grained Systems. O'Reilly Media, USA (2015)
77. Nurolahzade, M., Walker, R.J., Maurer, F.: An assessment of test-driven reuse: promises and pitfalls. In: In: International Conference on Software Reuse, pp. 227–252 (2013)
78. Oracle Practitioner Guide: Determining ROI of SOA through Reuse (2012)

79. Orrego, A., Mundy, G.: SRAE: an integrated framework for aiding in the verification and validation of legacy artifacts in NASA flight control systems. In: Proceedings 31st Annual International Computer Software and Applications Conference (COMPSAC). IEEE Computer Press (2007). http://doi.org/10.1109/COMPSAC.2007.199
80. Paulk, M.C.: How ISO 9001 compares with the CMM. IEEE Softw. **12**(1), 74–83 (1995). http://doi.org/10.1109/52.363163. See also https://resources.sei.cmu.edu/asset_files/TechnicalReport/1994_005_001_435267.pdf, and https://www.sei.cmu.edu/search.cfm#stq=cmm&stp=1
81. Poulin, J.S., Caruso, J.M., Hancock, D.R.: Business case for software reuse. IBM Syst. J. **32**(4), 567–594 (1993)
82. Pressman, R.S.: Software Engineering: A Practitioner's Approach, 9th edn. Palgrave Macmillan, London (2020)
83. Rapps, S., Weyuker, E.J.: Selecting software test data using data flow information. IEEE Trans. Softw. Eng. **11**(4), 367–375 (1985)
84. Royce, W.W.: Managing the development of large software systems: Concepts and techniques. Proc. IEEE WESCON **26**(1970), 1–9. Available in Proc. ICSE 9, Computer Society Press (1987)
85. Specht, J.: Terminology Proposal: Redundancy for Fault Tolerance. University of Duisburg—Essen available at http://www.ieee802.org/1/files/public/docs2013/new-tsn-specht-redundancy-terminology-20130115-v01.pdf
86. Sametinger, J.: Software Engineering with Reusable Components. Springer, Berlin (1997)
87. Sim, S.E., Gallardo-Valencia, R.E. (eds.): Finding Source Code on the Web for Remix and Reuse. Springer, Berlin (2013)
88. Tracz, W.: Confessions of a used-program salesman: lessons learned. In: Proceedings of the Symposium on Software Reusability, pp. 11–13 (1995)
89. Weißleder, S.: Influencing factors in model-based testing with UML state machines: report on an industrial cooperation. In: Proceedings of Model Driven Engineering Languages and Systems (MODELS). Lecture Notes in Computer Science, vol. 5795, pp. 211–225. Springer, Berlin (2009). http://doi.org/10.1007/978-3-642-04425-0_16
90. Whittaker, J.A., Poore, J.H.: Statistical testing for cleanroom software engineering. In: Proceedings of Twenty-Fifth Hawaii International Conference on System Sciences, vol. 2, pp. 428–436 (1992)
91. Wirth, N.: A plea for lean software. IEEE Comput. **28**(2), 64–68 (1995). https://doi.org/10.1109/2.348001

Chapter 2
Application Aspects and Organization of Dependable Software Reuse

The terms and primary concepts of reuse have been introduced, discussed, and exemplified at the end of the last chapter. The present chapter will review aspects of application, architecture and organization, and assurance and legal issues of reuse.

A broadly observed phenomenon is that, while planning substantial reuse of software components, software engineers are often overly optimistic concerning how much reusable functionality can be achieved. It cannot be stressed strongly enough that reuse is not an ultimate saver of costs, schedule, or dependability. Even COTS (commercial-off-the-shelf) deployment often satisfies only less than 40% of the functionality of an industrial application [8, 9, 18, 20, 21].

The reason for this discrepancy is twofold. First, the costs of the adaptation of the component to be reused might come higher than expected. Second, the test and follow-on costs to ensure the required level of dependability can prove to be more challenging than foreseen. Therefore, it is important to build components from scratch for reuse. Also, the system that is supposed to accept reusable components must be prepared to be built by reuse. Thus, the architecture of both, that is, the component built-*for*-reuse and the system built-*by*-reuse, must be appropriately designed.

To avoid dependability, environmental, and productivity risks, reusable components need to be of high quality. Each time a low-quality component is reused, the poor quality and associated risks would consequently be reproduced. To avoid those risks, the development of dependable, environment-friendly and reusable components requires considerably extra effort.

Dependability methodologies of reuse include application aspects and the organization of the reuse. Pre-store and pre-use characteristics of reusable components are to be met and the cases built-for- reuse or built-by-reuse should be distinguished. In the following sections of this chapter, these important characteristics of reuse, including energy saving and ecologic aspects, will be discussed before starting with the technical aspects in the next chapter.

2.1 Systems Architecture and Software Reuse

The architecture of a system commits the structure to combine and integrate its elements and their features, and determines the relationships of those elements to each other and interdependencies between them and the environment [15]. The architecture of a system is of decisive importance for its design and evolution, including reuse characteristics of its components.

Some logical/topologic structures are hierarchical, centralized (star-form), or decentralized (network-form) as graphically schematized in Fig. 2.1. Architectural elements can be event-, state-, or service-oriented. Relations between graph elements are often defined as *consists-of*, *uses-services-of* or *neighbored*.

It is important for reuse that the software architecture allows a precise design and specification of interfaces and their dependability and other critical features so that it enables evaluation, selection, acquisition, and integration of reusable components into the receiving system.

It is evident that the architecture is the key to software reuse; its form and preciseness affect the features of the system.

2.1.1 Built-for-Reuse, Built-by-Reuse

Apart from functional requirements, the addressing of critical non-functional requirements, that is, dependability and quality, is also important. Neglecting these aspects certainly results in schedule and cost impacts, and invokes severe safety and security risks. Therefore, besides the functional and interface requirements, non-functional requirements are then also clearly to be fulfilled; for example, glue code and wrappers are to be carefully planned, specified, designed, implemented, and tested (Sect. 1.5.3).

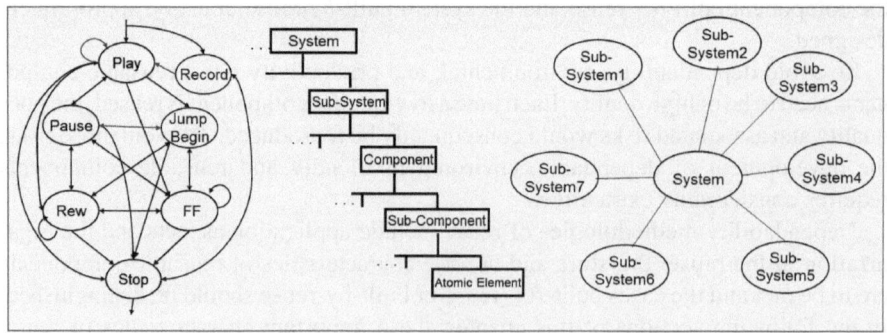

Fig. 2.1 Typical structures: decentralized (network-form), hierarchical, and centralized (star-form)

Software-by-reuse is the use of existing applications or their components to build new applications. It is widely accepted and convenient to consider software reusability from the following viewpoints:

- Built-*for*-reuse aims at planned development of reusable components.
- Built-*by*-reuse aims at planned development of systems using reusable components.

Both of these viewpoints focus on characteristics of reusability that are to be checked before storing the component and before reusing this component in a new product.

The following recommendations address not only internal reuse, they can easily be adopted, as well, for external reuse (in case you forgot these reuse terms, see Sect. 1.5).

2.1.2 **Pre-store *Characteristics of Reusability***

Before storing a component for reusability, the following characteristics should be checked to determine whether and how to use it in other systems (Fig. 2.2 see also Sect. 5.3.2).

- *Universality* is defined over the range of the *functionality*, enabling reusability in a large class of domains and contexts. It requires the following sub-characteristics:

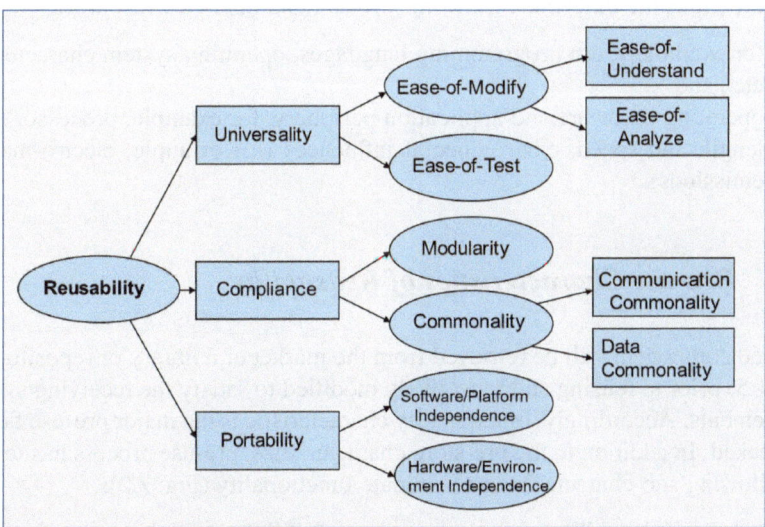

Fig. 2.2 Characteristics of reuse (based on [16])

- *Ease-of-modify* requires the availability of the source code and an appropriate documentation of the component. Its most important factors are

  ease-of-understanding and ease-of-analyze.

- *Ease-of-test* requires availability of appropriate *test criteria* to generate test cases, and *test oracles* to justify the decision whether or not the modification has achieved its goals, required by both reusable components and new system's specifications.

Remember that the universality feature is likely to be in conflict with dependability that requires it be tailor-made, in accordance with the individual structure and environment of the component.

- *Interoperability* is defined in terms of the ability to communicate with other systems, to adapt and communicate, and requires following the features:

  - *Modularity* is an architectural property of software being composed of units that are functionally independent on each other in the sense that

    a change in one component has minimal impact on other components.

  - *Compliance* requires the component follow existing standards and de-facto standards, that is, state-of-the-art rules methodology or best practice techniques (see Sect. 1.3.3.14).

    Compliance concerns primarily *interfaces* and *data structures* used.

- *Portability* is defined by the ability of software to operate on different platforms and requires the independence from software and hardware resources:

  - for example, from programming languages, operating system characteristics, etc., and
  - operating hardware and application periphery, for example, processor's word length and speed, environmental influences (for example, electro-magnetic emissions).

2.1.3 Pre-use *Characteristics of Reusability*

A stored component will be retrieved from the market or a library or repository (see Sect. 1.5) prior to reusing and most likely modified to satisfy the receiving system's requirements. Accordingly, functionality characteristic is the major pre-use factor to be checked. In addition to the pre-store characteristics, pre-use process has to check the following sub-characteristics to validate functionality (Fig. 2.2):

- *Suitability* is the ability of the component to fulfill the receiving system's requirements and the expectation it will produce the results (outputs and behavior) of the new system.
- *Accuracy* is the precision of the results expected from the reused component.

- Just like pre-store and pre-use, *compliance* of the reused component is a key feature to comply with certain standards the composite system has to follow, in addition to the standards followed by pre-store process (Sect. 2.1.2).

2.1.4 *Built-for-Reuse*

Figure 2.3 summarizes the concept of building components that are planned to be reused. This concept consists of the following, partly interactive phases [6]:

- *Development* phase is, at first glance, similar to the development phase of common, one-way components. However, there are substantial differences to ensure the reusability that will be discussed in the following section. Therefore, the developed component will be viewed solely as a *potentially* reusable component, not yet ready for operation.
- *Pre-Store* phase checks the reusability of the potentially reusable component, based on the reusability characteristics discussed in the last section. If it passes

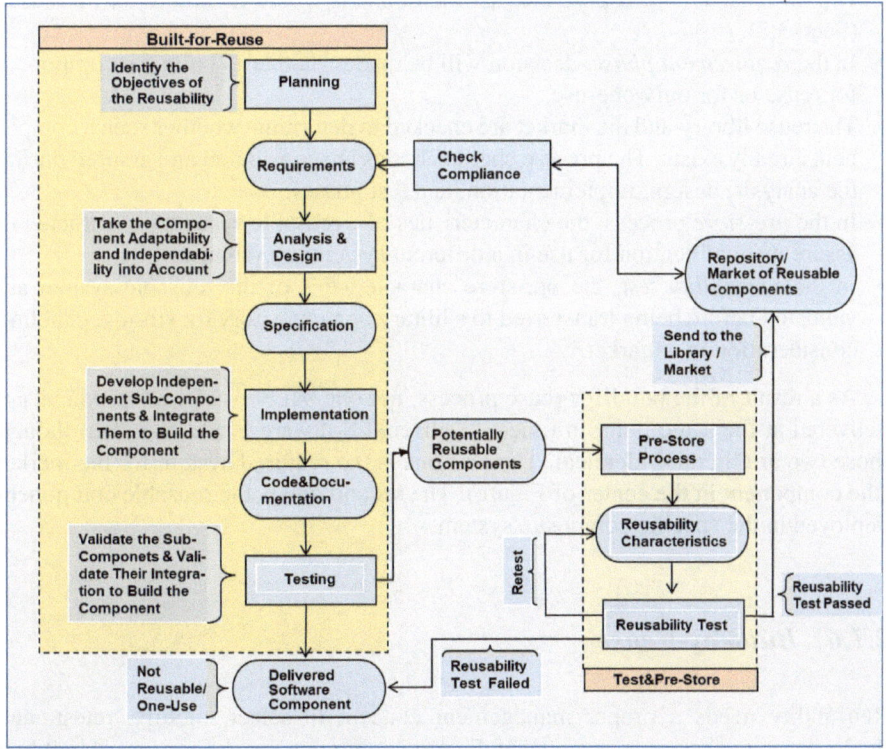

Fig. 2.3 Built-for-reuse (based on [6, 16])

the checks, this component constitutes a reusable component. Otherwise, it is a one-way component.

- *Marketing* the reusable component is the next phase that interacts with the development phase of another reusable component. Developers of reusable components have to closely observe the open market of reusable components that will be stored in the company-owned repository.

2.1.5 *Stages of the Development Phase of* Reusable Components

The stages of the development phase of reusable components differ considerably from those of the common software development and entail the following activities (Fig. 2.3):

- In the *planning phase,* the objectives of the reusability to enable component extraction are defined. The component to be reused is identified.
- *Domain analysis* provides precise explanation when/where/how the component will be reused, and, if possible, list the domains where it should not be reused (Sect. 3.2).
- In the *requirement phase*, decision will be made whether to build the component for reuse or for only one-use.
- The reuse library and the market are checked to determine whether such a component already exists. The pre-use characteristics are considered and assured during the analysis, design, implementation, and test phases.
- In the *pre-store* process, the characteristics of a reusable system are extracted to assure its qualification for use in a different system or systems.
- In the *reusability test,* the pre-store characteristics of the reusable system are validated before being transferred to a library or a repository for storing, enabling consideration in a market.

As a result of the build-for-reuse process, not one but two software products are delivered at the same point in time—"Delivered Software Product"—even though those two are literally identical. The first one is the required system for the market (the component in the center of Figure). The second one is the reusable component deployed in the specific composite system.

2.1.6 *Built-by-Reuse*

Reusability needs a proper management concept to select, modify, retest, and deploy/operate a reusable component. Figure 2.4 summarizes the concept of building systems out of reusable components. This concept consists of the following, partly interactive phases:

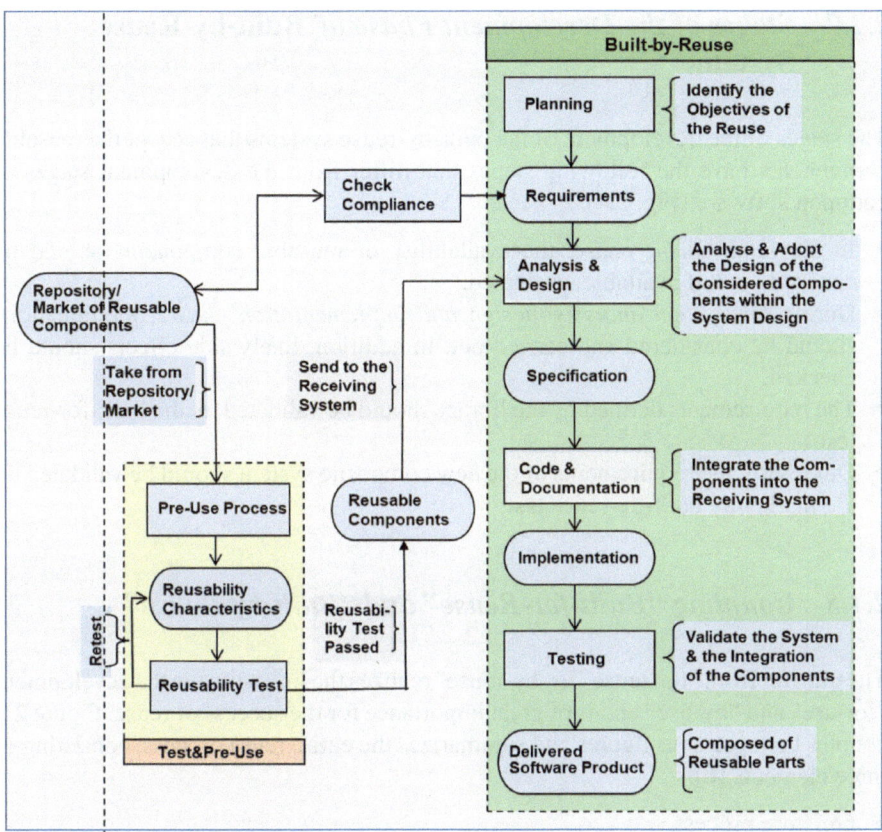

Fig. 2.4 Built-by-reuse (based on [6, 16])

- *Requirement* for a specified reusable component sent by the receiving system to be built by reusable components.
- *Compliance Check* to find potential components and the selection of one of them out of a repository or market, and checking its reusability characteristics. If the selected component is not appropriate, select another and retest. If eventually an appropriate one found, send it to the receiving system.
- *Development* of the receiving system.

2.1.7 Stages of the Development Phase of Built-by-Reuse System

The stages of the development of the built-by-reuse systems that accept the reusable components have the following stages that differ from the development stages of common software (Fig. 2.4).

- In the *requirement* phase, the availability of reusable components should be checked, and, if available, evaluated.
- During the *system analysis, design and implementation phases,* modifications should be considered and carried-out. In addition, likely side effects should be checked.
- The requirements defined by the library should be validated by the built-for-reuse test.
- Otherwise, the requirements of the new composite system should be validated on the market by built-by-reuse test.

2.1.8 Coupling "Built-for-Reuse" and "Built-by-Reuse"

The transfer from "for-reuse" to "by-reuse" realizes the link between the development "for-use" and "by-use" and is of great importance for the success of reuse. Figure 2.5 couples both previous figures and summarizes the entire reuse process, consisting of three elements [6]:

- pre-store process,

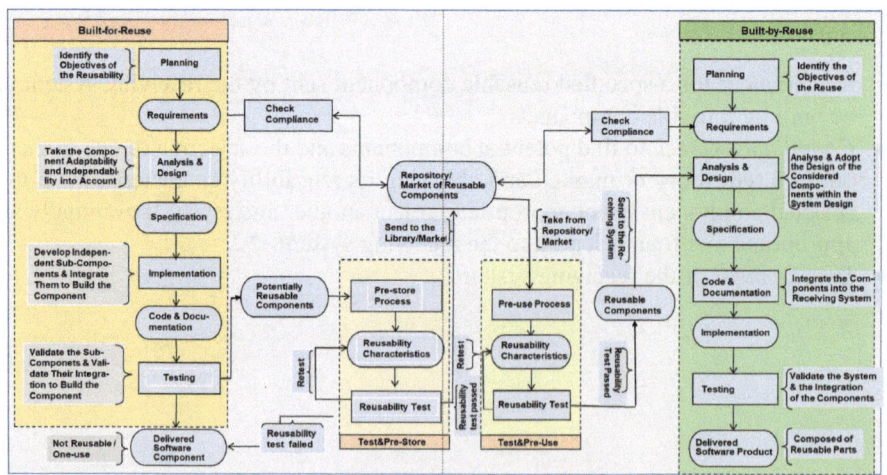

Fig. 2.5 Built-for-and-by-reuse (based on [6, 16])

- pre-use process, and
- storage of reusable components.

Figure 2.5 reveals also the main steps of the reuse process, as well as its requirements:

Understand—analyze—modify—test

As explained in Sect. 2.1.2,

ease-of-understanding, ease-of-analysis, ease-of-modification, and ease-of-testing

are the key factors to *ease*-of-reuse of a component. They enable the selection and analysis of components to be reused in a new system.

2.1.9 Some Remarks

Note that during its life cycle, a reusable component encounters two types of modifications and thus two types of tests:

- The first to satisfy and assure the *characteristics* required by the library and/or the market, and
- a second type of modification and test are then necessary prior to its *deployment* after having been removed from storage.

Note also that the first modification and test aim at *generalization* of the component, whereas the second one aims at *specialization* of this component to satisfy the requirements of a new system in realizing another application in another domain.

2.2 Dependability and Ecological Aspects Reuse

Hardware reuse, as well as software reuse, whether involving home-grown or COTS components, certainly promises lower cost, better dependability, thus providing a decrease in risk, increase of productivity, and, consequently, considerable potential for a less stressful development process [10, 16].

During the last decades, much research progress has been achieved in technically mastering software reuse by industrial best practices. However, software reuse has proven to be complex and steadily evolving with the progress of software engineering so that it needs appropriate methods and techniques for dependability assessment and assurance.

The applicability of the conventional dependability techniques for the analysis and evaluation of software reuse should be considered carefully. During the last decades, specific methods provided effective and broadly well-understood and, therefore, accepted solutions for dependability assurance, especially concerning functionality and testing of software. A selective amount of them is appropriate for effective reuse dependability analysis and evaluation, applicable to both reusable components and composite systems. Indeed, many factors influence the dependability performance in the life cycle, including the early stages and implementation and integration phases.

A common recommendation for software reuse is to constraint the reusable software components to perform only one function completely. This restriction is designed to ease the implementation, deployment, and maintenance of reusable components and composite systems that contain such components. Furthermore, deviation from this restriction could have an adverse effect on dependability due to the possibility of errors introduced into the software during implementation or maintenance. However, the concept "one component—one function" is a severe constraint that limits the scope of software reuse and, therefore, has difficulty being accepted by the industrial and commercial software development and marketing, where rather universally deployable components are more attractive.

Constructive methods and approaches are well-understood to help avoid faults while producing reusable software, which is the best way of fault handling. Nevertheless, analytic methods are also necessary to detect and eliminate faults that could not be avoided.

Existing analytical approaches use either source code (if available), leading to white-box testing, or software specification and user profile, leading to black-box testing for testing and test case generation. Grey-box testing combines both approaches, which is rare.

As we discussed in the Sects. 1.3.3.11–1.3.3.13, measures of the effectiveness of a test set in fault revealing, *coverage-oriented adequacy criteria* use the ratio of the portion of the specification or code that is covered by the given test set, while testing to the uncovered portion.

EXAMPLES are: $C0$ test, $C1$ test, dd-test, du-test (white-box testing), or cause-effect analysis and operational profile analysis (black-box testing). The belief is, the higher the degree of test coverage, the lower the risk of having critical software artefacts that have not been sifted through.

Formal, meaning sound mathematical, methods are recommended, such as model checking, model-based testing, and formal proofs of programs, to avoid singularity of case-oriented testing.

Finally, reliability growth models statistically evaluate software based on test data recorded during testing, more precisely, the number and time intervals of failures triggered by test cases [5]. Thus, with some effort, it is also possible to determine the reliability of software reuse [7].

When considering the dependability of reuse, it is essential to include the operation and maintenance stages in the analysis of composite systems, such as legacy or heritage software.

Dependability performance and dependability of service of reuse should be continuously monitored, analyzed, and evaluated by the methods revisited in this section.

2.2.1 Validation, Re-validation and Cost Aspects of Software Reuse

The most common form of reuse is using software developed for one-use in a new application, which is *accidental* reuse. One of the major objectives of the present book is to warn managers that this kind of unplanned reuse can be a potential minefield because it can cause the inheritance of all the problems of the pre-existing software in the reaping of only a few of its benefits.

Many managers, while planning for software reuse, forget that both the reused component <u>and</u> the composite system are to be tested in the new domain. Experience reports say that developing reusable software can cost 60% more than one-use software, whereby a good portion of additional costs goes to testing [8, 9, 18, 20, 21].

Note that the software reuse involves redesign, reimplementation, and re-testing. Redesign arises if the existing functionality does not fulfill the requirements of the new task, since it requires reworking to realize the new function, and, prior to this, it necessitates reverse engineering to understand its current functionality.

The design change leads to reimplementation. Thus, re-testing (as a kind of regression testing) is necessary to validate the functionality of the reused software in the new domain to determine whether or not redesign and reimplementation are needed.

Re-implementation requires re-coding, code review, and unit testing [11, 14]. Re-testing activities can be clustered in the following groups:

- test re-planning,
- test procedures to be altered,
- re-integration testing,
- re-release and re-acceptance testing,
- test drivers/simulators to be altered,
- test reports to be rewritten.

The following undesirable events/situations, mostly caused by managerial misjudgment, can negatively influence the dependability of software reuse:

- failing to select the right component or to wrongly favor the wrong selection criteria;
- failing to justify and adjust the need for and/or extent of the modification of the selected component to fulfill operational or application requirements;
- failing to justify and adjust the need for and/or extension of the maintenance of the selected component during operational stage.

2.2.2 Redesign

Modification of architectural design entails:

- identification (finding out) of architectural design part(s)/element(s) to be modified:

- realization of the modification:
- re-validation of the entire architectural design.

A detailed redesign requires the identification of the component design part(s) to be modified, realization of the modification, and re-validation of the entire design.

Reverse engineering helps to detect the part(s) to be modified, which are not familiar to the developers, and aids in the understanding, modification, and re-validation of the entire component.

Re-documentation also requires the detection of the part(s) to be modified, and the modification and re-validation of the entire document.

The following fundamental facts influence dependability, especially reliability when using commercially available components, for example, COTS components for software development.

- Very often no source code is available, thus there is no way to correct a detected fault.
- This is a great restriction that prohibits application of the most widely used reliability models ("reliability growth models" (AIAA R-013-1992 [5], IEEE 1633 [14]) that require perfect correction of detected faults.
- If source code is available: Note that COTS software is no longer COTS after its source code has been modified to correct a fault detected because the COTS supplier no longer maintains the documentation and source code (just as electronics equipment warranties are no longer valid after a seal is broken).
- Furthermore, the modifications can violate the original software design. From then on, modified COTS software is to be handled as an accidental reuse.

2.2.3 Assumptions and Rules for Improving Software Reuse Dependability

Commercially available software components for reuse, for example, COTS software, address common needs.

2.2.3.1 Folklores

Arguments for COTS software often induce the following assumptions, which should be, however, seriously questioned.

- "COTS software contains fewer faults than the one-use ones." The reality is that this kind of software is also made by ordinary developers and is as likely to be subject to having bugs as any other kind of software.
- "System integrators know exactly the functionality and interfaces of COTS." The reality is that COTS software is ordinary software with ordinary software

documentation. So this kind of software also behaves regularly and requires a learning curve.

- "Glue code" and "wrappers" are "easy to write." The reality is that they interface with other people's software, meaning to get them running properly and functioning correctly can be a very tedious and costly process.
- "Composite system will meet user requirement." The reality is that almost always additional effort is needed.

2.2.3.2 Real Rules

Realistic rules stemming from experiences with real-life projects might aid proper reuse and improve the dependability of reuse:

- *Minimize* the use of reusable components which

  - use "combos," that is, combine and perform many non-trivial functions (all of which need to be learned, trained, and maintained), and
  - do not have clearly-defined interfaces.

- *Maximize* the use of reusable components, which

  - perform clearly-defined functions,
  - have clearly-defined interfaces, all with easy-to-understand, predictable inputs and outputs,
  - have a visible, intuitive architecture that can be identified and easily understood, and
  - have been on the market for some time, with equivalent alternates available at competitive sources (then look for likely industrial standards).

2.2.3.3 Limitations Due to Incompatibilities

Incompatibility of a software to networks or external systems might cause limitations. These cases can violate dependability requirements, or it might not be economical to enforce compatibility.

2.2.4 Dependability, Energy Consumption, and Ecology

The emission of CO_2 due to IT has been found to be equivalent to the aviation sector. On the other hand, a good portion of IT emission is, again, due to IT! In earlier decades, studies on the energy efficiency of IT focused on the hardware part, although the software application is responsible for executing tasks. In other words, software directly affects the hardware energy consumption [4].

Moving to cloud computing is an effective way of reducing energy consumption and carbon emissions by 30% or more compared to running the same applications on individual infrastructure [19]. Again, we immediately come to software as a major actor in energy consumption as it also controls and affects cloud. Another good idea for energy saving is to not solely concentrate on the CPU, but rather spread the attention to other components, as well, such as memory, disk, and network.

These are convincing arguments that the dependability of a reusable software is not alone a criterion for the fitness of reuse. In a hardware product, consisting of new and old components, new and old software usually have to cooperate. This requires many measures from a combined design to a combined testing.

The reused SW might have been tested to be free of failures, but the current common requirement of low energy consumption can cause a practice-proven, aged software to no longer be used. This situation is more common than thought. An example of critical, energy-consuming software is embedded software, such as wireless sensor networks, in which most applications do not have the sensors plugged in. Thus, sensors are powered by the batteries they carry. To keep the network alive as long as possible, it is very important to conserve energy while the network is functioning. For this purpose, energy-efficient algorithms are necessary. This situation is similar for software that is embedded in equipment such as mobile phones. Another example is software that controls technical processes invoked by green-house gases emissions and waste and pollutants production. Environmental aspects are important for refurbishing. IT has to become green!

A tradeoff analysis is in most cases necessary between energy efficiency, sustainability, and dependability requirements. An example is workload balancing for life-cycle extension of the power supply. It is then very important for the dependability of the entire system to have a software that conserves energy to exploit all the possible power-saving features of the hardware and controlled devices and processes.

There are many other ways to reach a better energy consumption goal; for example, by sparingly using battery charging commands, avoiding excessive transport of large amounts of data, or banning obsolete software elements not capable of operating in an energy-efficient way. It is obvious that a reusable software that is not energy-efficient makes the situation worse each time it runs. Therefore, software reuse should be critically checked concerning energy-saving aspects and made "lean" wherever possible. For this purpose, knowledge about the operational conditions of the hardware components is necessary and helpful. Checklists for energy saving measures are included in Appendices C and D of Part II at the end of the book.

Aspects of a closely related area, *Ecodesign* of HW in combination with SW, are discussed in Part II. Usually, if HW is planned for reuse, the remanufacturer might not systematically check the features of the SW embedded in this HW. Techniques and tools of Green IT might help to select the best environmentally compatible SW. Especially for "as-new" components, which are not new but qualified as new (see Sect. 6.2), the customer expects a modern energy-saving consumption, which also depends on the standards applied to the development. If the component is actually not as-new, some restrictions might have not been validated. So, a detailed planning of all requirements of reuse, including strategies for design of HW in combination with SW, is necessary (cf. Fig. 1 in Prologue).

Besides all these requirements, it is crucial to keep an eye on the costs of reuse that are limited by the rules of the market and management. They will be decisive to which kind of products are worthwhile to be refurbished and how.

2.3 Software Reuse Assurance and Legal Aspects

Aspects of validation and the qualification of components to be reused for build-for-reuse and build-by-reuse were already discussed in Terms, and Definitions of Reuse. Software components that are to be reused should be qualified and the qualification process should be documented; for example, in accordance with ISO/IEC 12207.

It is important that the context and domain of reuse and operational and/or embedding hardware are identified. Dependability and safety requirements and potential exceptions should be specified; likely conflicts between those requirements are to be identified and solved.

2.3.1 Built-for-Reuse—Validation and Qualification of Components to Be Reused

The manufacturer of reusable components and the reusing party should together specify the functional properties of these components wherever possible and should identify the quality targets; for example, in accordance with ISO/IEC 12207. Design, redesign, test, and re-test issues are to be defined.

Any modification and additional software, including wrapper and glue code necessitated by reuse, might influence dependability, safety, and/or energy efficiency. They are to be considered and checked.

Conformance to and compatibility with hardware and overall system requirements are to be validated.

2.3.1.1 Validation and Qualification

The manufacturer of reusable components and the reusing party should together specify and perform the validation and qualification process wherever possible and should document the test and re-test results; for example, in accordance with ISO/IEC 12207 [12]. These documents should be included in the validation and qualification documentation.

For external reuse, for example, COTS software, where the reusing party is not always known, the validation and qualification process should be carried out by the manufacturer of the reusable component. Qualification of any other kind of reuse should be carried out by the reusing party.

Recommendations concerning organization, characteristics of reusability and validation process, test and re-test criteria, as explained in Sect. 1.4, Terms, and Definitions of Reuse, and Sect. 2.2, should be taken into account.

2.3.1.2 Assessment of Quantifiable Quality Targets

The manufacturer of reusable components or the reusing party should ensure that quantifiable quality targets are met; for example, reliability measures, complexity metrics, and test coverage measures.

2.3.2 Built-by Reuse—Validation and Qualification of the Receiving System

Before integration of the reusable component, the receiving system should have been qualified and the qualification process documented; for example, in accordance with ISO/IEC 12207, IEC 61508-3 and IEC 61508-4 [17].

The manufacturer of the receiving system and of the reusable component should together specify the functional properties of these components wherever possible and should identify the quality targets; for example, in accordance with ISO/IEC 12207. Design issues should be defined.

Architectural issues, such as redundancy for realizing fault tolerance, to increase the availability, and/or other aspects to meet the quality targets, should be considered during design and implementation.

2.3.2.1 Validation and Qualification

The manufacturer of the receiving system should specify and perform the validation process and should document the validation results; for example, in accordance with ISO/IEC 12207 IEC 61508-3 and IEC 61508-4 [17].

Qualification documents of reused components should be included in the qualification documentation of the composite system. Quality targets should be validated.

The composite system should be qualified by the reusing party, considering the documentation of the components manufacturer in the light of the context and domain foreseen for the reuse.

Recommendations concerning organization, characteristics of reusability and validation process, test and re-test criteria, as explained in the previous sections, should be taken into account.

2.3.2.2   Assessment of Quantifiable Quality Targets

The manufacturer of the composite system should ensure that quantifiable quality targets are met, including reliability measures, complexity metrics, and test coverage measures.

Key Points, Exercises, Recommended Further Reading, References

Key Points

- Architecture of a system to be built by reuse is crucial; it necessitates components to be built in.
- Characteristics of a component to be reused is crucial.
- Reusable components need to be tested before storage and retested before usage.
- The development process of components to be reused differs from the development of one-way components.
- The development process of systems to be built by reusable components differs from the development of conventional systems.
- Validation, re-validation, and reliability of software reuse each have their own rules, technically and legally.

Exercises

1. Which architectural form is appropriate for reusability?
2. What are the differences between the pre-store and pre-use characteristics of reusable components?
3. Are there any specific features of systems built by reuse?
4. What are the differences between the development of reusable components and one-way components?
5. What are the differences between the development of systems built by reuse and conventional systems?
6. Name the differences between the validation assurance of reusable components and those of conventional systems.
7. List the differences in the legal aspects of reusable components and conventional systems.

Recommended Further Reading

1. Ezran, M., Morisio, M., Tully, C.: Practical Software Reuse. Springer Practitioner Series (2002)
2. IEC 62628:2012 guidance on software aspects of dependability
3. Petersen, K., Badampudi, D., Syed, M.A.S., Wnuk, K., Gorschek, T., Papatheocharous, E., Axelsson, J., Sentilles, S., Crnkovic, I., Cicchetti, A.: Choosing component origins for software intensive systems: in-house, COTS, OSS or outsourcing?—a case survey. IEEE Trans. Softw. Eng. **44** (2018). https://doi.org/10.1109/TSE.2017.2677909

References

4. Acar, H., Benfenatki, H., Gelas, J.-P., da Silva, C.F., Alptekin, G., Benharkat, A.-N., Parisa Ghodous, P.: Software greenability: a case study of cloud-based business applications provisioning. In: Proceedings of IEEE 11th International Conference on Cloud Computing (CLOUD), pp. 875–878 (2018). https://doi.org/10.1109/CLOUD.2018.00125.hal-01887065
5. AIAA R-013-1992: Recommended practice: software. Reliability (1992)
6. Badareen, A.B., Selamat, M.H., Jabar, M.A., Din, J., Turaev, S.: Reusable software component life cycle. Int. J. Comput. **5**(2), 191–199 (2011). Available at http://www.naun.org/journals/computers/19-863.pdf
7. Dimov, A., Punnekkat, A.: On the estimation of software reliability of component-based dependable distributed systems. In: Proceedings QoSA-SOQUA 2005, LNCS 3712, pp. 171–187. Springer, Berlin
8. Frakes, W.B., Isoda, S., Success factors of systematic reuse. IEEE Softw. **11**(5), 14–19 (1994). Available at http://ieeexplore.ieee.org/stamp/stamp.jsp?tp=&arnumber=311045
9. Hossain, S.: Rework and reuse effects in software economy. Glob. J. Comput. Sci. Technol. C Softw. Data Eng. **1**(4) (2018)
10. IEC 62309: Dependability of products containing reused parts—requirements for functionality and test (2004)
11. IEC 62628:2012 Guidance on software aspects of dependability
12. ISO/IEC 12207: 2017 systems and software engineering—software life cycle processes
13. IEEE 1633: 2016, Recommended practice for software reliability
14. IEEE 1517-2010: IEEE standard for information technology, system and software life cycle processes, reuse processes (2010)
15. ISO/IEC/IEEE 42010:2011: Systems and software engineering—architecture description
16. IEC/PAS 62814-2012: Dependability of Software Products Containing Reusable Components—Guidance for Functionality and Tests (withdrawn, however available)
17. IEC 61508:2010: Functional safety of electrical/electronic/programmable electronic safety-related systems: Part 1: General requirements, Part 2: Requirements for electrical/electronic/programmable electronic safety-related systems, Part 3: Software requirements, Part 3-1: Software requirements—reuse of pre-existing software elements to implement all or part of a safety function (2016), Part 4: Definitions and abbreviations
18. Mili, A., Chmiel, S.F., Gottumukkala, R., Zhang, L.: Managing software reuse economics: an integrated ROI-based model. Ann. Softw. Eng. **11**, 175–218 (2001)
19. Microsoft, Accenture and WSP environment & energy study. https://news.microsoft.com/2010/11/04/microsoft-accenture-and-wsp-environment-energy-study-shows-significant-energy-and-carbon-emissions-reduction-potential-from-cloud-computing/
20. Oracle practitioner guide, determining ROI of SOA through reuse (2012)
21. Poulin, J.S., Caruso, J.M., Hancock, D.R.: Business case for software reuse. IBM Syst. J. **32**(4), 567–594 (1993)

Chapter 3
Reusability-Driven Software Development

The last two chapters introduced the terminology of software engineering and software reuse as used in this book. Furthermore, principles of dependability and features of dependable software have been introduced. Application aspects of reusability based on these introductory elements have been discussed.

Now it is time to find out how software is to be constructed to enable a seamless reusability. Accordingly, this chapter discusses aspects of development and packaging of reusable software.

3.1 Reusable Components—Their Constituents and Packaging

Preparing and making a component easy to reuse require not only a high-quality, but also an outlook, in other words, a package that motivates to obtain and deploy it. This section discusses principles of implementation and packaging of reusable components.

3.1.1 Fixed and Variable Parts of Components

Reusable components consist of two main parts [23]. One is the *fixed* part that is kept the same in all applications. The other one is the *variable* part that will realizes the specific application. The development of the fixed part requires a thorough analysis of the applications targeted to be realized (*domain*) reusing the component under consideration (CUC). This is also called *domain analysis*. The counterpart is the *application analysis* to realize the specific applications using the CUC (Sects. 3.1–3.3). Accordingly, CUC includes the following elements as a package (Fig. 3.1):

F. Belli and F. Quella, *A Holistic View of Software and Hardware Reuse*,
Studies in Systems, Decision and Control 315,
https://doi.org/10.1007/978-3-030-72261-6_3

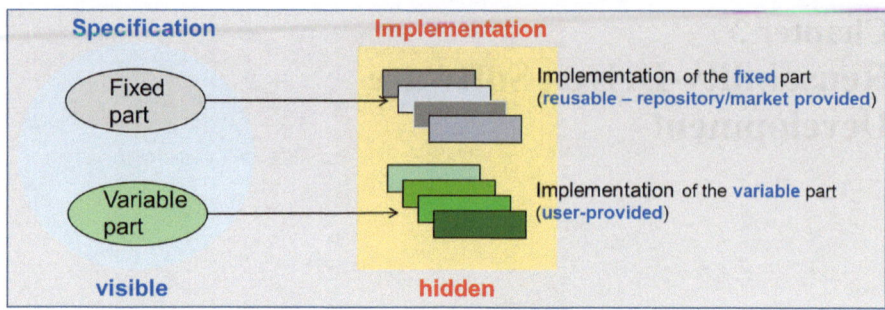

Fig. 3.1 Fixed and variable parts of components (based on [25])

- *Specification*, which is visible to the user (usually, the developer of the reusable component), contains

  – a fixed part that is reusable for many variations, and
  – a variable part that realizes the particular use that has to be supplied by the user, and

- *Implementation*, which is hidden.

3.1.1.1 How to Separate Fixed and Variable Parts from Each Other

The following principles help with developing reusable components:

- Keep the fixed part as large as possible (this will become a reuse *asset*).
- Complementarily, keep the variable part as small as possible (this will become the work to be done when adapting the asset).
- Before packaging, identify the variable part that will be proceeded, dependent on the application domain.
- Complementarily, identify the fixed part during the specification of any functionality.

  EXAMPLE: Design a car platform for handling the following variations [21]:

- four-cylinder, manual transmission, front-wheel drive, four-door sedan,
- four-cylinder, manual transmission, all-wheel drive, four-door sedan, and
- four-cylinder, automatic-transmission, all-wheel drive, four-door sedan.

Domain Analysis:

Car: power bloc (to move the car) + **chassis** (to carry everything and be moved) one fixed part: chassis, three different variable parts: power blocs.

Refining the analysis leads to

power bloc: engine, transmission, powertrain

and to another fixed part:

engine (to power all models);

Thus, in addition to **one type of chassis**, develop only

one type of engine and
two transmission and **powertrain models**

to power all three car models.

3.1.1.2 Three Aspects of Separation

Refining the domain can increase the level of *granularity* that, in turn, can lead to more fixed parts.

However, take care! Too high granularity leads to parts that are too small being fixed, which can lead to excessive adaptation effort for reusing this component. This is the first and the most important aspect of reusability, given that disproportionate effort for reuse would discourage potential customers.

The second aspect concerns the packaging that leads to:

- one part with *all* of the variable part and *as little* of the fixed part *as possible*, and
- another part with the *remainder* of the fixed part and *no* variable part.

The third aspect requires a good balance between *generality* and *adaptation effort*.

3.1.1.3 Generality of and Adaptation Effort for Reusable Components

In Sect. 1.4, different kinds of repositories have been introduced: *vertical* repository that is in the same domain and *horizontal* repository that is different domains.

A reusable component can be realized in such a way that it can be reused in many different applications. It is then a *universal* component (the reader might find this designation exaggerated). Contrarily, a reusable component can be realized in such a way that it can be reused only in a specific family of applications. Accordingly, it is to be expected that the effort to adapt a universal component to a specific application will be greater than that of a specific component designed to a specific kind of application.

Figure 3.2 exemplifies this aspect, attempting to relatively quantify the effort needed for adaptation. Application 1 needs no adaptation effort. 65% of its implementation can be realized using vertical repository, while the rest (35%) needs units delivered by horizontal repository. Application 2, on the other hand, needs adaptation effort (30%) [25].

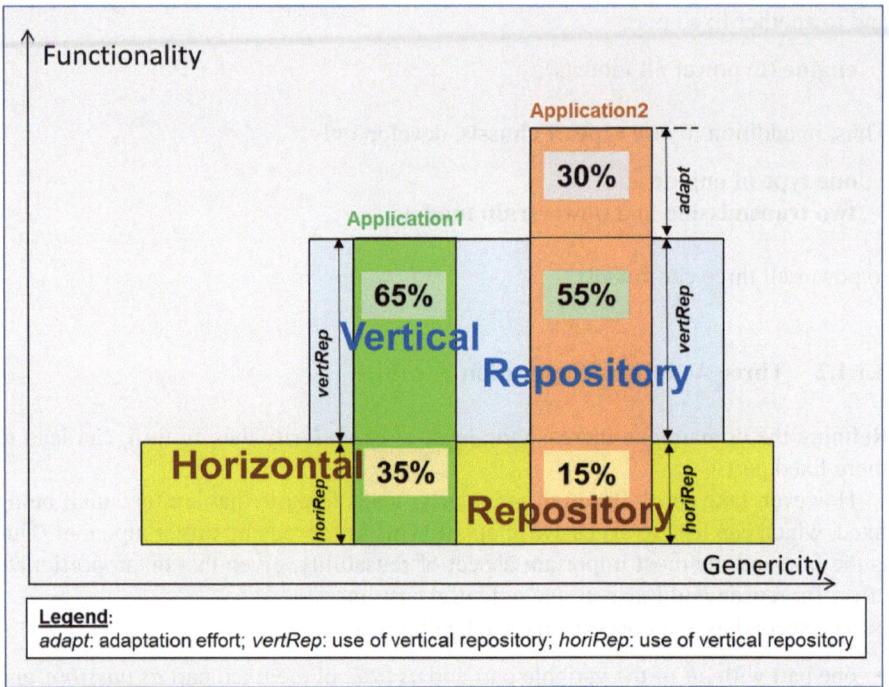

Fig. 3.2 Repositories—horizontal versus vertical (based on [25])

3.1.2 Packaging

After the fixed and variable parts of reusable components have been separated, they must be, figuratively speaking, prepared for shipping, that is, marketing. To improve their marketability, an attractive packaging is needed for their visible and invisible parts [26].

Packaging includes all activities to prepare the component for attracting potential users and for shipping. A good package clearly specifies the domain and the usage of the component and also provides information on how *not* to use of it, that is, its limitations, special case handling, and likely misusage.

3.1.2.1   External Packaging

This part of the packaging supports selection and analysis of a component to be reused. The selection requires information about a reusable asset to compare with the developer's needs. This is called *indexing*, which is a subject of application analysis (Sect. 4.1).

To sum up:

- Usage-based indexing "speaks" the user's language, for example, when inter-changeable algorithms are desired.
- Computation-based indexing needs the user to think like an implementer of the asset, for example, about message patterns between components of the design patterns.

3.1.2.2 Internal Packaging

This part of the packaging supports the development of a component to be reused. This requires information about reusable assets that are supposed to fulfill the needs of many developers. This process is a subject of domain engineering (Sect. 3.2, [26]).

EXAMPLE: Design a network application for handling messages transmitted using *different* encryption algorithms (that is, type of encryption as the *variable* element).

- Task: Package this variability in a way that it isolates the encryption algorithm in a component that is as small as possible.
- Exemplary implementations [13]:

  - Procedural programming: A single function

    String encrypt (**Message** m)

  - Object-oriented programming: encryption embodied in a member function

    String Message :: encrypt ()

To implement different encryption algorithms, either

edit the member function encrypt, or
create a subclass of **Message** in which encrypt is to be to re-defined, or
delegate actual encryption to another object Encrypter that embodies the encryption algorithm.
In C++: Declare encrypt a virtual method.

3.2 Domain Engineering (DE)

Domain Engineering (DE) covers development aspects of reusable components, while Application Engineering (AE) handles their deployment (Fig. 3.3; REBOOT methodology, [15]). Accordingly, DE "assists" the activities of component building. These activities can be grouped in the following categories:

- classification of applications and their realization by reusable components,
- reengineering of components for adapting to an application and for ease-of-reuse, and
- qualification of the component for a dependable reuse by V&V techniques.

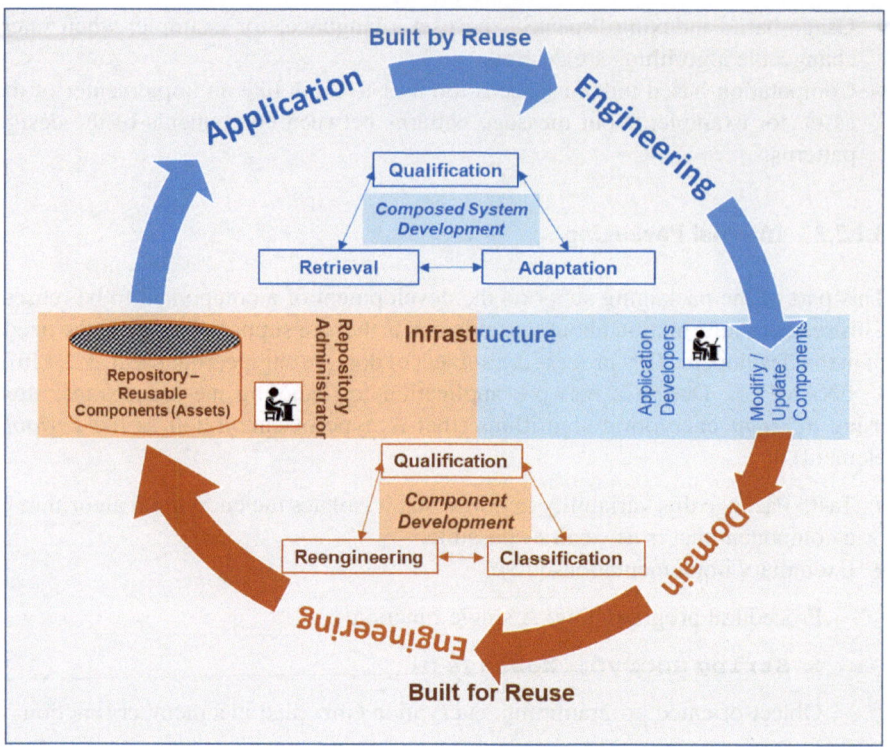

Fig. 3.3 Domain engineering and application engineering (based on [15])

Accordingly, AE assists to retrieve, adapt, and evaluate a potentially reusable component, or components, for a specific application.

Last but not least, a common infrastructure of DE and AE, realized by a base of reusable components (assets) as candidates for reuse, which are possibly to be modified and/or updated before reuse. A base (reuse) administrator usually controls the process. The circular format of Fig. 3.3 suggests a component that has been modified to become a new asset, that is, to be considered as another reusable component (REBOOT methodology, see [15]).

DE is interpreted and realized in two alternative ways:

- DE is a major, *stand-alone* activity to produce assets to be used in application engineering (AE) projects. This view of DE might be considered by companies to be too expensive and risky.
- DE as a *side show* activity of AE to initiate an asset production if and only if there is a need to fulfill an application-specific need. Thus, there is no need to view DE as a discipline on its own.

This book follows the "stand-alone" view.

3.2.1 Activities of Domain Engineering and Application Engineering

As already mentioned, DE includes the *analysis* of typical applications that are expected to lead to reusable components, which will then be collected in a repository of reusable assets. This is called *Domain Analysis*, which follows the step determination of the range of the domain, that is, its delimitation. The goal is the implementation of reusable components.

On the other hand, AE analyzes the requirements of an application and looks for reusable components to realize this specific application. The selected candidates might need to be adapted, which is also a kind of implementation (Figs. 3.3 and 3.4; [15, 26]).

Reusable components can be developed "from scratch," or they can be "inherited" from legacy software (Fig. 3.5, see also Sect. 1.4.3). The first case entails a common, *forward* engineering-oriented development process known from practice; maybe model-based, and leads to reusable components via an architectural design.

In the second case, a legacy application necessitates *reverse* engineering techniques that leads, via architectural design, to models, which are then supposed to be processed as in the first case to realize the component.

Another way, maybe a "quick-and-dirty" one, can be taken using *reengineering* techniques. This can be achieved directly from the legacy system, or indirectly via an architectural design.

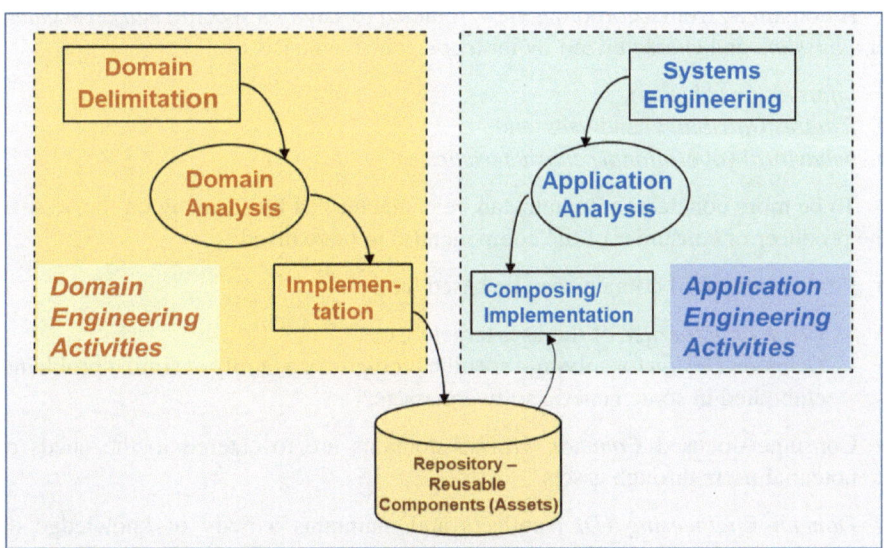

Fig. 3.4 Producer–consumer development model for reuse (based on [25])

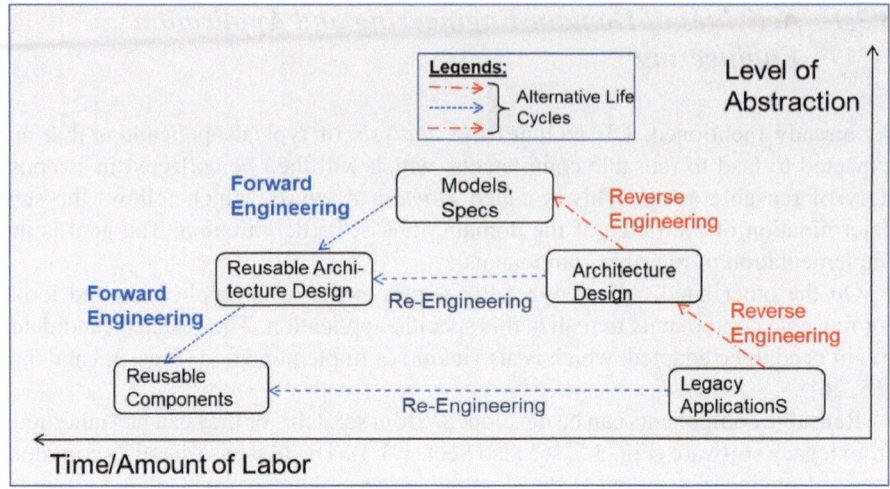

Fig. 3.5 Where the reusable components come from (based on [25])

3.2.2 Domain and Domain Analysis

A *domain* is an area of knowledge and activity characterized by a family of related systems. It is defined by a set of concepts and terminology unambiguously used and well-understood by expert practitioners in this specific area.

A domain is, from a corporate view, founded to satisfy a specific *market segment* or a *mission*, and characterized by its *scope*, which consists of:

- *information* (*objects*),
- *features* (*attributes*) and *uses*, and
- *behavioral* (*operational*) *characteristics*.

To be more concrete, a domain can be characterized by focusing on the view of the producer or consumer of the component(s) to be realized:

- Producer-focused domains can be based on

  - *Common Expertise* of the user targeted, or
  - *Common Design* for solving a specific problem, or a family of similar problems, embodied in some generic software assets.

- Consumer-focused *Common Market* domains are to catered to the needs of potential users through assets.

Domain Engineering (*DE*) collects and maintains a body of knowledge on activities to establish a technical and organizational infrastructure for effectively developing a family of applications within a given domain.

Domain Analysis (DA) is the process of capturing, analyzing, and modeling information about applications in a domain, specifically, common characteristics and reasons for variations.

DA addresses following issues.

- *Technical*: How to find, or produce, widely useful components, and how to package them in a way that they are easily retrieved, adapted, and (re)used (*packaging*, Sect. 1.2).
- *Organizational*: How to integrate them into the running development process of the accepting system in such a way that does not slow down the process but optimizes the use of the organization's resources.
- *Techno-Organizational*: How to organize these activities into a systematic and controllable process.

3.2.3 Activities, Issues and Models of Domain Analysis

Activities of DA include:

- *identifying* domain and its stakeholders,
- *defining* a *glossary* of terms for the domain in consideration,
- *documenting* domain *assumptions* and technical *risks*,
- *identifying problems* within the scope of the domain and their *variations*,
- *identifying legacy* system artifacts that reflect current deployed applications and implementing *functionality* required in the domain applications, and
- *identify commonalities* and *variabilities* in the family of applications lying in the domain.

Issues and *models* of DA include:

- *terms* specific to the domain, and their semantics and pragmatics for characterizing elements and relationships within a family of related applications;
- *commonalities* as services and functionalities (often related to the user requirements) shared across various applications in the same domain;
- *variabilities* to be identified and defined by how each application differs from another within the same domain;
- *rules and constraints* on structuring and implementing the applications;
- *boundaries* of separate applications and their environment, which define their stimuli, events, input/outputs, etc.;
- behavioral and qualitative *requirements*:
- *decision models* to determine what is *inside* the domain, and what is *outside*;
- *documentation* of the *issues* on problems identified during DA, potential solutions and ideas.

3.2.4 Over- and Underscoping of Domain Analysis

If DA is kept too wide, perhaps overloaded with many functions and responsibilities, the risk of *overscoping* arises, with the produced models being too general and targeting as many applications as possible. This would lead to overly costly implementation of assets.

Underscoping arises if the domain artifacts developed are too narrow, that is, not specific enough to be easily retrieved.

Scoping is a critical activity that requires a great deal of experience and skill to make the appropriate trade-offs.

3.3 Domain Analysis (DA) Methods

The objective of DA is the development of domain models. This requires abstracting from the irrelevant aspects of the component under consideration and focusing on the relevant ones. There are two ways of modeling:

- *Black-Box* abstraction from application-oriented (users') needs leads to *requirements* models; for example, as practiced by feature-oriented analysis.
- *White-Box* abstraction from implementation-oriented (programmer's) needs leads to *architectural* models; for example, as practiced by object-oriented (OO) development.

3.3.1 Black-Box Abstraction

A *feature* is a characteristic, end-user-visible behavior of a system [3]. This technique concentrates on the external, user-oriented features of the domain. There are several techniques that can be borrowed from requirements engineering, including *Joint Object-Oriented Domain Analysis (JODA)*, *Organizational Domain Modeling (ODM)*, and *Feature-Oriented Domain Analysis (FODA)* [1, 2]. The last one, FODA, is very relevant to reuse and will be reviewed in the next section as an example.

3.3.1.1 Feature-Oriented Domain Analysis (FODA)

Feature-Oriented Domain Analysis (FODA) is one of the popular techniques introduced by Kang et al. [2], and focuses on identifying features that characterize a domain as explained in Sect. 3.2.

The objective of FODA is to create a domain model representing a family of systems, which can then be refined and specialized into the particular desired system within the domain.

Applications in a domain provide several capabilities that FODA technique models as features in three steps:

- *Context Analysis* scopes the domain, that is, determines its boundaries by identifying the relationship between the domain applications and the external elements. The outcomes of this step are:

  - *Context Diagrams* relating the domain to the environment, and
  - *Structure Diagrams* relating the domain to other domains.

- *Domain Modeling* identifies commonalities and variabilities and consists of:

  - *Feature Analysis* focusing on the end-user's view of the configurable requirements and candidate systems within the domain, and
  - *Information and Operational Analysis* enabling customers to select from configurable requirements to specify a final system.

- *Architecture Modeling* creates a common solution structure.

Context Diagrams

Context Diagrams in FODA show data flows between:

- a generalized application within the domain, and
- the other entities and abstractions with which this application communicates.

Note that the variability of the data flows across the domain boundary must be specified, for example, either by a set of diagrams or text describing the differences.

The example in Fig. 3.6 depicts a context diagram for the Army Movement Control Domain [9]. The arrows represent the information received or generated by the window manager. The closed boxes represent the set of sources and sinks of information. The open-ended boxes represent the databases with which the window manager must interact.

Structure Diagrams

A structure diagram shows the position of the object under consideration within the application's entire domain. For example, the window manager (Fig. 3.7, [22]) is shown in bold outline in the upper left corner.

A structure diagram is used to clearly separate the logical concept of the object under consideration from many other related items with which it is often confused. For example, a window manager is not a graphical user interface (GUI), although it may provide the "feel" portion of a GUI's "look and feel."

A second kind of structure diagrams determines the dependencies of the components of the object, thus its internal structure. As an example, Fig. 3.8 sketches the elements of an imaginary car model. The lines stand for the relation "consist of" between the upper elements and lower ones. Note that the identifiers of

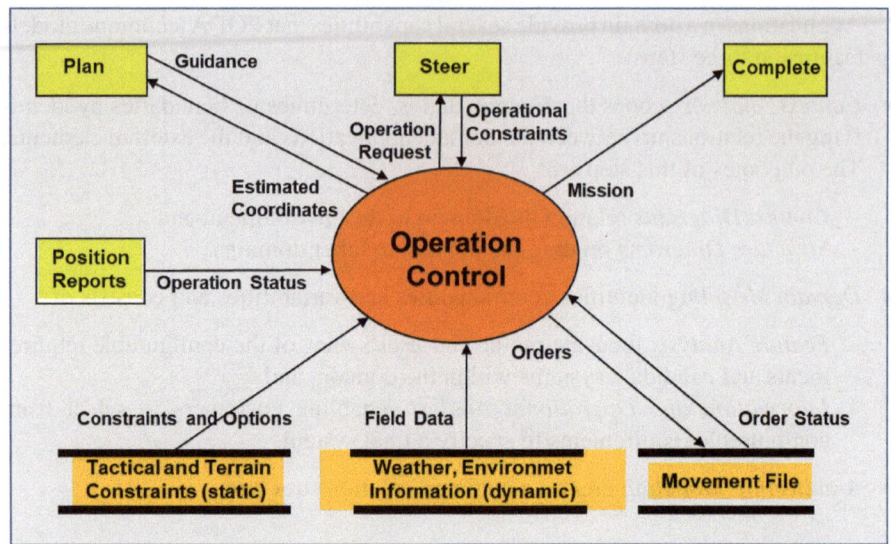

Fig. 3.6 Example for a context diagram (based on [22])

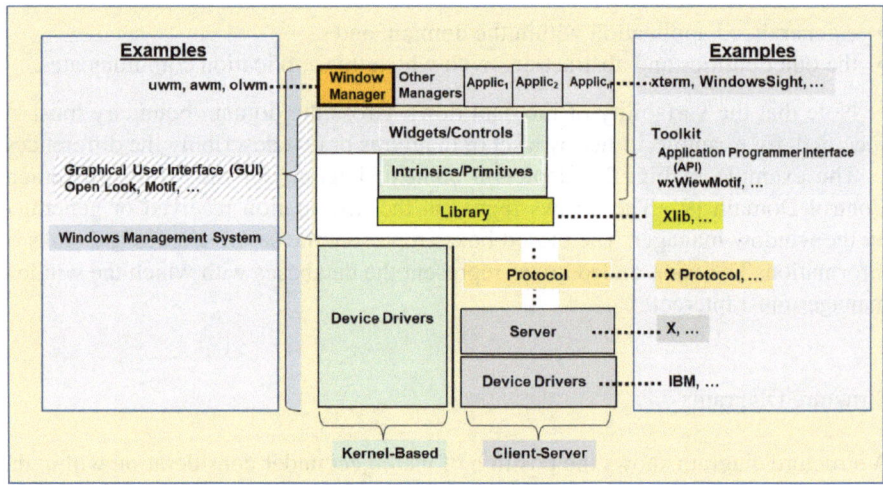

Fig. 3.7 Example for an architectural (external) structure diagram (based on [22])

elements reflect their hierarchical position, as the naming concept of configuration management suggests (Sect. 1.3.1). The variability of some elements is represented by shadowed rectangles.

3.3.1.2 Feature Diagrams and Similarity Modeling

Feature Diagrams (or *Feature Models*), synonymously called *feature models*, are tree-like models representing parent–child relationships, enriched by Boolean attributes (similar to AND-OR- Diagrams, or Fault Diagrams, [7]).

As an example, Fig. 3.9 depicts a simple, not necessarily smart, family of mobile phones that are mandatorily supposed to have the ability to call and, at the same time, display the number called. The number can be typed in or selected from a phone book or from a call log. These alternatives can be available from only one of them to all at the same time. The display, on the other hand, can be either black-and-white or color, but not both.

Note that the leaves of the tree show the features a product can have, that is, each phone must *mandatorily* have the ability to call and display by *selecting one or more alternative*(s) from the variety of features represented in the feature diagram (symbolized as dark half-moon). For example, the user of the uppermost left phone can only dial the phone number using the numeric pad, while the owner of the right-most device comfortably has all three alternatives for dialing. Finally, the leftmost phone and the phone in the middle have a black-and-white display, while the right-most, comfortable one has a color display, whereby the choice of one alternative excludes the other (symbolized as white half-moon).

A feature diagram can also be used to define a new product type that is similar to the existing ones, but differing in some ways by not having exactly the same features; for example, to satisfy expected needs and preferences of the market. The new phone at the bottom left side of Fig. 3.9, for example, is supposed to have a 0.3 degree of similarity with the simplest phone and a 0.9 with the most comfortable one [17, 20]. The similarity degree can be determined by subjective perceptions and opinions, but also market analyzes.

Feature diagrams are used in product line engineering for variability modeling (see Sect. 4.2.2). They are subject to configuration management (Sect. 1.3.1) and thus changes in their structure are to be systematically documented. Additionally, any change must be analyzed to assure consistency and avoid redundancy in product families derivable from those feature diagrams [10].

3.3.2 White-Box Abstraction—Architectural Usability and Usefulness of Assets

The objective of the white-box abstraction is to meet the needs of implementation-oriented programming, which is supposed to lead to an *architectural* model. Layered models are broadly accepted for reusability aspects.

Inspired by the "Open Systems Interconnection (OSI)" notion, a layered model needs to possess the following features (Fig. 3.10, [25], see ISO/IEC 7498 [18]):

- Each layer is characterized by its well-defined service(s).

- This service is offered to the next layer.
- Contrarily, each layer produces its service(s) using the service(s) of its next-lower-level layer.
- The top most layer interacts with the (end) user, which is usually human.

3.3.2.1 Architectural Variability

Analyzing object-oriented (OO) source code can be used to create an architecture. The architectural variability depends on the following features (see also Fig. 3.9):

- *Component variability*: many similar variants of one component.
- *Connector variability:* to interoperate with the accepting system (see Sect. 1.4).
- *Configuration variability* is given by the presence/absence of

    – Components that can be (see also the example given in Fig. 3.8) either:

    > *Core* (mandatory) components, for example, Sound Source, or
    > *Optional* components, for example, Purchase Reminder.

    – Component-to-component connectors (links).

In the example depicted in Fig. 3.11, a CD reader needs, in contrast to a CD Player, the service of a MP3 decoder in addition to the services of a sound source, which, in turn, takes the services of a Player.

Identification of component variants is based on its provided and required services. Thereby, dependencies among components and functional constraints can cause an additional problem. A new technique, Formal Concept Analysis, can help here and is explained in the next section.

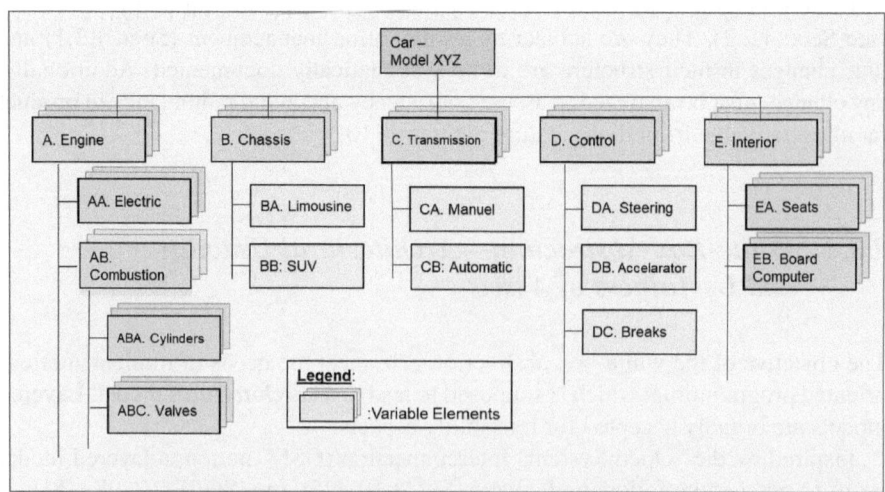

Fig. 3.8 Example for an internal structure diagram

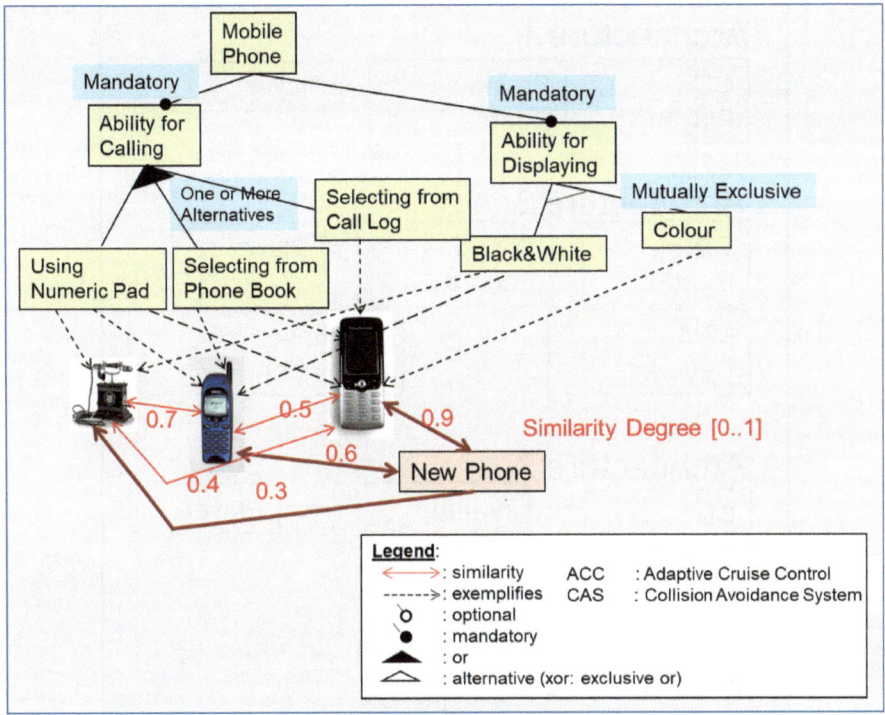

Fig. 3.9 Feature diagram for developing a family of mobile phones (based on [20])

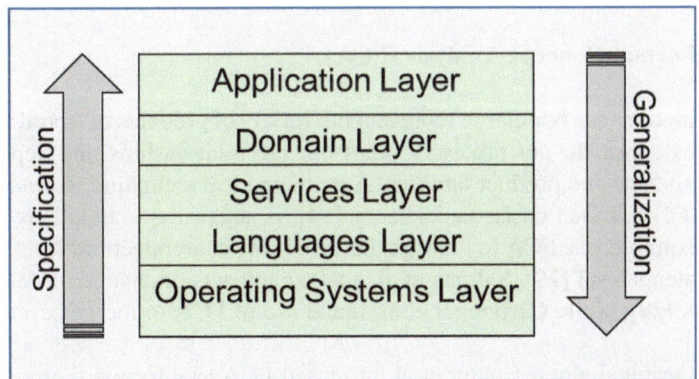

Fig. 3.10 A layered architecture of assets (based on [25]; different interpretations of the symbols of the legend are possible)

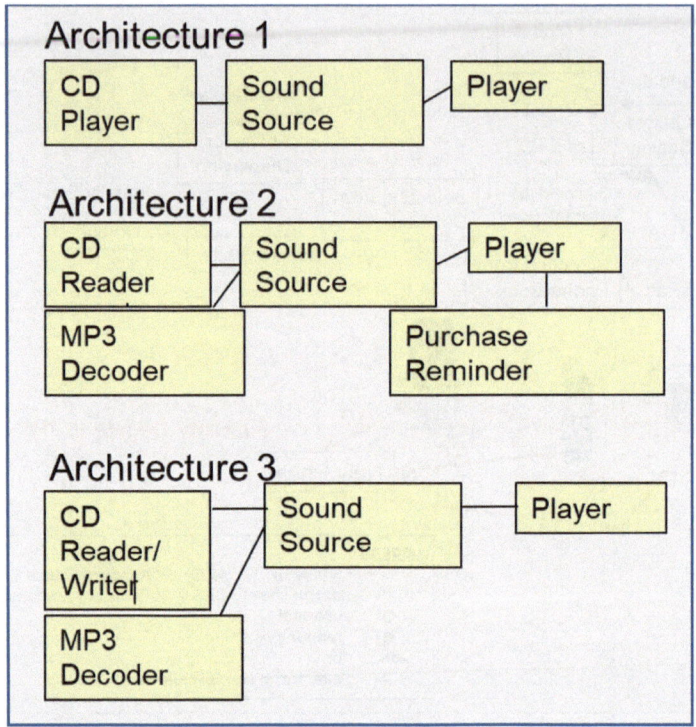

Fig. 3.11 Different types of architectures (based on [29])

3.3.2.2   Formal Concept Analysis (FCA)

Feature Diagrams are popular to represent the variety of products of a product family. Nevertheless, they do not precisely point out the relationships and dependencies between products and product families. A mathematical technique, *Formal Concept Analysis (FCA)*, based on Lattice Theory [14], is suggested to help here. Shatnavi et al., for example, use FCA to manage the variability at architectural level instead of at requirements level [29]. Salman et al. analyze impacts of changes at feature level using FCA [29], while Carbonnel et al. make use of FCA to merge several feature models [5].

In their original work, Ganter et al. proposed FCA as a formal representation to analyze data on relationships between a set O of *objects* described by a set A of *attributes* [14].

Maximal groups of objects sharing the same attributes (*Formal Concepts*) are extracted and hierarchically organized into a graph (*Concept Lattice*). So, each formal concept *FC* consists of:

- *Extent E* of the concept *FC*: Set of the objects covered by the concept;
- *Intent I* of the concept FC: Set of the attributes shared by the objects belonging to this extent.

Concepts can be linked through *sub-concept* and *super-concept* relationships, resulting in a lattice as a partially ordered structure.

A is a *sub-concept* of the *super-concept B*, if the extent of the concept *B* includes the extent of the concept *A*, and the intent of the concept *A* includes the intent of the concept *B*.

A *formal context*: The input of a formal concept is given by a *Formal Context*, $K = (O;A;R)$ with O (Objects), A (Attributes), and $R \subseteq O \times A$, with X as Cartesian (cross) product.

Thereby, R is a binary relation between O and A, indicating a set of attributes held by each object.

Summarizing the introduced notions leads to the following sets and relations:

A *formal concept FC = (E,I)* is an ordered pair with O: Objects, A: Attributes; E: Extent with $E \subseteq O$; I: Intent with $I \subseteq A$, if and only if E consists solely of objects sharing all attributes in I, and I consists of only attributes shared by all objects in E.

EXAMPLE (based on [29], https://en.wikipedia.org/wiki/Formal_concept_analysis): Analyzing the following bodies of waters as objects and their states as attributes can lead to a formal concept:

O={river,sea,reservoir,channel,lake},   A={natural,artificial,stagnant, running,inland,maritime,constant}

The Formal Context is: $K = (O,A,R)$ with O and A as given above and

R={*(river,natural), (river,running), (river,inland), (river,const), (sea,natural), (sea,stagnant), (sea,maritime), (sea,constant), (reservoir,artificial), (reservoir,stagnant), (reservoir,inland), (reservoir,constant), (channel,running), (channel,inland), (channel,constant), (lake,natural), (lake,stagnant), (lake,inland), (lake,constant)*}

Formal Context (FC) of this example can be represented by Table 3.1.

Table 3.1 Formal context of the example (based on [29])

Formal context (FC) of the example

Ext/Int	Nat	Art	Stag	Run	Inl	Mari	Cons
River	×			×	×		×
Sea	×		×			×	×
Reser		×	×		×		×
Chann				×	×		×
Lake	×		×		×		×

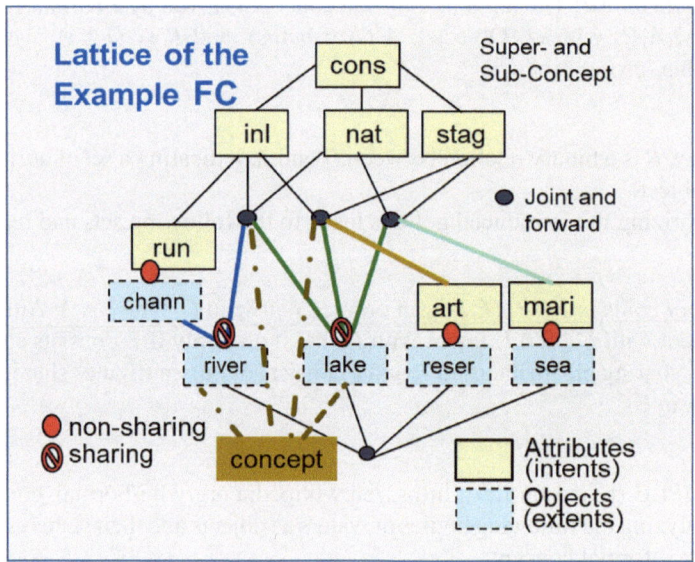

Fig. 3.12 Formal concept (FC) of the example (based on [29])

In accordance with its definition above, the (only) formal concept is then given by Fig. 3.12.

$FC=(\{river,lake\}, \{natural,inland,constant\})$

The next section explains the use of the formal concept notion.

3.3.2.3 Reverse Engineering to Identify Architectural Variability

It is obvious that the configuration variability is related to the set of *mandatory* (core) components and *optional* components and links. This fact results in the following clauses enabling the use of the formal concept analysis to design architectural configurations (see also Fig. 3.13):

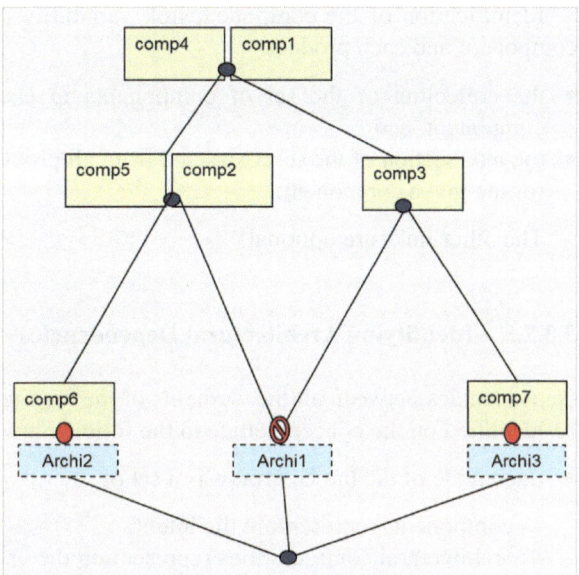

Fig. 3.13 Architectural variability represented as the lattice of a formal concept (FC) (based on [29])

- Each *object* of the formal concept lattice represents an *architecture*.
- Each *attribute* of the lattice represents a *member* component.
- *Common* attributes are grouped into the root (comp4, comp1).
- *Non-root* concepts represent the *variable* ones (*comp5*, *comp2* are present in Arch2 and Arch1, but not in Arch3).

The next section further expands and explains the variety concept.

3.3.2.4 Identifying Component-Link Variability

A connection between two components represents an *abstraction* of a group of *method invocation*, *access attribute*, or *inheritance* links between *classes* composing these components. Thereby, a component may have links with a set of components in:

- one product, and/or
- another product.

Thus, *link variability* relates to *component variability* in that:

- A link *A-B* refers to selecting the components *A* and *B* in the architecture.
- Considering a *core link* is based on the occurrence of the linked components, but *not* on the occurrence in the architecture of products.
- Therefore, a *core link* requires a link occur in the architecture configuration and is selected for composing the product.

Identification of the component link variability requires for each architectural component and each product:

- the collecting of the set of components in each product connected to this component, and
- the intersection of the sets extracted from all products, determining all core links for the given component.

The other links are optional.

3.3.2.5   Identifying Architectural Dependencies

Dependencies between all the elements of the architecture, such as constraints, can be identified on the concept lattice in the following ways:

- Each node of the lattice groups is a set of

  - components representing the intent,
  - architectural configurations representing the extent.

- Configurations are paths starting from their concepts to the lattice concept root, based on super-/sub-concept relationship. A path contains an ordered list (tuple) of nodes based on their hierarchical distribution.

Finally, the dependencies between each pair of nodes can be identified by traversing the lattice:

- "required": *comp6* requires *comp5* and *comp2*; root node(s) is/are to be selected.
- "exclusive or": *comp6* and *comp7* excludes each other, that is, no object exists containing both.
- "and": bi-directional version of the "required" constrained; for example, *comp5* and *comp2* are in "and" dependency.
- "(inclusive) or": either one is accepted; for example, "exclusive or" for *comp6* and *comp7* is ignored.

3.4   Application Engineering

While domain engineering attempts to support production of attractive, versatile reusable components, application engineering's objective is to select best-fit components for the receiving system. All of these efforts aim at improving the time-to-market, reducing costs, and, not trivially, keeping the quality high.

Even informal, ad-hoc reuse of code, developed anywhere else, was a modest start that usually costed almost no extra effort. The follow-on research activities led to the state-of-the art we have now, based on domain-specific organization and application-oriented, reuse-driven development.

Section 5.3.1 discusses the efforts of reuse maturity research and development activities in the course of investments, experience, and time, along with the aimed/yielded improvement of time-to-market, and quality and cost of these efforts.

3.4.1 Reminder: External Packaging

Recall that external packaging was supposed to support the *selection* for reusable components and their *analysis* for reuse in the specific case under consideration. Selection requires information about a reusable asset to satisfy the developer's needs (which is called *indexing*).

Indexing information can be:

- *textual*, that is, a list of terms out of a controlled, unambiguous vocabulary, or
- a *formal/semi-formal* specification of the functionality of the asset.

The style of indexing can be:

- *usage-based*, describing potential uses of the asset the way the developer would expect, and
- *computation-based*, describing the intrinsic, logic features of the asset.

EXAMPLE: The sorting of a list can be indexed in the following alternative ways:

- Usage-based, textual: the asset can be programmed to *rank* employees *subject to* their salaries, students to their grades, sale-people to their sales, etc.
- Computation-based, formal: the asset can be programmed to *get* an input set S and for all $s \in S$, and a user-supplied function $f()$ *reorganizes* it the way that

$$f(s) \geq f(successor(s)).$$

3.4.2 Acquisition of Assets

Acquiring reusable assets can be:

- *Composite*, that is, attempting to reuse and integrate all available assets to develop the new system, or
- *Generative*, that is, taking as input a set of specifications and generating and outputting the artifacts to be integrated in the new system.

However, current approaches in the practice are often a hybrid mixture of the both.

An *application generator* is, ideally speaking, a tool, or a chain of tools, that inputs a set of specifications and outputs the code of an application within an implementation language, for example, via generative programming [6]. So, those generators are *meta*-compilers, whose inputs are abstract specifications or even programs that are:

- incomplete and to be completed using a set of domain-dependent reasonable defaults,
- partially or totally non-procedural, that is, declarative, graphical, formal.

These generators should be:

- user-friendly, usable by non-professional programmers,
- supportive for fast prototyping, and
- performable to generate components with tangibly less effort.

Metaprogramming is developing with an application generator that can be constructed in following steps [8, 24]:

1. Recognizing domains
2. Defining domain boundaries
3. Defining an underlying (abstract) model as specification
4. Defining the variant and invariant parts
5. Defining the specification input method
6. Defining the products

Last but not least, while acquiring reusable components, the question arises: *Build-or-Buy*?

A popular rule of thumb recommends that an organization should build an asset in-house if and only if:

- the asset corresponds to company's core business,
- is not available on the market, and
- the company has an advantage in terms of technology, quality and/or cost, based on its own experiences.

If any one of these conditions is not fulfilled, purchasing the asset could be the better choice.

3.4.3   Analysis and Assessment of Assets for Reuse

Acquisition of reusable components, or assets, requires information about:

- How the asset under consideration performs its function or fulfills its role; for example, which algorithm is used, its complexity, and needed swapping space. This is information of *intrinsic/structural*, or *white-box* fashion.
- How to use this asset, for example, values of performance measures and interface characteristics. This is information of *extrinsic/attributive*, or *black-box* fashion.

In other words, intrinsic (internal) packaging contains information about the implementation of the usage interface for:

- customization for adapting to the problem at hand, and
- instantiation to integrate the reusable component into the system at hand.

Accordingly, extrinsic (external) packaging forms accompanying documentation to support:

- selection of the reusable assets, and
- analysis of these selected assets—how to use/reuse and customize them.

Usage information must include the following (not limited to those):

- *contextual bindings*, describing the necessary settings, for example, connecting the asset to a database driver;
- *adaptation procedure*, describing the operations needed for custom-tailoring the asset (they can be vague, and thus difficult to settle up, or very easy, for example, setting some parameters);
- *instantiation/assembly*, describing the operation of the asset once it is integrated within its host environment and assembled with other components, for example, for creating its instances and how to connect them;
- *platform/infrastructure*, describing the environmental requirements, for example, a software bus, a service;
- example(s) for usage.

3.4.4 Domain Versus Application Requirements

A common problem that frequently arises is DA having to distinguish between the requirements common to all applications and those that are specific to a particular application because this can help trace the requirements of software assets used in the development across applications in the domain. Moreover, this trace could help to find out whether or not the user's needs have been satisfied.

EXAMPLE: (Boeing, [4]) A noise generator used to simulate the noise of an engine in a flight simulator. The DA requires:

- noise generation in every system, and
- that the generated noise has to depend on engine conditions and atmospheric conditions.

Now, *generally*, it may be said that the generated noise has a pitch, intensity, and a duration.

And: Instantiating this requirement in specific applications would say something about these values. But what values? Who knows them? Who cares that they will be measured and recorded? These questions needing clarification by DE.

The next two sections consider the common aspects of DE and AE.

3.4.5 Requirements Engineering in the Context of DE

In the context of DE, several problems arise to be analyzed and solved using techniques of requirements engineering:

- Challenge 1: Manage various application requirements as a *function* of common domain requirements and specific application requirements.
- Challenge 2: Manage requirements in the *commonality/variability* context and the relationship between domain requirements and specific application requirements.
- Challenge 3: Manage requirements *traceability* along the domain and application engineering lifecycle.

3.4.6 An Example of Requirement Traceability in the Context of DE

Domain analysis of a variety of products delivers the domain requirements, which are necessary to specify and develop assets. Analyzing the legacy system, if existing, for example, by requirements abstraction, can also lead to such domain requirements (Fig. 3.14).

Once domain requirements are available, the asset development can start, considering mechanisms concerning assets and customization due to requirements of

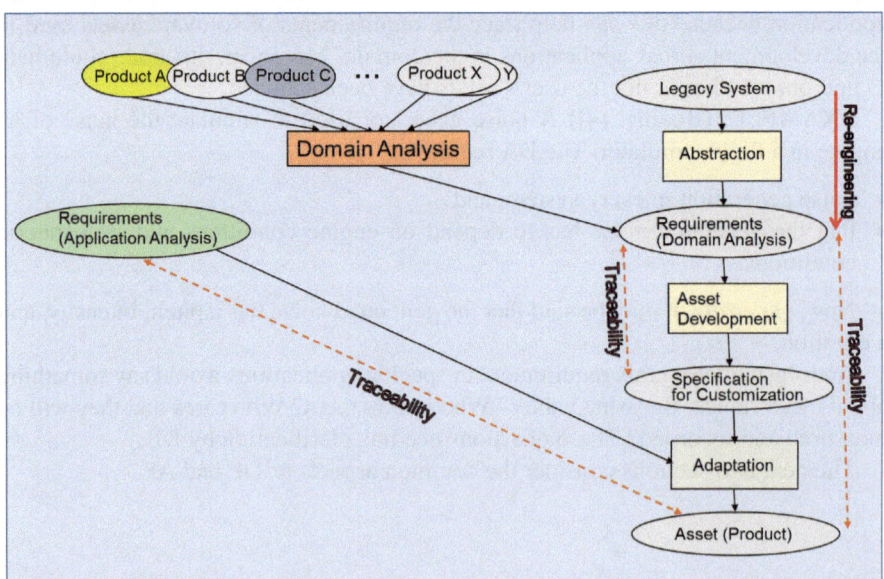

Fig. 3.14 Traceability of requirements (based on [25])

the specific applications. The traceability of these requirements is a necessary instrumentalism for the validation of the developed asset.

3.5   Reusability Organization—Managerial Aspects

Before concluding this Chapter, this section discusses briefly the managerial aspects of reuse, namely, its styles and organizations forms.

It is evident that both business managers and software managers closely cooperate to make a reuse initiative successful. They clearly need to have maximum commitment to, and involvement in, the reuse initiative. They also must decide on the allocation of reuse-related responsibilities to "produce reuse" [11, 16, 27].

In this context, a "reuse producer" is supposed to coordinate the efforts for realizing reuse and, in the end, contributes the most to produce reusable components. Moreover, this reuse producer is also the promoter of the application of those assets at different organizational levels. The "reuse consumer," on the other hand, are application groups who design and develop products with reusable components.

Fafchamps [12] identified at Hewlett-Packard different models of these producer/consumer relationships that will be discussed in the following subsections (see also Sect. B.8.1 of Appendix B).

3.5.1   Lone *and* Nested *Reuse Producer*

The simplest form of the reusability organization relies on a *Lone Reuse Producer*, who coordinates the reuse activities among the project teams. This producer directly reports to the next-level manager, thus being on the same managerial level as the heads of the project teams involved (Fig. 3.15, left, [12, 19, 25, 28]).

A *Nested Reuse Producer* does not directly operate with the project teams but with the sub-reuse producer assigned to these teams. Accordingly, this organization is more expensive, having greater potential to lead to a more successful reusability (Fig. 3.15, right).

3.5.2   *Reuse Producer* Pool

More investment is necessary if each project possess its own reuse team instead of a single reuse producer. This team reports to the reuse manager assigned to the project. Note that the project manager and reuse manager are on the same managerial level, directly reporting to the company's next managerial level.

The reuse teams are gathered in a *Reuse Producer Pool* (Fig. 3.16). This enables to assign multiple projects to a reduce team staff in order to reduce the costs.

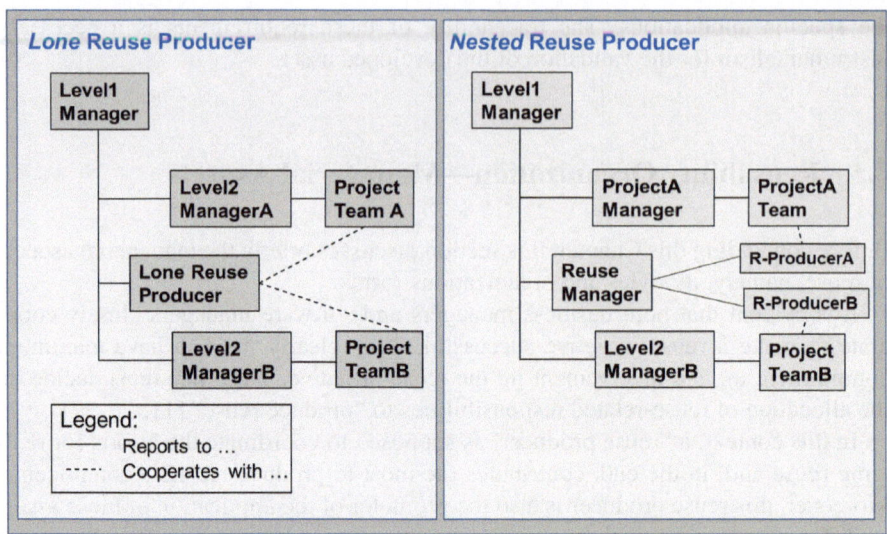

Fig. 3.15 Different types of reuse producers (based on [25])

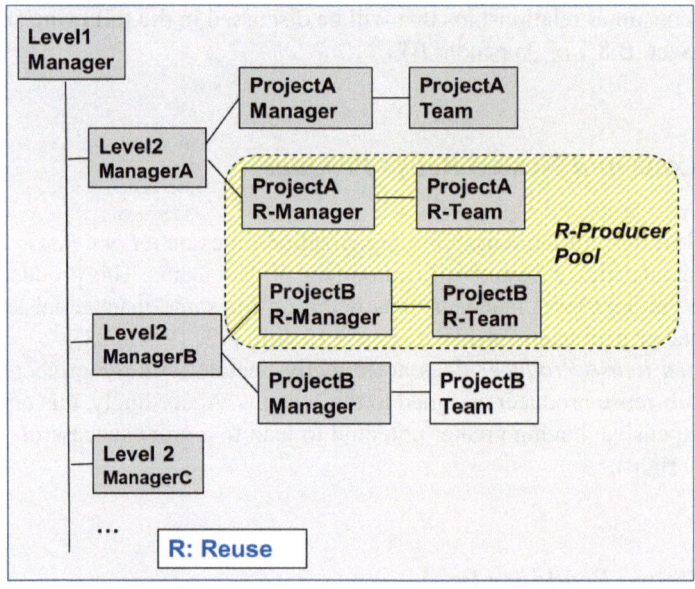

Fig. 3.16 Reuse pool (based on [25])

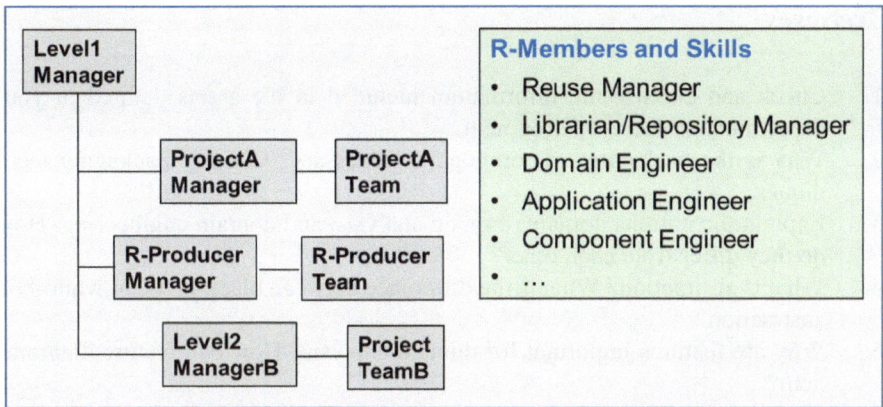

Fig. 3.17 Reuse producer team and its members (based on [25])

3.5.3 Reuse Producer Team

Instead of a pool, reduce producers, including their head, can be gathered in a department. Reuse producers can then be following the need concerned on a case-by-case basis (Fig. 3.17, [12, 28]).

Key Points, Exercises, Recommended Further Reading, References

Key Points

- Architecture of a system to be built by reuse is of crucial importance.
- Characteristics of a component to be reused is also of crucial importance.
- Domain Engineering (DE) covers development aspects of reusable components, while Application Engineering (AE) handles their deployment.
- A proper internal and external packaging of components identifying DE and AE is necessary to enable ease of use and ease of selection, respectively.
- Abstraction is a key method of domain analysis (DA) to develop reusable components.
- Feature-oriented DA (FODA), as a black-box abstraction, concerns the external, user-oriented characteristics of reusable components; white-box abstraction has architectural aspects.
- Reusability needs an appropriate organization that can be realized in different forms, in accordance with the structure and size of the company.

Exercises

1. Check and classify the information included in the assets defined in your department or social environment.
2. Why is the packaging of components necessary? Classify packaging techniques.
3. Explain the notions domain, domain analysis, and domain engineering. How do they differ from each other?
4. What is abstraction? What is the difference between black-box and white-box abstraction?
5. Why are features important for domain analysis? How can feature diagrams help?
6. How many different phones can be developed following the feature diagram of the example in Fig. 3.9?
7. Explain formal concept analysis. Why is it a white-box technique?
8. Is formal concept analysis a black-box technique?
9. Apply formal concept analysis to a simple product family, for example, mobile phones.
10. What is the objective of application engineering? What are its subjects?
11. Why is requirements traceability useful for application engineering?
12. Name the different forms of a reusability organization.
13. Which reusability organization structure would be appropriate for a small-size software house?

Recommended Further Reading

1. Jatain, A., Goel, S.: Comparison of domain analysis methods in software reuse. Int. J. Inf. Technol. Knowl. Manage. **2**(2), 347–352 (2009)
2. Kang, K., Cohen, S., Hess, J., Nowak, W., Peterson, S.: Feature-Oriented Domain Analysis (FODA) Feasibility Study (Report). Software Engineering Institute, Carnegie Mellon University (1990). Available at http://www.floppybunny.org/robin/web/virtualclassroom/chap12/s4/articles/foda_1990.pdf

References

3. Apel, S., Batory, D., Kästner, Ch., Saake, G.: Feature-Oriented Software Product Lines. Springer, Berlin (2013)
4. Bass, L., Campbell, G., Clemens, O., Northrop, L., Smith, D.: Third Product Line Practice Workshop. Technical Report CMU/SEI-99-TR-003, ESC-TR-99-03 (1999)
5. Carbonnel, J., Huchard, M., Miralles, A., Nebut, C.: Feature model composition assisted by formal concept analysis. In: Proceedings of ENASE: Evaluation of Novel Approaches to Software Engineering, pp. 27–37 (2017)

6. Czarnecki, K., Eisenecker, U.: Generative Programming: Methods, Tools, and Applications. Addison-Wesley, Boston (2000)
7. Czarnecki, K., Wasowski, A.: Feature diagrams and logics: there and back again. In: 11th International Software Product Line Conference (SPLC 2007), pp. 23–34 (2007)
8. Cleaveland, C.T.: Building application generators. IEEE Softw. 25–33 (1988)
9. Cohen, S.G., Stanley, Jr., J.L., Peterson, A.S., Krut, Jr., R.W., Application of Feature-Oriented Domain Analysis to the Army Movement Control Domain, Technical Report CMU/SEI-91-TR-028, ESD-91-TR-028, 1992
10. Dintzner, N., Kulesza, U., Deursen, A.V., Pinzger, M.: Evaluating feature change impact on multi-product line configurations using partial information. In: International Conference on Software Reuse, pp. 1–16 (2015)
11. Ezran, M., Morisio, M., Tully, C.: Practical software reuse. In: Chapter "Two Major Case Histories". Springer Practitioner Series, pp. 155–169 (2002)
12. Fafchamps, D.: Organizational factors and reuse. IEEE Softw. **11**(5), 31–41 (1994). https://doi.org/10.1109/52.311049
13. Gamma, E., Helm, R., Johnson, R.E., Vlissides, J.: Design Patterns: Elements of Reusable Object-Oriented Software. Addison-Wesley, Reading, MA (1995)
14. Ganter, B., Stumme, G., Wille, R.: Formal Concept Analysis: Foundations and Applications. Springer, Berlin (2005)
15. Hallsteinsen, S., Paci, M.: Experiences in Software Evolution and Reuse—Twelve Real World Projects. Springer Science & Business Media, Berlin (1997)
16. Hooper, J.W., Chester, R.O.: Software Reuse Guidelines and Methods. Springer, Berlin (1991)
17. Henard, C., Papadakis, M., Perrouiny, G., Klein, J., Le Traon, Y., Assessing Software Product Line Testing via Model-based Mutation: An Application to Similarity Testing, IEEE Sixth International Conference on Software Testing, Verification and Validation Workshops, Luxembourg, pp. 188-197, 2013, doi: 10.1109/ICSTW.2013.30
18. ISO/IEC 7498: Information technology. Open Systems Interconnection. Basic Reference Model: The Basic Model (1994)
19. Jacobson, I, Griss, M., Jonsson, P.: Software Reuse—Architecture, Process and Organization for Business Success. ACM Press, Addison Wesley Longman, Boston (1997)
20. Kaindl, H., Mannion, M.: A feature-similarity model for product line engineering. In: International Conference on Software Reuse, pp. 34–41 (2015). https://doi.org/10.1007/978-3-319-14130-5_3
21. Karlsson, E.A., Brantestam, J.: Generic reuse development processes. In: E.A. Karlsson (ed.) Software Reuse: A Holistic Approach. Wiley, Hoboken, pp. 253–270 (1995)
22. Kang, K., Cohen, S., Hess, J., Nowak, W., Peterson, S.: Feature-Oriented Domain Analysis (FODA) Feasibility Study (Report). Software Engineering Institute, Carnegie Mellon University (1990)
23. Krueger, C.W.: Software reuse. ACM Comp. Surv. **24**(2), 131–183 (1992)
24. Levy, L.S.: A metaprogramming method and its economic justification. IEEE Trans. Softw. Eng. **SE-12**(2), 272–277 (1986)
25. Mili, H., Mili, A., Yacoub, S., Addy, E.: Reuse-Based Software Engineering, Techniques, Organization, and Controls. Wiley, Hoboken (2002)
26. Noback, M.: Principles of Package Design: Creating Reusable Software Components Technology. Springer, Berlin (2018)
27. Reifer, D.J.: Practical Software Reuse—Strategies for Introducing Reuse Concepts in Your Organization. Wiley, Hoboken (1997)
28. Sametinger, J.: Software Engineering with Reusable Components. Springer, Berlin (1997)
29. Shatnawi, A., Seriai, A., Sahraoui, H.: Recovering architectural variability of a family of product variants. In: Proceedings ICSR, LNCS 8919. Springer, Berlin, pp. 17–33 (2015)
30. Salman, H.E., Seriai, A.-D., Dony, Ch.: Feature-level change impact analysis using formal concept analysis. Int. J. Softw. Eng. Knowl. Eng. **25**(1), 69–92 (2015)
31. Schaefer, I., Stamelos, I. (eds.): Software reuse for dynamic systems in the cloud and beyond. In: 14th International Conference on Software Reuse (ICSR), Lecture Notes in Computer Science, vol. 8919. Springer, Berlin (2015)

Chapter 4
Software Reuse Technologies

Different technical, managerial, and organizational methods are available for developing reusable software components. The best-known ones are Component-Based Software Engineering (CBSE), Commercial-Off-the-Shelf-Software (COTS)–Based Development, and Product-Line Engineering (PLE). However, these methods are not orthogonal, that is, they have a lot of issues in common. The common terminology and techniques they share have already been introduced in the previous chapters. Therefore, this chapter will briefly review these methods, while pointing out their differences.

Needless to say, the methods and models presented here can also be applied to developing families of products other than software.

4.1 Component-Based Software Engineering (CBSE)

Component-Based Software Engineering (CBSE) is one of the most popular and widely accepted reuse technologies. It is even viewed as a special area of software engineering for the developing *of*, and *with*, reusable assets. These assets have a special packaging, forming them to components, *ideally* as a component that can be easily identified, put to market by the producer, retrieved by potential customers, and (re)used to compose new products. Thus, the *ideal* component (not yet existing in large amounts) is self-contained and requires no customization, in other words, it is "plug and play," providing well-defined services to the applications in which it is integrate [19]; see also Sect. 1.4.3 for a precise definition of *component*.

© The Author(s), under exclusive license to Springer Nature Switzerland AG 2021   113
F. Belli and F. Quella, *A Holistic View of Software and Hardware Reuse*,
Studies in Systems, Decision and Control 315,
https://doi.org/10.1007/978-3-030-72261-6_4

4.1.1 Techniques of CBSE

To recap the features mentioned above, reusable software components are self-contained and clearly identifiable pieces of software that describe and/or perform specific functions, have clear interfaces, appropriate documentation, and a defined reuse status.

A component's status indicates its realization, for example, implemented (binary code) or specification (formal or informal). Thus, not only executable units are components, but also specs (specifications) or test suites can form reusable components.

Implementation environments (programming platforms) are not the subject of a component and need not be of a specific paradigm, for example, object-oriented (OO). Nevertheless, platforms supporting OO programming (OOP) are popular and have many advantages (see Sect. 1.5.5).

Using the notions of OOP, components can be assumed to be typically larger than classes and are implemented as a set of classes, one of which realizes the component interface. Several languages are available for component description, for example, MIL (Module Interface Languages), IDL (Interface Definition Languages) and ADL (Architecture Definition Languages).

Techniques preferred in CBSE are:

- *abstraction*, for example, metaprogramming, design patterns [24, 52] and
- *composition*, for example, event-based composition, simulated reflection [41, 42].

CBSE is centered around integration with the receiving system through:

- standardized interfaces, and
- middleware (computational infrastructure) as a separate software layer.

to mediate the interaction; for example, CORBS, Java RMI, CJB. Thus, interoperability with legacy and heritage systems poses challenges since those might have been implemented before such standards had been known.

Generally speaking, Component-based (CB) development is lucrative if the domain possesses a dynamic component market with broad offerings and competitive prices. This requires that accepted and standardized models be available to abstract the components, that is, to easily explain and to model their structure and function.

4.1.1.1 The Structure of a Component

Figure 4.1 sketches constituents of a component to be integrated into the receiving system, including its interfaces to interoperate with other components [44].

Internals of a component are its *implementation*, which is not visible to outside parties, contrary to its visible *specification* (see Fig. 4.2, [32]). Its *interfaces*, on the other hand, enable its reuse. They are of the *horizontal* type to communicate, and thus interoperate, with other components. The *vertical* interface of a component enables

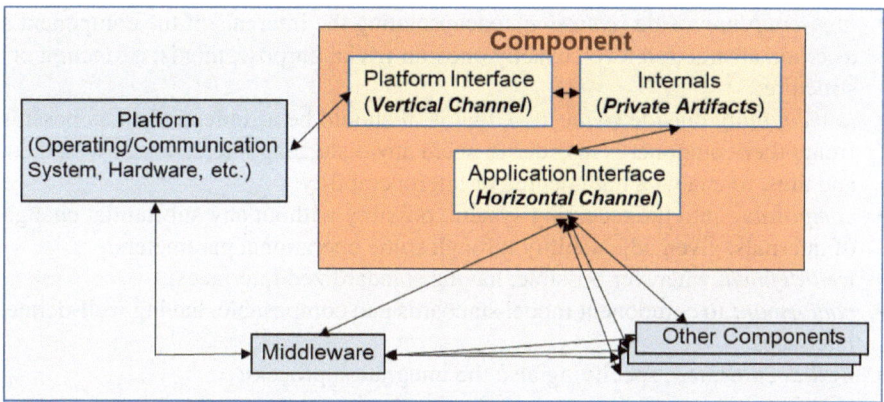

Fig. 4.1   A component and its interfaces (based on [44])

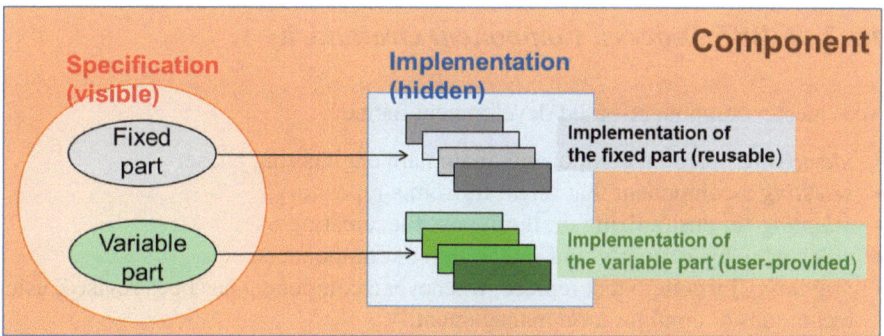

Fig. 4.2   Constituents of a component: Variable and fixed parts (based on [32])

its technical conformance with the reuse environment, such as operating system, system hardware.

The interfaces of a reusable component enable the user, who is usually a programmer, to implement the variable part.

4.1.1.2   Desirable, Non-functional Features of Components

A component's functional features relate to its usage. On the other hand, non-functional features, aside from general non-functional features, such as safety or security, are concerned with *how well* this component fulfills its expected functionality (see Sect. 1.2). The following rather reuse-specific features are greatly desirable for a component to possess. Thus, it is expected to be:

- *high-coupling* inside (*cohesive*), encapsulating the internals of the component at a certain abstraction level, exactly one non-trivial purpose, that is, a function or a structure;
- *low-coupling* outside (*adhesive*), that is, it should be as independent as possible from other components in order to avoid any disturbing interferences with them, and thus, to enable a high degree of interoperability:
- *composable* into the receiving system, possibly without any substantial changes of internals, given adjustability through some operational parameters;
- *well-defined*, wherever possible, having standardized interfaces;
- *conformant* to component model standards and composable, having well-defined interfaces;
- *well-documented*, specifying also the integration process;
- *certified* by a supplier or a trustable third party or a project-independent department of the company.

4.1.2 CBSE Process, Component Granularity

Activities in component-based development include:

- *identifying* potential components in domain engineering phase;
- *selecting* a component that might fit from a repository;
- *adapting* for customizing or, if not possible, creating one;
- *composing* by assembly-and-integration of components;
- *upgrading* by component replace, whenever a component has been revised, using techniques of configuration management.

Components break down systems and development processes, leading to units that are of different *granularity*, that is, extent of functionality, resources (costs, runtime, memory) or distribution.

A component's granularity dimensions can be defined for different levels and environments, for example at:

- *business level* as units of dispute, for tracing faults and corresponding responsibilities;
- *management level* for controlling resources;
- *abstraction level* for encapsulating details that define functionality, resources, structure, or state;
- *compilation* as individually units that are composable;
- *distribution level that* defines remoteness of the units.

4.1.2.1   Commercial-Off-The-Shelf-Software (COTS)–Based Development

As a CBSE-related reuse technique, COTS–Based Development is the process of building software applications with commercially available software components.

A COTS software is (see also Sect. 1.4.3 for a definition of COTS, and Sects. 2.2.2 and 2.2.3):

- sold, leased, or licensed to the general public;
- offered by a vendor (or a group of vendors) who has created it and is usually responsible for its maintenance and upgrades;
- delivered without its source code so that the buyers, lessees, and licensed can use it as a black box;
- available in multiple identical copies within the same version on the market.

The expected advantages of using COTS products is that it can bring gain in:

- cost, operational quality, functionality/specificity, and
- reduced time-to-market, maintenance overhead.

4.1.2.2   COTS and CBSE

Both COTS and CBSE are compositional approaches to software development. Thus, COTS development is a special case of CBSE. However, COTS development necessitates development by third parties, leading to the following differences:

- white-box (CBSE) vs. black-box (COTS) reuse,
- copyright privileges of CBSE,
- maintenance advantage of CBSE (user does not depend on vendor), and
- retrieval criteria that might influence the frequency and scope of the usability and reusability.

4.1.2.3   Lifecycles for COTS Development

While life cycles of CBSE are similar to the ones known from ordinary software development (Sect. 1.3.2),COTS development has significantly different ones. The most significant differences are summarized below:

- Design: bottom-up (CBSE) versus top-down (COTS)
- Selection of COTS components: subject to quality, fitness, and interaction
- Integration of COTS components: matching of interface(s), functionality, inter-component communication.

4.2   Product-Line Engineering (PLE)

Product lines (PL), also called *product family* or *business unit*, form a suite of end-user products

- of similar use and similar features, also called „look-and-feel" of a family,
- but fulfilling different requirements, and
- having different pricing.

Typical examples of PLs are depicted in Fig. 4.3. The systems represented are different, but obviously belong to the same family. The products (first, second, and third rows) are prepared in accordance with customers' individual expectations and, last but not least, at varying costs.

In software engineering, a *software product line* (*SPL*) is a set of software-intensive systems that share a common, managed set of features, which satisfy the specific needs of a particular market segment or mission, and that are developed from a common set of core assets, that is, platforms, in a prescribed way [46]; see [37, 49] for further examples. An SPL approach enables the generating of the product-line members' code from a common product-line infrastructure instead of having to develop them independently from scratch.

Fig. 4.3   Some product and system families (lines)

Product-Line Engineering (PLE) attempts to develop similar products in the same domain, explicitly modeling commonalities and differences within the products of a line.

The basic technical means of PLE can be summarized as follows:

- *Domain Analysis* to identify the commonalities and variations among the PL members
- *Software Architecture* as the skeletal (infra)structure adopted to all PL members
- *Development Process*, consisting of Domain Engineering and Application Engineering

PLE combines the advantages of different techniques developed in the course of software reuse research and practice.

4.2.1 Building Blocks, Platform, Product Family

PLE identifies blocks of products that are different but belong to the same PL. Usually, some features are common to a subset of the products of the PL under consideration. These commonalities define a *feature platform*. A *product family* is the set of products that are defined based on the same platform. Figure 4.4 depicts these relations. In this example, features are represented as blocks (*feature blocks*) that the members of the family can (shaded) or cannot (white) share.

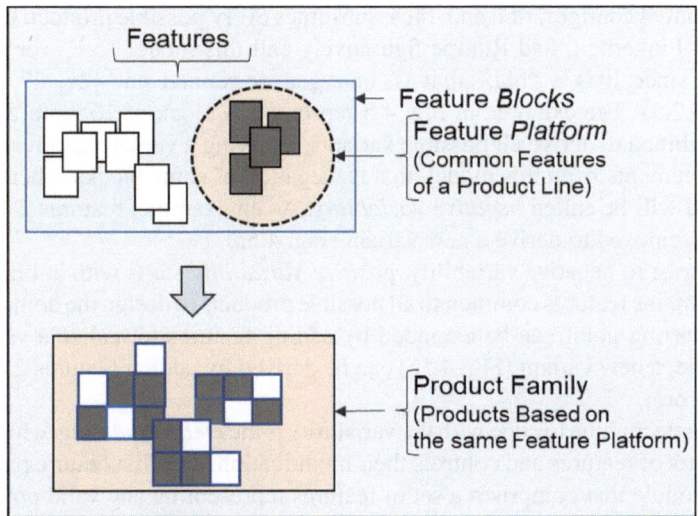

Fig. 4.4 Building a product family (based on [53])

EXAMPLE: The product platform of the hamburger family depicted in Fig. 4.3 are the upper and bottom halves of the rolls that are included in each and every of the family member.

4.2.2   Variability Modeling

As Seidl et al. mentions [51], there are various notations for variability models; for example, feature models [20, 31], decision models [45], orthogonal variability models (*OVM*s) [49], and variability specifications (*V-Spec*s) of the Common Variability Language (*CVL*) [30, 28] as used in generative programming, and variability management [54].

Eichelberger et al. reviews some models concerning their expressiveness vs. analyzability [22]. A collection of open-source plug-ins for the Eclipse is available in *FeatureIDE* that supports several aspects for feature-oriented software development that are comparable to industrial tools, for example, feature modeling, type checking, and testing [43].

One of the most favored modeling techniques uses feature models, also called feature diagrams, that represent all features available in a PL. A *variant* is an instance of this model, possessing some of these features.

Variability of the members of a PL can be presented in the following ways:

- One method is the presentation of all features that can be encountered in any product. This leads to a model that contains every element that is used in at least one product configuration and, thus, subsumes every possible product. Grönniger, Krahn, Pinkernell, and Rumpe figuratively call this model *150%* (or *complete*) *model* since it is a "big," that is, enlarged or refined one [26, 27, 52]; (see Sect. 4.2.5). The example in Fig. 4.5 represents a block of 16 features that can be combined to derive all possible variants. Deriving a variant requires removing some elements from this model, that is, negation of some blocks. Therefore, this method will be called *negative variability*. As an example, features 2, 8, and 16 can be removed to derive a new variant (Fig. 4.5a).
- In contrast to negative variability, *positive Variability* starts with a minimal core that contains features common to all possible products to design the domain model. This starting point can be extended by adding features to realize a variant. For example, a new variant (Fig. 4.5b) can be derived by adding features 2, 8, and 16 to the core.
- Another technique to cope with the variability is the *Delta Modeling*, which defines two parts of features and controls their modification. The first feature part forms a core module that comprises a set of features representing any valid product. The second part is a set of delta modules, which specify changes that will be applied to the core module. These changes can be either the construction (add) or destruction (remove) of features, that is, applying positive and/or negative variability. Delta modeling has also been successfully applied to product line testing [40].

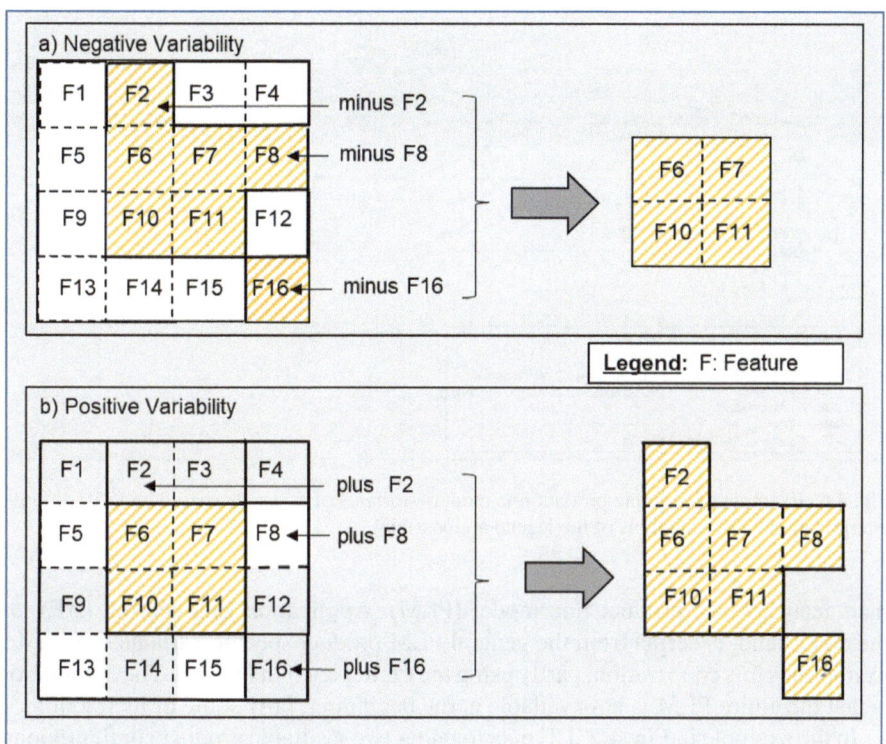

Fig. 4.5 Negative and positive variability (based on [26, 34])

4.2.3 Product Line Model Versus Variant Model

Feature diagrams stress modeling all features of a PL, as domain engineering has to do (see Sect. 3.3.1.2). In other words, all and every variant of the product line must be deductible from this complete model. Application engineering, on the other hand, has to model each and every individual product, that is, any variant should be visible based on its own features that are deducted (extracted) from the feature diagram.

Figure 4.6 depicts a PL model of a carmaker's family of models (see also Fig. 3.9 that was focused on the structure of this PL). Engine, chassis, transmission, and control units are mandatory. Optional accessories (*extras*), such as assistant systems, are available at additional cost. A symbol for the relation "requires" indicates dependencies of one feature on one or more others. For example, the feature "Comfort" necessitates the inclusion of the features "ACC (Adaptive Cruise Control)," "Collision Avoiding" System (CAS)," and "Leather Seats." Furthermore, "Comfort" excludes "Sport" chassis, which, on the other hand, excludes "ACC."

Now, different types of vehicles can be deduced from this PL model.

Construction of PL models belongs to activities typical of Domain Engineering (DE), which collects models of product variants and anaylzes them in order to unify

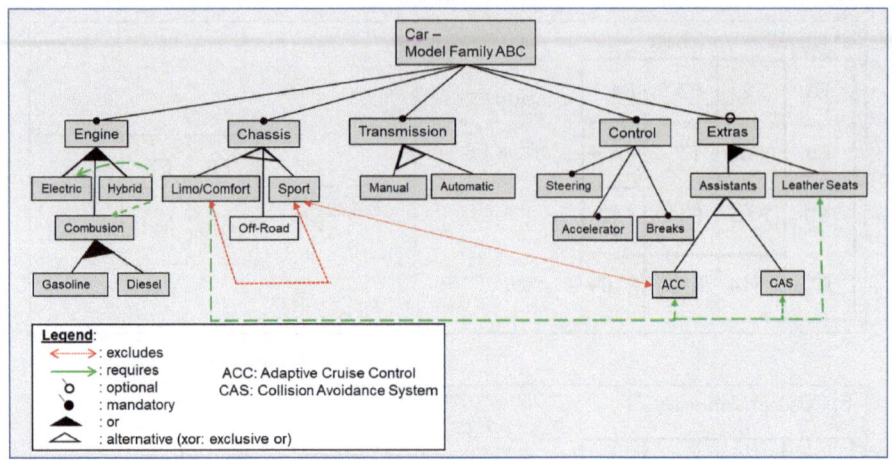

Fig. 4.6 Example of a simple product line model—variants of a vehicle (based on [27]; different interpretations of the symbols of the legend are possible)

their features in a product line model (PLM). Application Engineering (AE), on the other hand, excerpts from the general PLM product specific variants. Figure 4.7 exemplifies this cooperation, partly using the PL depicted in Fig. 4.6. The assumption is that the entire PLM is not available at the beginning, only some of its fractions.

In the example in Fig. 4.7, DE puts together two available product configurations, P1 and P2, to construct a PLM. Note that this is the first step; this starting PLM will be iteratively, step by step, extended by considering and analyzing more product models. Thus, DE works in an *inductive* manner, from specific to general. AE can then deduce and configure models of specific product variants from this PLM, extracting corresponding features.

The next subsections explain the deduction process in detail.

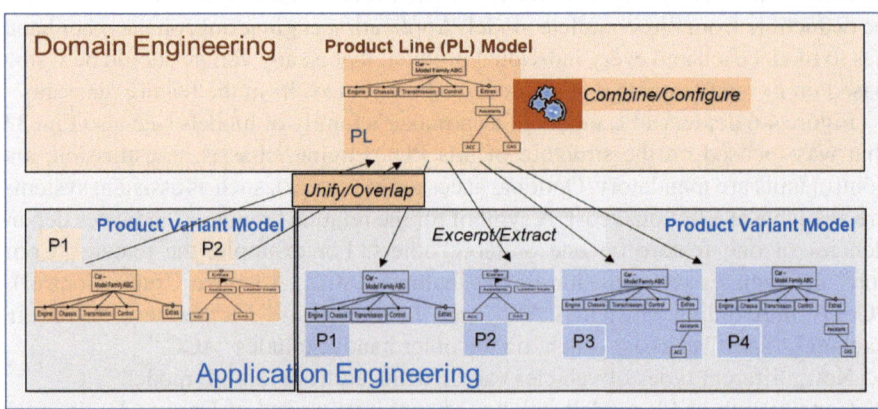

Fig. 4.7 Product line model and derivable variants

4.2.4 Deriving a Valid Model

Figure 4.8 shows the features of a simple online shop. A catalog of offered goods, a mechanism for the payment of obtained goods, and a security system form the obligatory parts of the modeled PL. A mechanism for a comfortable search of goods is optional. Payment, on the other hand, can be realized in three alternative ways, whereby using a credit card entails a high security procedure.

Figure 4.9 sketches the derivation of a payment case that can be observed in an online shop.

The starting feature model (Fig. 4.8) is supposed to specify all derivable variants. Therefore, it will be called *complete*, or 150%.

Once the 150% model has been constructed, application engineering can derive the models of variants as individual products. Figure 4.9 shows an example based on the 150% feature model (Fig. 4.8) that describes the structure of a variant having no search mechanism and that operates with a credit card. This variant realizes a payment by a credit card under high security handling, given that credit card payment requires this kind of handling.

The model depicted in Fig. 4.9 is a *100% feature model* and describes the structure of the variant under consideration, exactly and entirely, and nothing else.

Figuratively speaking, a 100% model is supposed to be a "small" and somehow narrow one compared to the "big" 150% model [27], ([16, 51, 52], and [17, 35]; see also Sect. 4.2.2 and [39]). These percentage notions are borrowed from modeling in life sciences, primarily bio-medicine.

Feature models are easy to learn and work with, as well as useful and powerful for specifying large PLs and complex relationships between features. Even this simple scheme leads to 20 variants, varying the features "payment" and "search," leading to different security options and search policies, respectively.

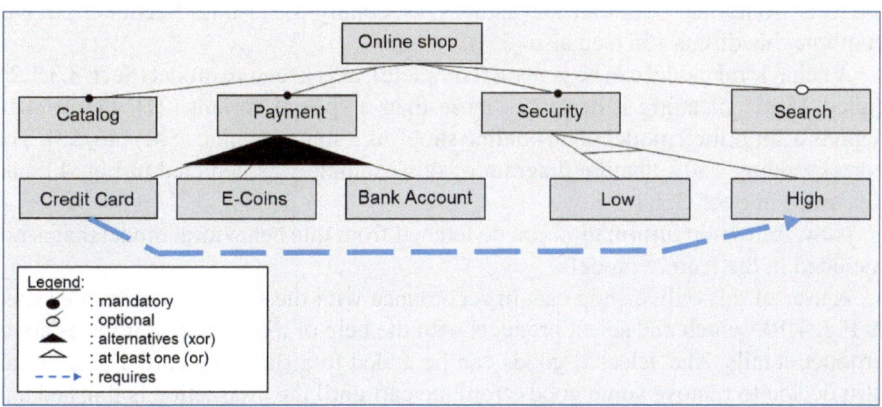

Fig. 4.8 Online shop PL (from [53], based on https://en.wikipedia.org/wiki/Feature_model; different interpretations of the symbols of the legend are possible)

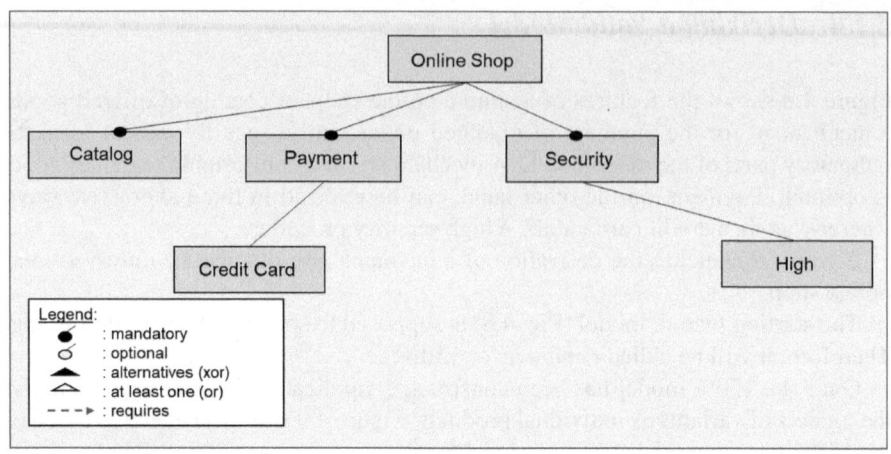

Fig. 4.9 A valid variant of the online shop PL, based on payment (based on [53])

4.2.5   Behavioral Model of a Product Family ("150% Model")

The variability, especially the variation points of a product family, can be represented simply by a feature model. This model does not, however, include any other, notwithstanding significant aspects of software engineering, such as requirements and architecture of the product family and its variants. A *behavioral* model can help here by representing these aspects, especially the usage of the SPL, taking also its environment into account, including the interactions with the potential users and the systems to which this SPL can be connected. So, the modeling of the behavior of the SPL is an important task of the domain engineering and completes the information included in the feature diagrams. Combining these two models during the design of the SPL would add both their advantages, especially for testing. Section 5.2.5 will continue this discussion (see also [21]).

A behavioral model can be general (complete), as is a feature model (Sect. 3.3.1.2), called 150%, meaning it describes more than a special variant [27]. Figure 4.10 depicts a simplified model of an "online shop" as a state machine (SM) [46, 53]. The corresponding 150% feature diagram of this example was depicted in Fig. 4.8 and discussed in Sect. 3.2.4.

Now, follow-up information can be fetched from this behavioral model that is not included in the feature model.

A user of this online shop can, in accordance with the state machine represented in Fig. 4.10, search and select products with the help of a catalog that informs about product details. The selected goods can be added to a shopping cart. The user can also decide to remove some goods from the cart until the transaction is finished and the user is ready to check out. The payment can be made by either a bank transfer, e-coins, or a credit card. The system then validates the payment by checking user data, bank data, etc. If this information is valid, the order will be processed and

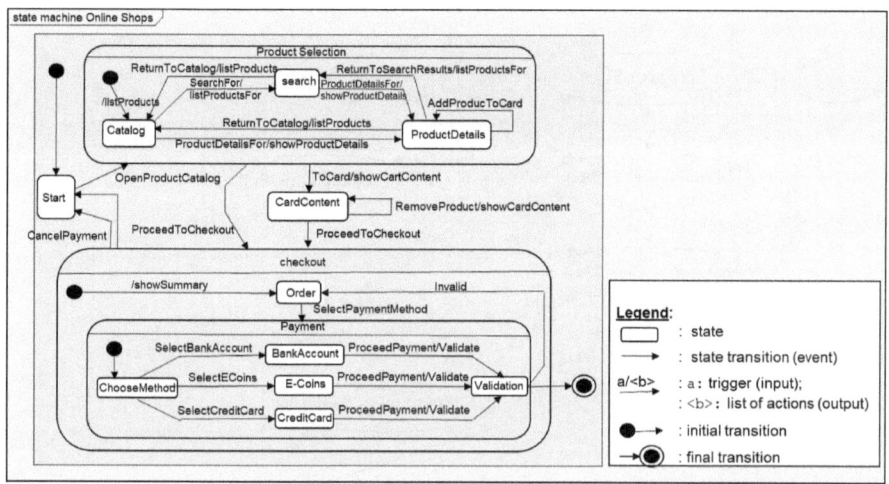

Fig. 4.10 State machine of the product family of online shop, based on payment (based on [53])

confirmed to the user. Otherwise, the user can select another method of the payment. In the end, the user will be forwarded to the initial stage.

Another option to model the behavior of a SPL is given by event sequence diagrams (ESG) that support test-driven design and validation strategy (Sect. 1.3.3.11.5, and see Appendix A for a brief introduction to ESG concept, and [12] for more details). As the name indicates, the focus of this strategy is on events because testing is an *event-centric* activity [13]. Test engineers usually observe events that are noticeable and perceptible in the sense that an unambiguous decision on a test having been "passed" or "failed" can be made (Sect. 1.3.3.5).

The nodes of an ESG correspond to the state transitions of a state machine (SM). Thus, the ESG in Fig. 4.11 represents exactly the same constellation of events observable in a simplified online shop as the SM in Fig. 4.10. Note that some trivial transitions of SM, for example, "ReturnToCatalog," are not included as double arrows in the corresponding ESG. Therefore, the number of transitions of an SM can differ from the number of events of the corresponding ESG. Different forms of SM are available in the literature; Belli and Hollmann suggested a basic format also used in this book [11].

The ESG strategy has several advantages. First, an ESG has less syntactical elements than a state machine, which makes its learning and usage simple. This, in turn, decreases the likeliness of design faults. All this helps to save costs. Actually, an ESG has only one kind of node that semantically represents the events connected by directed arcs. Second, and more important, an ESG, represented by a directed graph, enables to utilize the notions of, and results from, the "good old" graph theory, for example, for test case generation and test optimization (see Sect. A.7 of Appendix A and Sect. 5.2.3). Thus, the rich repertoire of algorithms and strategies developed over the centuries in the course of graph theory, such as to solve the

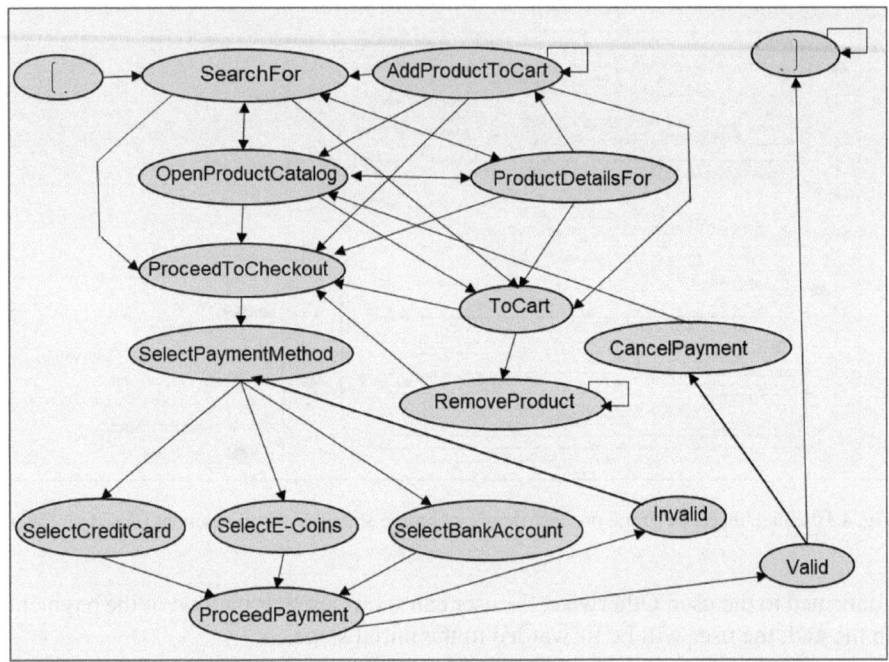

Fig. 4.11 Event sequence graph model of the product family represented in Fig. 4.9 (based on [53])

traveling salesman problem or the Chinese postman problem, are relevant and can be used for automatizing and optimizing test processes [7, 14, 23].

Figure 4.11 depicts the ESG corresponding to the state machine given in Fig. 4.10. The node noted by the bracket "[" is a pseudo-event and indicates the starting point of the process modeled by this ESG; its end is noted by the bracket "]," which is also a pseudo-event.

Note that both state machines and event sequence graphs support modeling only sequential processes. Petri nets, as suggested by Heuer, Budnik, Lauenroth, and Pohl [29] and by Gorgonio and Perkusich [25], can be used to overcome this constraint and to model also concurrent processes (see Appendix B for a brief introduction).

4.2.6 Extracting the Behavior of an Individual Product via Negative Variability

The attentive reader might have already noticed the problem with 150% behavioral model, despite being represented as a state machine (Fig. 4.10) or an ESG (Fig. 4.11). As already mentioned, all potential variants are expected to be derivable from this 150%, complete model. This is true, but neither model tells everything about how this derivation is to be carried out. This problem stems, in this example, from the fact that the event `search` in this example is optional, that is, it can be excluded. This is a variation point that cannot be included in the 150% model.

A method introduced by Lackner, Schmidt, and Weißleder combines feature diagrams with UML state machines to utilize the merits of both and to overcome their limitations [33, 34, 53]. This book favors ESG because of the reasons mentioned in Sect. 4.2.5.

The additional information gained by coupling feature diagrams and behavioral diagrams is useful, as discussed in this and the preceding section, not only for better understanding and modeling the SPL and easing its realization, but also for testing, as will be explained in Sects. 5.1 and 5.2.

Figure 4.12 illustrates the coupling of a feature diagram (Fig. 4.8) and an ESG. Here, only an excerpt out of the model (Fig. 4.11) is used in this figure to keep the example simple.

In this example, the feature "Credit Card" will be mapped to the event "Select Credit Card." Keep in mind that using a credit card entails a high security handling of the payment process. This can be realized by a configuration function of the SPL specification that assigns a value to every feature of the feature model, denoting the presence ("TRUE") or absence ("FALSE") of the mapped features. One can view this mechanism as "guards" known from programming.

Remember also that this type of modeling is called "150%," including all features. The features not included in a variant under consideration will be removed or "negated" in the sense of negative variability.

Figure 4.13 gives an example of the mapping of the feature model in Fig. 4.8 to the ESG in Fig. 4.11, and depicts the resulting 100% behavioral model without Search and Operating with Payment by Credit Card (also as 100% model). Exploitation of this combination is discussed in Sect. 5.2.5

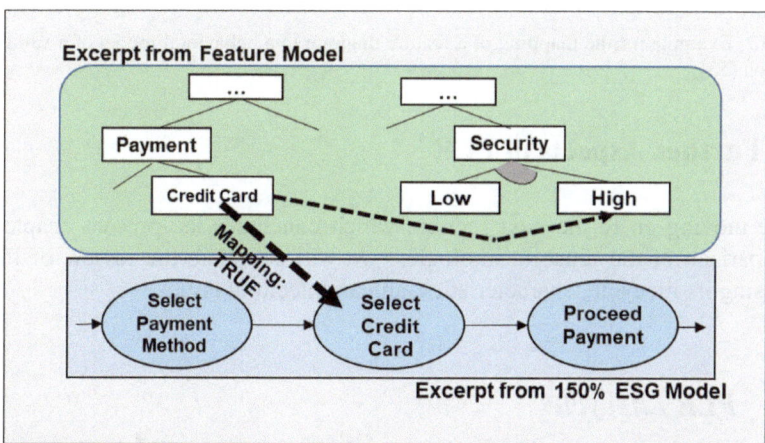

Fig. 4.12 SPL design with negative variability and behavioral Augmentation Using ESG Model (based on [34])

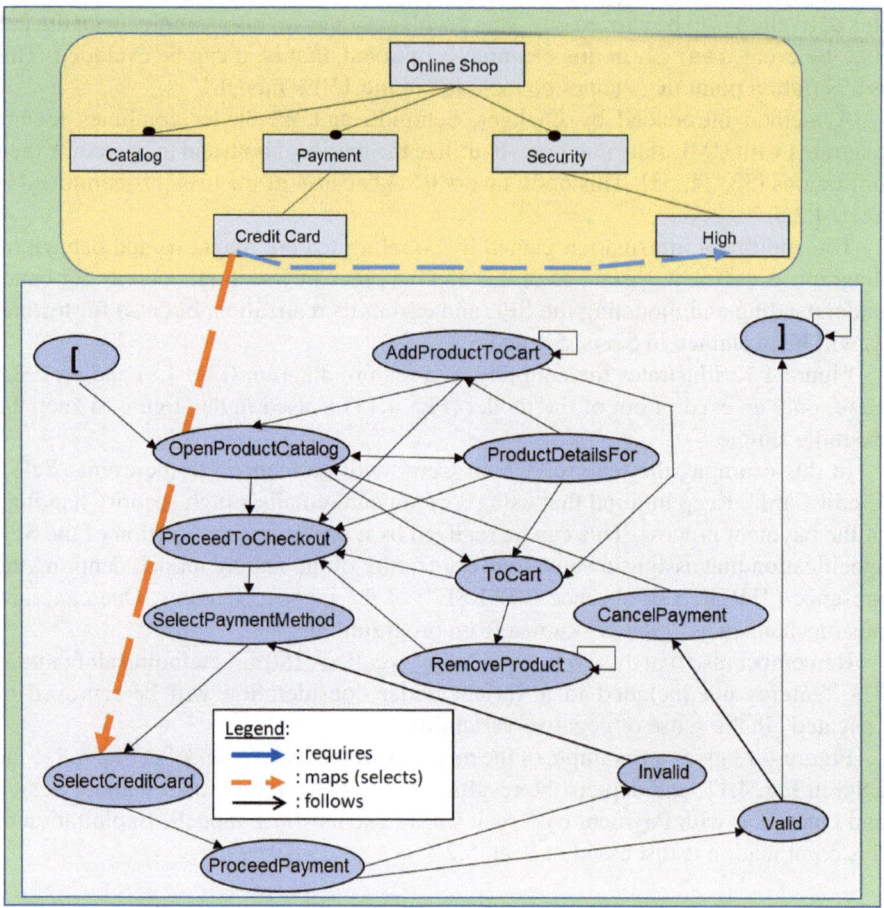

Fig. 4.13 Example for the mapping of a feature diagram to a behavioral model of a valid variant (based on [53])

4.3   Further Aspects of PLE

Before moving on to the next section, which concludes the present chapter with a comparison of the reuse technologies, we will complete the review of PLE by discussing its lifecycle, characteristics, and architectural issues.

4.3.1   PLE Lifecycle

As stressed several times in the previous sections, Product Line Engineering (PLE) consists of two main phases: Domain Engineering (DE) and Application Engineering

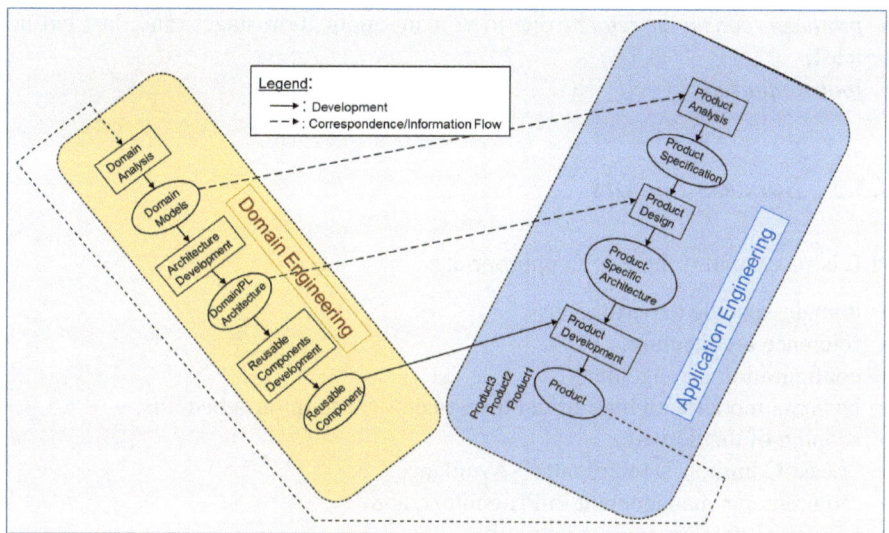

Fig. 4.14 PLE lifecycle (based on [6, 44])

(AE). The objective of DE is to produce reusable components that necessitate development of a domain architecture, considering also the needs of the AE. AE, on the bottom side, designs the products, develops an architecture for each product, and selects reusable components offered by DE. Figure 4.14 depicts this process ([6, 37], [AAYA02]), resembling the V-model of the software process (Sect. 1.3.2.5). The sequence of activities of the DE are placed into the left leg of the model, from top to bottom. The activities of the AE are located in the right leg, also from top to bottom. Note the direction of the development flow; the sequence of the activities of the AE starts at the top, same as DE does. The asset that will be produced by DE will be transferred to AE, close to the bottom. Both legs are also interconnected via information flow. Note also the concept that the DE can also consider the final, composed product as a new asset, as indicated by the dotted connection between the product (AE) and domain analysis (DE).

4.3.2 PLE Characteristics

The features of PLE are significant compared with other software reuse technologies because PLE is:

- *architecture-based* since a reference architecture is incorporated early on;
- *economically-driven* rather than technically-driven;
- *domain-specific*, *reuse-based*, and *process-driven*, that is, adaptable to the specific needs of the domain;

- *producer–consumer-related* due to domain-application stages, and, last but not least,
- *tool-supported*.

4.3.3   Success Factors

PLE is successful if there is an appropriate:

- domain-specific expertise,
- reference architecture,
- configuration management,
- business model with high initial investments for long-term benefits,
- scoping of the domain,
- "Least-Common-Denominator"-Avoidance,
- requirements management in PL-context, and
- a separate domain engineering unit.

4.3.4   PL Architectures

A *system architecture* is defined by the *kind* and *constellation* of the system (architectural) elements (Sect. 2.1).

Elements of a PL architecture consist of *components* and *middleware*, including connectors and glue code (Sect. 1.5.3). The latter can be as simple as a procedure call yet be as elaborate as client-server-protocols as a links between distributed databases.

The difference between the software architecture in general and the product line architecture (PLA) is that PLE has a *domain* (*reference*) architecture common to all members of the PL under consideration.

A PLA contains the *framework* and *context* for developing reusable assets.

The more robust and highly flexible and customizable the PLA is, the better the PL is.

PLA defines and manages the fixed and common parts of PL, in close cooperation with configuration management.

4.3.5   Some Commercial PLE Approaches and an Example

Some well-known, successful commercial reference approaches include (see also [8]):

- SYNTHESIS introduced by Software Productivity Consortium (SPC) [51, 44, 52].
- Product Line Framework introduced by Software Engineering Institute (SEI) [37].

- FAST (Family-oriented abstraction, specification, and translation) by Lucent Technologies [38].
- PuLSE, a product line software engineering methodology by Fraunhofer Institute for Experimental Software [10, 50].

SYNTHESIS technology is a good example of a typical PLE approach in a range of "opportunistic mode" for an ad-hoc application (see "accidental reuse," Sect. 1.4.3), and a comprehensive "leveraged mode" for systematic reuse as recommended by the authors of this book. The "Leveraged mode" is discussed in the next sections.

Synthesis views the construction of versions of a product family not as successive modifications to previous versions of similar programs (see Sect. 1.3.1, Configuration Management), but as instances of a family of related systems having similar descriptions as a consequence of common properties, which can be represented by an abstract model, for example, a feature model. The product family under consideration forms a business area that is viewed as a domain. SYNTHESIS consists of the following integrated and iterative subprocesses:

- Domain Engineering (DE) for generating

  - standardized work products, including code and documentation (Fig. 4.15), and
  - a template for application engineering process.

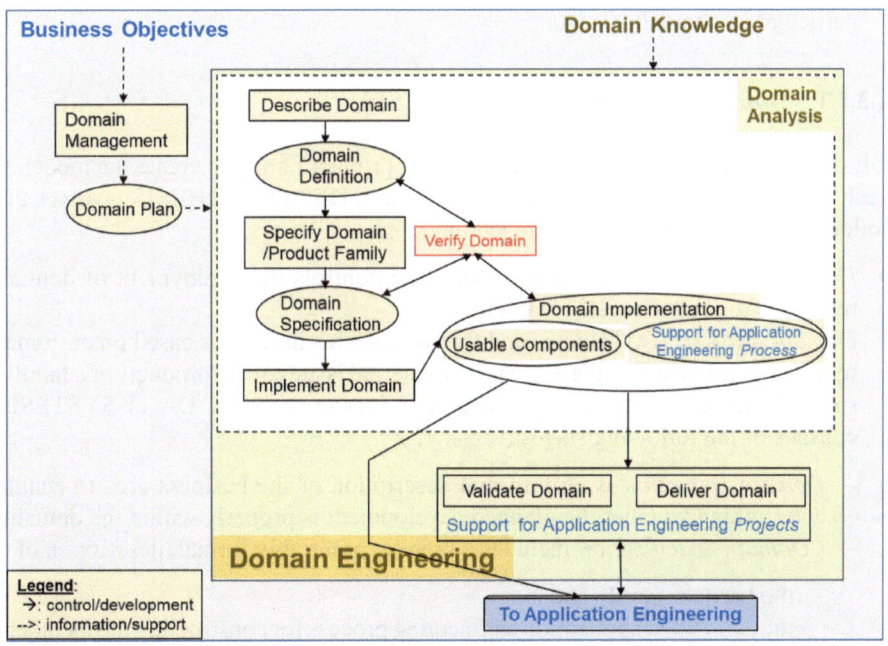

Fig. 4.15 Domain engineering in SYNTHESIS (based on [44, 52])

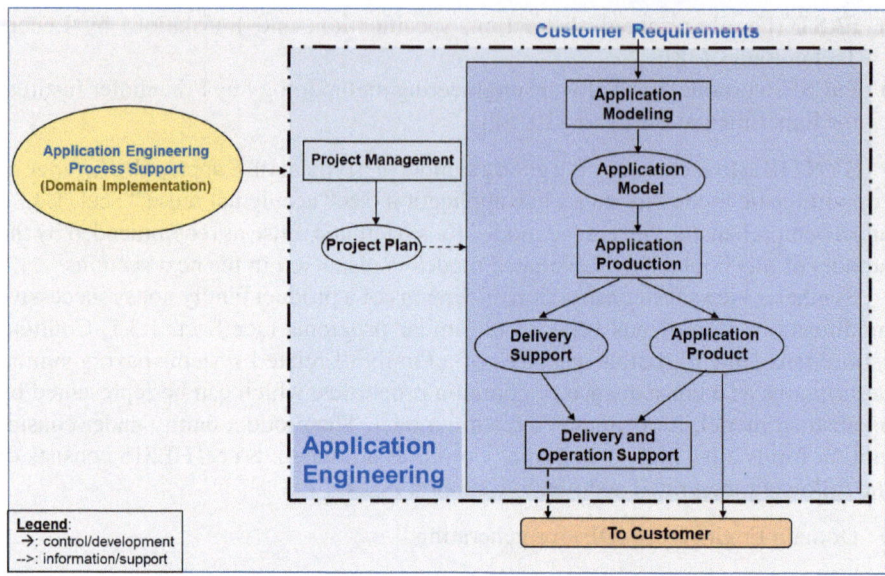

Fig. 4.16 Application engineering in SYNTHESIS (based on [44, 52])

- Application Engineering (AE) with emphasize on requirements for describing a particular system (Fig. 4.16).

4.3.5.1 Domain Engineering in SYNTHESIS

DE provides AE the architecture model of the product family to create the model of each and any product in the scope of the domain. DE in SYNTHESIS involves the following activities (in the "Leveraged mode"):

- *Domain Management* plans, monitors, and controls the deployment of domain resources to produce reusable components.
- *Domain Analysis (DA)* studies and formalizes a business area based on customer requirements that affect the external and internal issues of the products of a family, such as the form, structure, and content of those products. DA of SYSTESIS consists of the following stages (Fig. 4.15):

  - *Domain Definition* is an informal description of the business area to enable determining whether the planned development is properly within the domain.
  - - *Domain Specification* includes a precise, preferably formal, description of

    the product family and
    the associated application engineering process for constructing the members of the family.

- *Domain Verification* is supposed to ensure the correctness, consistency, and completeness of the work products of the domain engineering.
- *Domain Implementation* realizes the production of the entities as described in the domain specification that are

  adaptable components, including code, documentation, and support for V&V, and
  procedures to realize the deliverable work products of the application engineering process, that is, to develop the product, and
  process support to define the policies and procedures to enable the application engineering to develop a work product and provide tools for supporting this process.

- Project Support validates the domain implementation and releases it for the application engineering's use, supporting its activities in developing individual products in the following steps:

  - *Domain Validation* evaluates and certifies the effectiveness and quality of domain implementation.
  - *Domain Delivery* to an application engineering project aimed at realizing the production of the members of the family, with all necessary support and assistance.

4.3.5.2 Application Engineering in SYNTHESIS

The objective of the application engineering process is to develop a product that satisfies customer requirements in a particular business area and consists of the following activities (Fig. 4.16):

- *Project Management* plans, monitors, and controls the deployment of project resources to deliver a product.
- *Application Modeling* models and specifies a required product based on customer requirements, considering also technical and financial constraints.
- *Application Production* develops a standardized product and the associated delivery supporting items in compliance with the application model.
- *Support of Delivery and Operation* of the product, including the evaluation of its effectiveness, finalizes the activities of the application engineering.

4.4 Comparison of CBSE, COTS, and PLE Along Some Dimensions

Before concluding this present chapter on "Software Reuse Technologies", Table 4.1 in the following discusses the three mainstreams of reuse, which are component-

Table 4.1 Comparison of CBSE, COTS, and PLE along the dimensions organization, technical aspects, and economical aspects (based on [44])

Aspect	Factor/form	CBSE	COTS-based development	PLE
Organizational	Primary interactions	Component and application engineers, repository/library retrieval	Third Party, mostly COTS vendor	Domain engineer
	Primary reuse skils	Component and application engineers repository/library assistant	Reuse manager	Domain engineer, system architecture
Technical	Integration	Compositional/generative	Compositional only	Composition into Domain Architecture
	Constraints	Components comply with component model	Components comply to application constraints	Components comply to domain architecture constraints
	Support responsibility	Application engineer	Component vendor	Domain engineer
	Packaging	Binary source code, modifiable	Binary black-box components	source code, reference architecture, middleware infrastructure, guidelines
	Retrieval	Public/domain libraries	Commerical Advertisement	Domain libraries
	Implementation technology	Defined by component model	Not controllable	Defined by reference architecture
Economical	Component Acquisition	Adapt	Buy	Instantiate
	Market scope	ROI-driven from an application developper's view	ROI-driven by component provider's view	ROI-Driven by domain engineer's view
	Economical motivations	Common component models	Common market	Common design and architecture

based software engineering (CBSE), commercial-off-the-shelf (COTS) and product line engineering (PLE). Aspects the discussion considers are *Organizational, Technical Direction*, and *Economical*. These aspects are not necessarily orthogonal; they commonly make use of some techniques and notions. For example, the "Integration"

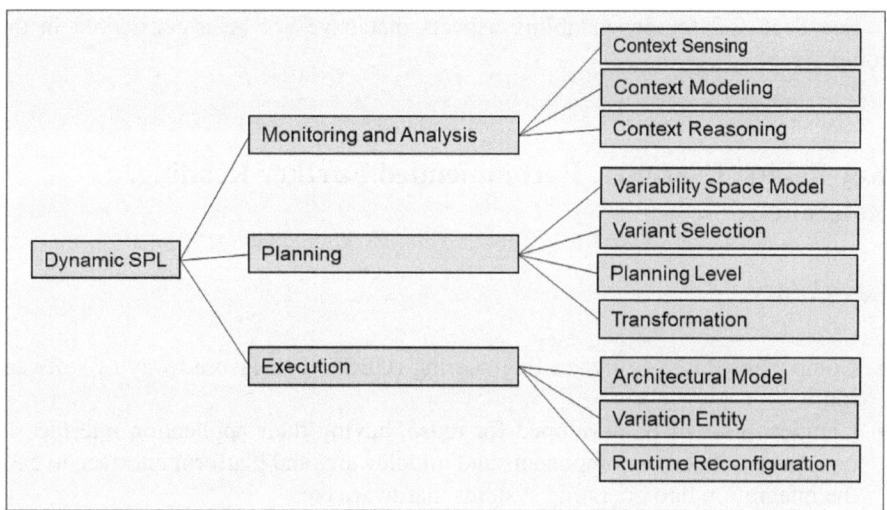

Fig. 4.17 Dynamic SPLE (based on [9])

factor of the "Technical Aspect" can be "compositional" in all three reuse technologies. Nevertheless, they differ tremendously in many important issues so that a decision to adapt a reuse technology has managerial and significant technical and financial consequences, since a switch from one to another can be difficult and costly. Even a combination, for example, CBSE and COTS, must be thoroughly and well thought out (Table 4.1).

During recent years, SPLE has been favored and deepened in research and practice. Bencomo et al., for example, introduced *Dynamic SPL (DSPL)*, where the variability of the products is captured at the domain engineering phase and the best variant for a specific operating environment and user requirements is selected during the application engineering phase [15]. Bashari et al. reviewed and refined this concept by summarizing the activities in following dimensions [9] (Fig. 4.17).

- *Monitoring and Analysis* aims at modeling the domain context by focusing on relevant properties and the semantic relations between them. These activities require sensing those properties and their likely changes and generating alternatives supported by a knowledge-base and a reasoning mechanism.
- *Planning* activities attempt to make the application engineering superfluous by using the results of the monitoring and analysis to define the space of the variability. Depending on the requirements, this dimension selects appropriate variants at the level considered, which are the feature level or architecture model. Additionally, a transformer converts one level to another.
- Finally, the *execution* step is supposed to dynamically form the configuration at runtime, based on the architectural model and entity required.

See Sect. 2.2 for dependability aspects that have not been considered in the comparison.

Key Points, Exercises, Recommended Further Reading, References

Key Points

- Component-Based Software Engineering (CBSE) is a favored way of software reuse.
- Components will be developed for reuse, having their application interface to cooperate with other components and middleware, and platform interface to ease the integration into operating systems, hardware etc.
- Components can be of different granularity to ease the integration into the composed system; this can, however, increase their size and decrease their performance (run time, memory requirement).
- Commercial-Off-the-Shelf-Software (COTS)–Based Development is a special case of CBSE. It is customer-oriented.
- Thus, lifecycles of COTS Development and CBSE are different.
- The concept of product line engineering (PLE) is nowadays the most favored technique of reuse; it is almost used as a synonym for reuse.
- PLE combines most of the known reuse techniques.
- Model-based PLE describes the product family by a general ("150%" or "complete") model, while the behavior of each product (as a variant of the product family) will be modeled by an individual ("100%") model that instantiates the general model for a specific use case.
- The lifecycle of PLE consists of domain engineering (DE) and application engineering (AE).
- Accordingly, an appropriately carried out DE and AE are of decisive importance for the success of PLE.
- A comparison of CBSE, COTS–Based Development, and PLE reveals differences in organization, technical, and economical issues.

Exercises

1. Explain vertical and horizontal repository (or library), and explain when which one is more appropriate. Discuss the situation with an example from your environment.
2. Explain DA activities with an example from your environment. Discuss the situation (positives/negatives).

3. What are the issues and models of a working domain? Do you have one in your environment?
4. Learn more on MIL, IDL, ADL, etc., and discuss which one could be appropriate for which kinds of applications.
5. Discuss whether the company/department where you work or were apprenticed uses CBSE for reuse (or should use CBSE, in case it is not yet under consideration).
6. Can you identify the feature(s) above in components used by the company/department where you work or were apprenticed? Discuss how you can make certain the feature(s) is/are realized in the future.
7. Identify the common feature(s) of the components sold by the company/department where you work or were apprenticed.
8. Discuss whether the company/department in which you work or were apprenticed uses PLE for reuse (or should use PLE, in case it is not yet under consideration).
9. Discuss the architecture of a PL produced in the company/department in which you work or were apprenticed.
10. What is the difference between the lifecycles of CBSE and COTS development?
11. Does PLE have specific characteristics that differ from those of CBSE and COTS development?
12. When and how can PLE be successful?
13. Explain the main difference between software architecture in general and product line architecture (PLA).
14. Do you know any successful commercial PLE approaches?

Recommended Further Reading

1. Apel, S., Batory, D., Kästner, Ch., Saake, G.: Feature-Oriented Software Product Lines. Springer, Berlin Heidelberg (2013)
2. Binder, R.V.: Testing Object-Oriented Systems: Models, Patterns, and Tools. Addison-Wesley (2006)
3. Clements, P., Northrop, L.: Software Product Lines—Practices and Patterns. SEI Series in Software Engineering. Addison-Wesley (2002)
4. Pohl, K., Günter Böckle, G., Linden, F.v.d.: Software Product Line Engineering: Foundations, Principles, and Techniques. Springer (2005)
5. Sommerville, I.: Software Engineering, 10 e. Pearson (2015)

References

6. Addy, E.A.: A framework for performing verification and validation in reuse-based software engineering. Ann. Softw. Eng. **5**, 279–292 (1998). https://doi.org/10.1023/A:1018968222862

7. Aho, A.V., Dahbura, A.T., Lee, D., Uyar, M.Ü.: An optimization technique for protocol confor-mance test generation based on UIO sequences and rural Chinese postman tours. IEEE Trans Commun **39**, 1604–1615 (1991)
8. de Almeida, Alvaro, Lucredio, D., Garcia, V.C., de Lemos Meira, S.R.: A survey on software reuse processes. In: Proceedings of the IEEE International Conference on Information Reuse and Integration (IRI), pp. 66–71 (2005). https://doi.org/10.1109/IRI-05.2005.1506451
9. Bashari, M., Bagheri, E., Du, W.: Dynamic software product line engineering: a reference framework. Int. J. Software Eng. Knowl. Eng. **27**(2), 191–234 (2017). https://doi.org/10.1142/S0218194017500085
10. Bayer, J., Flege, O., Knauber, P., Laqua, R., Muthig, D., Schmid, K., Widen, T., DeBaud', J.-M.: PuLSE: a methodology to develop software product lines. In: Proceedings of ACM Symposium on Software Reusability, pp. 122–131 (1999)
11. Belli, F., Hollmann, A.: Test generation and minimization with "basic" statecharts. In: Proceed-ings of the ACM symposium on Applied Computing SAC '08, pp. 718–723 (2008). https://doi.org/10.1145/1363686.1363856
12. Belli, F., Budnik, Ch.J., White, L.: Event-based modelling, analysis and testing of user inter-actions: approach and case study. In: Software Testing, Verification and Reliability, pp. 3–32 (2006)
13. Belli, F., Beyazit, M., Memon, A.: Testing is an event-centric activity. In: Proceedings of the International IEEE Conference on Software Security and Reliability, pp. 198-206 (2012)
14. Belli, F., Budnik, Ch.B., White, L.: Event based modelling, analysis and testing of user inter-actions: approach and case study. Software Testing, Verification and Reliability **16**(1), 3–32 (2006)
15. Bencomo, N., Hallsteinsen, S., Almeida, E.: A view of the landscape of dynamic software product lines. Computer **45**(10), 36–41 (2012)
16. Cichos, H., Oster, S., Lochau, M., and Schürr, A.: Model-based coverage-driven test suite gener-ation for software product lines. In: Proceedings of the MODELS 2011, LNCS 6981, pp. 425–439. Springer (2011); Extended Version of Model-based Coverage-Driven Test Suite Gener-ation for Software Product Lines, Informatik-Bericht Nr. 2011–07, Technische Universität Braunschweig (2011)
17. Cichos, H., Lochau, M., Oster, S., and Schürr, A.: Reduktion von Testsuiten für Software-Produktlinien. In: Proceedings of the Software Engineering, Gesellschaft für Informatik e.V., pp. 143–154 (2012)
18. Cohen, S.G., Stanley, Jr., J.L., Peterson, A.S., Krut, Jr., R.W.: Application of Feature-Oriented Domain Analysis to the Army Movement Control Domain. Technical Report CMU/SEI-91-TR-028, ESD-91-TR-028, 1992
19. Cox, B.J.: Planning the software revolution. IEEE Softw. **7**(6), 25–35 (1990)
20. Czarnecki, K., Eisenecker, U.: Generative Programming: Methods, Tools, and Applications. Addison-Wesley (2000)
21. Devroey, X.: Behavioural model-based testing of software product lines. Ph.D. Thesis, University of Namur, PReCISE Research Center (2017)
22. Eichelberger, H., Kröher, Ch., Schmid, K.: An analysis of variability modeling concepts: expres-siveness vs. analyzability. In: Proceedins of the 13th International Conference on Software Reuse—ICSR, pp. 32–48 (2013)
23. El-Fakih, K., Hierons, R.M., Turker, U.C.: K-branching UIO sequences for partially specified observable non-deterministic FSMs. IEEE Trans. Software Eng. (2019). https://doi.org/10.1109/TSE.2019.2911076
24. von Gamma, E., Helm, R., Johnson, R.E., Vlissides, J.: Design Patterns. Elements of Reusable Object-Oriented Software. Addison-Wesley Professional Computing Series (1994)
25. Gorgonia K.C., Perkusich A.: Adaptation of coloured petri nets models of software artifacts for reuse. In: Proceedings of the 8th International Conference—Software Reuse: Methods, Techniques, and Tools (ICSR), pp. 240–254 (2004)
26. Groher, I., Voelter, M.: Expressing feature-based variability in structural models. In: Proceed-ings of the Workshop on Managing Variability for Software Product Lines at SPLC (2007).

Available at http://citeseerx.ist.psu.edu/viewdoc/download?doi=10.1.1.571.593&rep=rep1& type=pdf

27. Grönniger, H., Krahn, H., Pinkernell, C., Rumpe, B.: Modeling variants of automotive systems using views. In: Proceedings of the Modellierungs-Workshop MBEFF, Berlin, Informatik-Bericht 2008–01 (2008). Available at www.serwth.de/publications

28. Haugen Ø., Møller-Pedersen B., Oldevik J, Olsen, G.K., Svendsen, A.: Adding standardized variability to domain specific languages. In: Proceedings of the International Software Product Line Conference (SPLC '08), pp. 139–148 (2008). https://doi.org/10.1109/SPLC.2008.25

29. Heuer, A., Stricker, V., Budnik, C.J., Konrad, S., Lauenroth, K., Pohl, K.: Defining variability in activity diagrams and petri nets. Sci. Comput. Program. **78**, 2414–2432 (2013)

30. James Blair, J., Batory, D.: A comparison of generative approaches: XVCL and GenVoca (2004). Available at https://www.cs.utexas.edu/ftp/predator/xvcl-compare.pdf

31. Kang, K., Cohen, S., Hess, J., Nowak, W., Peterson, S.: Feature-Oriented Domain Analysis (FODA) Feasibility Study (Report), Software Engineering Institute, Carnegie Mellon University (1990). Available at http://www.floppybunny.org/robin/web/virtualclassroom/chap12/s4/articles/foda_1990.pdf

32. Krueger, C.W.: Software reuse. ACM Comp. Surv. **24**(2), 131–183 (1992)

33. Lackner, H.: Domain-centered product line testing. Ph.D. Thesis, Humboldt-University, Berlin (2016)

34. Lackner, H., Schmidt, M.: Potential errors and test assessment in software product line engineering. In: Proceedings of the Tenth Workshop on Model-Based Testing, Springer, EPTCS 180, pp. 57–72 (2015). https://doi.org/10.4204/EPTCS.180.4

35. Lackner, H., Thomas, M., Wartenberg, F., Weißleder, S.: Model-based test design of product lines: raising test design to the product line level. In: Proceedings of the 7th International Conference on Software Testing, Verification, and Validation (ICST), pp. 51–60 (2014)

36. Linda, M., Northrop, L.M., Clement, P.C.: A Framework for Software Product Line Practice, Version 5.0, Software Engineering Institute (SEI), Carnegie Mellon University, 2012 (REV-03.18.2016); also available at https://resources.sei.cmu.edu/asset_files/WhitePaper/2012_019_001_495381.pdf

37. van der Linden, F.J, Schmid, K., Rommes, E.: Software Product Lines in Action. Springer (2007)

38. Lisboa, L.B., Li, J.J., Morreale, P., Heer, D., Weiss, D.M.: An evaluation to compare software product line decision model and feature model. In: Proceedings of 9th International Conference on Evaluation of Novel Approaches to Software Engineering (ENASE), pp. 1–8 (2014)

39. Lity, S., Nahrendorf, S., Thüm, T., Seidl, C., Schaefer, I.: 175% modeling for product-line evolution of domain artifacts. In: Proceedings of 12th International Workshop on Variability Modelling of Software-Intensive Systems (VAMOS), pp. 27–34 (2018). https://doi.org/10.1145/3168365.3168369

40. Lochau, M., Schaefer, I., Kamischke, J., Lity, S.: Incremental model-based testing of delta-oriented software product lines. In: Brucker, A.D., Julliand, J. (eds.) Tests and Proofs. TAP 2012, Lecture Notes in Computer Science, vol. 7305, pp. 67–82. Springer, Berlin. https://doi.org/10.1007/978-3-642-30473-6_7

41. Luckham, D.: The Power of Events. Addison Wesley (2005)

42. Maglio, P.P, Weske, M., Yang, J., Fantinato, M.: Service-oriented computing. In: Proceedings of the 8th International Conference (ICSOC) (2010)

43. Meinicke, J., Thüm, T., Schröter, R., Benduhn, F., Leich, T., Saake, G.: Mastering Software Variability with FeatureIDE. Springer (2017)

44. Mili, H., Mili, A., Yacoub, S., Addy, E.: Reuse-Based Software Engineering, Techniques, Organization, and Controls. Wiley (2002)

45. Muthig, D., Atkinson, C.: Model-driven product line architectures. In: Proceedings of the Software Product Lines (SPLC 2002), Lecture Notes in Computer Science, vol. 2379, pp. 110–129. Springer (2002)

46. Northrop, L.M.: SEI's software product lines tenets. IEEE Softw. **19**, 32–40 (2002)

47. Object Management Group (http://www.uml.org), Unified Modeling Language (UML®), Version 2.5.1, 2017
48. O'Connor, J., Mansour, C., Jerri Turner-Harris, J., Campbell, Jr., G.H., Reuse in command-and-control systems. IEEE Software **11**, 70–79 (1994). https://doi.org/10.1109/52.311065
49. Pohl, K., Günter Böckle, G., Linden, F.v.d.: Software Product Line Engineering: Foundations, Principles, and Techniques. Springer (2005)
50. Schmid, K., John, I., Kolb, R., Meier, G.: Introducing the PuLSE approach to an embedded system population at Testo AG. In: Proceedings of 27th International Conference on Software Engineering, (ICSE), pp. 544–552 (2005). https://doi.org/10.1109/ICSE.2005.1553600
51. Seidl, C., Wille, D., Schaefer, I.: Software reuse: from cloned variants to managed software product lines. In: Automotive Systems and Software Engineering: State of the Art and Future Trends, pp. 77–108. Springer International Publishing (2019)
52. Software Productivity Consortium (SPC): Reuse-Driven Software Process Guidebook, Herndon, VA (1993)
53. Weißleder, S., Lackner, H.: Top-down and bottom-up approach for model-based testing of product lines. In: Proceedings MBT 2013, EPTCS 111, pp. 82–94. Springer (2013). https://doi.org/10.4204/EPTCS.111.7
54. Zschaler, S., Sánchez, P., Santos, J., Alférez, M., Rashid, A., Fuentes, L., Moreira, A., Araújo, J., Kulesza, U.: VML*—a family of languages for variability management in software product lines. In: Proceedings of the Second International Conference Software Language Engineering (SLE), pp. 82–102 (2009)

Chapter 5
Reuse and Testing

CBSE, COTS and PLE are widely well-known techniques of software reuse, whereby PLE is favored in the practice. A software product line (SPL) consists of products which share some features but differ in some others. A large variety of features, such as a multiplicative factor, generally leads to a great number of variants as alternative products of a SPL. Very often, it does not make much sense or is even infeasible to test all of these variants. A bottom-up strategy to overcome this problem is to form a group of variants that share a set of features and generate test cases for them at domain engineering level. The opposite, top-down strategy at application level, is to generate test cases directly for products that can be derived. Chapter V reviews those strategies and illustrates them using simple examples, with the concept of model-based testing playing an important role.

Before discussing strategies, test methods reviewed in Sect. 1.3 will be revisited and appropriate test techniques for reuse will be selected and discussed in the context of reuse.

A discussion of cost aspects and reuse metrics concludes in this Chapter and, thus, Part A.

5.1 Reuse-Specific Problems and Approaches to Testing

Validation of reusable components and systems that are built by reusable systems is an important issue before releasing them. This chapter will brush up and extend the validation methods discussed in Chap. 1, especially in Sect. 1.3, and select and expand on some of them, considering the needs of reuse.

© The Author(s), under exclusive license to Springer Nature Switzerland AG 2021   141
F. Belli and F. Quella, *A Holistic View of Software and Hardware Reuse*,
Studies in Systems, Decision and Control 315,
https://doi.org/10.1007/978-3-030-72261-6_5

5.1.1 Testing Context-Dependent and -Independent Parts

A system built by reusable components forms a composite system, which consists of:

- common (fixed) parts and
- variable parts (Sect. 5.2.1.1).

Fixed parts realize reusable components. Thus, testing systems with reusable parts requires testing:

- components,
- variable parts, and
- composite system as a whole after integration.

The requirements of an application are often not, or incompletely, known while producing the components. Therefore, components will primarily be tested independently, without considering the context of their (later) application. This fact stresses the importance of testing the composite system as a whole after integration and before release.

Principally, component testing corresponds to unit testing, and can thus be carried out with conventional unit testing techniques. Composite systems' testing, on the other hand, corresponds to integration testing and can be carried out with conventional integration testing techniques (Fig. 1.24).

5.1.2 A Brief Survey of Software Testing Methods

Fig. 5.1 is a reprint of Fig. 1.24 to remind of the methods discussed in Sect. 1.3.3. Each of the methods gathered in the left column is a legitimate candidate for testing reusable components and composite systems. However, two groups have been highlighted to identify the test methods preferred to be applied to reuse. These are diversity-based techniques, for example, mutation testing, and combinatorial testing, for example, pairwise testing. These techniques will be discussed in the following sections after a brief introductory discussion on PL-oriented testing.

An SPL-approach is expected to enable the generation of the product-line members' code from a common product-line infrastructure instead of having to develop them independently. In case the generated software is embedded in a large product line, additional elements will be integrated to develop the end products, necessitating a *system te*st to assure there has been a correct integration. In the end, the *acceptance* test covers the entire system [27, 42]. In this case, the *behavior* of the SUC also becomes a subject of the validation process (Fig. 4.2, see also Sects. 4.2.5 and 5.2.5; [22, 34, 77]).

Fragal et al. propose extending finite state machine (FFSM) to featured FSM (FFSM) for considering the behavior aspect while testing software product lines

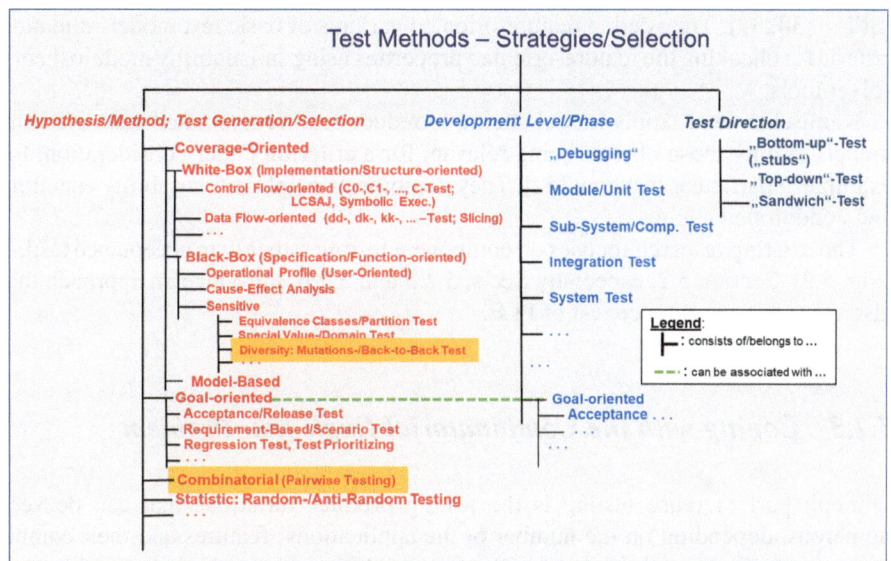

Fig. 5.1 Revisiting Fig. 1.24 for selecting test Methods for reuse

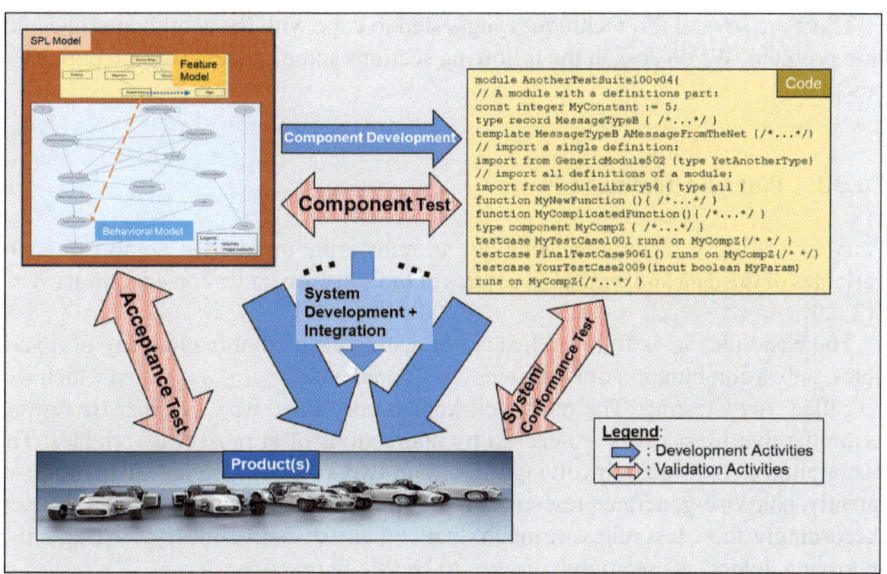

Fig. 5.2 Test activities for SPL

(SPLs) [38, 39]. They define feature-oriented variants of basic test model validation criteria for checking the feature-oriented properties using satisfiability modulo theory solver tools.

Kamischke et al. apply model slicing to reduce feature-annotated state machine models to only those objects being relevant for a criterion under consideration, for example, a particular test goal [52]. They combine principles of variability encoding and conditioned slicing.

The existing research focuses on component testing rather than acceptance testing (Fig. 5.2). Section 5.2, especially Sects. 5.2.5 and 5.2.6, introduce an approach that also considers acceptance test of PLE.

5.1.3   Coping with the Combinatorial Explosion Problem

Difficult part of reuse testing is the testing product variations that can be very numerous, depending on the number of the applications' features and their combinations (*combinatorial explosion*). Because of this problem, the individual testing of each variant of the product family is not always feasible and also does not make sense economically.

There are several test techniques suggested to cope with the combinatorial explosion problem. We discuss in the following sections some of them that are popular in practice.

5.1.3.1   Pairwise Testing

Pairwise testing is a popular technique to reduce the number of combinations of variables of the data and control structures of the software under consideration (SUC) [13, 29].

The basic idea is as follows. Instead of testing each combination, say of n variables, only a combination of i variables considered with $2 \leq i \leq n$ is tested which will be called i-*way* testing. The most well-known one is *two-way*, or *pairwise* testing, assuming that most faults are caused by interactions of at most two variables. The assumption is that faults involving more than two variables are less likely. Consequently, pairwise-generated test suites cover all combinations of only two variables. Accordingly, these test suites are much smaller than exhaustive ones, yet still effective in finding defects, as seemingly proven to be true in practice.

EXAMPLE: Consider a system S with three sets of input variables X, Y, and Z (Fig. 5.3, [13]). Assume that a function f maps each of the input variables, such that

$$f(X) = \{1, 2\}, \; f(Y) = \{Q, R\}, \; and \; f(Z) = \{5, 6\}.$$

"Brute force" combinatorial testing delivers $2 \times 2 \times 2 = 8$ test cases.

Table 5.1 Test (input) sets for the example in Fig. 5.3 [13]

Test input ID	Input X	Input Y	Input Z
T1	1	Q	5
T2	1	R	6
T3	2	Q	6
T4	2	R	5

Fig. 5.3 Example—a system S with three sets of input variables

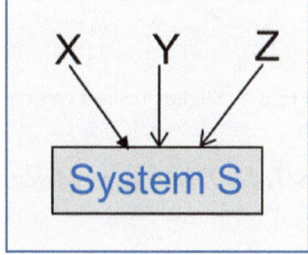

Table 5.1 summarizes the pairwise testing technique for the example given in Fig. 5.3 [13]. The test suite, consisting of the four test cases (instead of eight by brute-force), T1 to T4, covers all pairwise combinations of the variables of X, Y, and Z.

It is obvious that pairwise testing does not consider all combinations of all variables, that is, disregards many of them, including the ones that would enable to reveal critical faults. Nevertheless, the trade-off is attractive and explains the popularity of pairwise and n-way testing. Combination with model-based concepts help to further develop and refine this approach [67, 68].

5.1.3.2   Pairwise Testing Using a Feature Diagram (FD)

The feature diagram example given in Fig. 3.10 can be used to illustrate how to apply pairwise testing to reuse. Figure 5.4 abbreviates the id's of the features to avoid long identifiers (id's).

It can easily be seen that there are features in 3 groups of parties of 4, 2, and 3. They are as follows:

- 3 Groups: Ability for Calling/Displaying/SMS
- Party of Ability for Calling (4): NP/PB/CL/SB
- Party of Ability for Displaying (2): BW/CO
- Party of Ability for SMS (3): SG/SV/MP.

This leads to $4 \times 2 \times 3 = 24$ combinations. It is left to the reader to find out how many two-way (pairwise) test cases can be generated for the FD in Fig. 5.4 (see Exercises at the end of this section and [67]).

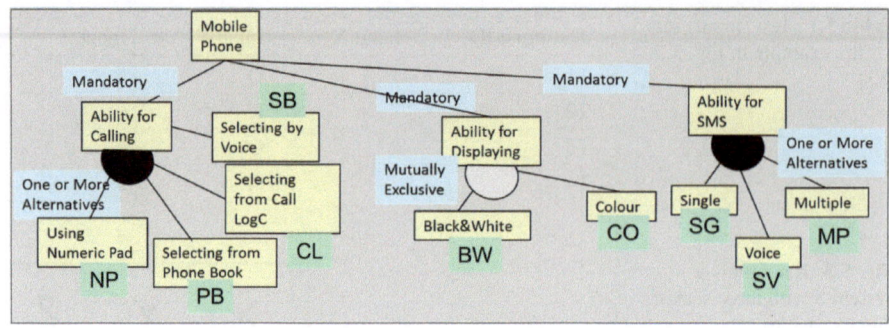

Fig. 5.4 Slightly revised version of the example in Fig. 3.10 (based on [51])

5.1.4   Diversity: Mutation/Back-to-Back Testing

A variant represents a mutant of the fixed software, in other words, a slightly modified version of the original software (Sect. 1.3.3.11.4). Thus, mutation analysis/testing techniques can be considered for reuse validation [45]. In this context, Back-to-Back test technique should also be considered (Sect. 1.3.3.11.2, [83]).

See also Sect. 2.2.6 for a mutation technique to test product lines.

5.1.5   Some Remarks

Remember that there are reuse assets other than programs, for example, tests or test specifications.

Testing during the domain engineering (DE) is context-independent since the context of the future application can hardly be foreseen. On the other hand, testing during the application engineering (AE) is strongly context-dependent, based on the application's features that are knowable. DE testing also differs from AE testing concerning following aspects that include a broad variety of assets:

- Requirements
- Goals
- Methods.

While DE testing enables the release of a component, AE testing triggers the acceptance of a selected component. Accordingly, the requirements in DE specify a family of products rather than describes a specific one, which is the subject of AE.

5.2 Model-Based Development and Validation

Model-based development is widely accepted for the construction and validating of systems that are expected to be easily modified and inexpensively maintained. These are features required also for reuse. This section focuses on model-based testing and discusses and expands on test techniques that are partly reviewed in Sect. 1.3.3.

A *model* is the representation of a system (or a process) with the focus on its relevant features, disregarding irrelevant details. Thus, a model is always more abstract than the modeled system, and the system itself is more concrete than its model. It is widely accepted that a *system* model comprises the relevant elements to produce a desired outcome, whereby a *process* model focuses on the activities of the involved actors, who contribute to the outcome and considering relevant aspects, for example, efficiency based on the actors' resource needs/consumptions. Generally, a system model is static, a process model is dynamic. This book uses the terms "system" and "process" interchangeably; the context determines which one is meant.

Modeling is useful and necessary to understand complex systems and processes. Engineering disciplines usually start with a model, a blue print, for example, for implementing (realizing, developing) a complex system.

Remember that methods and tools that are selected for implementing a system should support identification and evaluation of the software characteristics related to the test objectives.

Modeling in the large *abstractly* represents the major features of the system under consideration (SUC) *in toto*, that is, its architecture and how it communicates/interoperates with its environment. Modeling in the small, on the other hand, *concretely* represents the detailed features of the SUC, *refining* the abstract model, for example, the inner structure of its components, such as, data and control flow.

Formal models are mathematical representations, for instance, formulae, curves, graphs, etc. They allow quantifiable notation and calculus of the relevant characteristics of SUC using notions and methods based on a sound theory, for example, Boolean Algebra, Graph Theory. *Informal* models are usually qualitative; they are not much concerned about the precise measuring of the relevant characteristics. *Semiformal* models are in between. As we differentiate between *continuous* processes and *discrete* processes, we also differentiate between continuous models and discrete models, depending on observing the process (hypothetically) at any, uncountable point in time, or only at definite, countable points in time, respectively.

Model-based approaches have a long tradition in scientific-technical research; they are also well-established in all engineering disciplines. A *global* modeling for sustainability is even unanimously recommended to consider all relevant factors of environmental protection [53].

In analogy to "modeling in the large" versus "modeling in the small", test objectives can be viewed in two groups (see Sects. 1.3.3.6 and 1.3.3.11).

- *Testing in the small*: detection, localization, correction, masking, etc. of faults (*debugging*, that is, individual handling of coding faults, does not belong to the test process).

- *Testing in the large*: checking how far the test criteria selected prior to testing are fulfilled; for example, by determining:

  - path coverage, statement coverage, and
  - non-functional characteristics, for example, reliability or ease-of-use.

Model-based analysis and testing (MBT) differ considerably from traditional testing and will be reviewed in the next section. Formal models of MBT often come from automata theory (for example, final-state automata to model sequential systems and processes, see Appendix A), general net theory (for example, Petri nets to model concurrency, see Appendix B), decision tables to model causality, etc. Semi-formal models are Jackson diagrams, UML diagrams, etc. [56, 82, 86].

5.2.1   Model-Based Analysis, Test Case Generation, and Test Construction

Once the system under consideration (SUC) has been modeled, the very next step is to validate this model by:

- using formal methods/tools, for example, model checkers [28], and/or
- consulting end users/customer.

The validated model can then be used for validating the SUC concerning the user requirements and analyzing the system design—long before the implementation and testing start. Thus, a model is very useful, not only for testing. Therefore, the recommendation is:

Start always, and as early as possible with a model!

Nevertheless, and wherever feasible, the end user(s) should be involving in carrying out the analysis and discussed with them in detail. Experience and empiric studies show that in most cases these discussions reveal flaws in the model that can be explained by impairments in the requirements, that is, misunderstandings about what the user exactly needs and requires.

5.2.1.1   Model-Based Analysis and Testing Process

Based on the validated model, test construction includes the following steps (see also Sect. 1.3.3):

- Select a test *adequacy criterion*, or a set of criteria to justify the test termination.
- Transform the criteria into *coverage measure(s)*.
- Construct test cases and form *test suites*.
- Test using these test suites until the criteria have been fulfilled, or the test budget is exhausted (*test termination*).

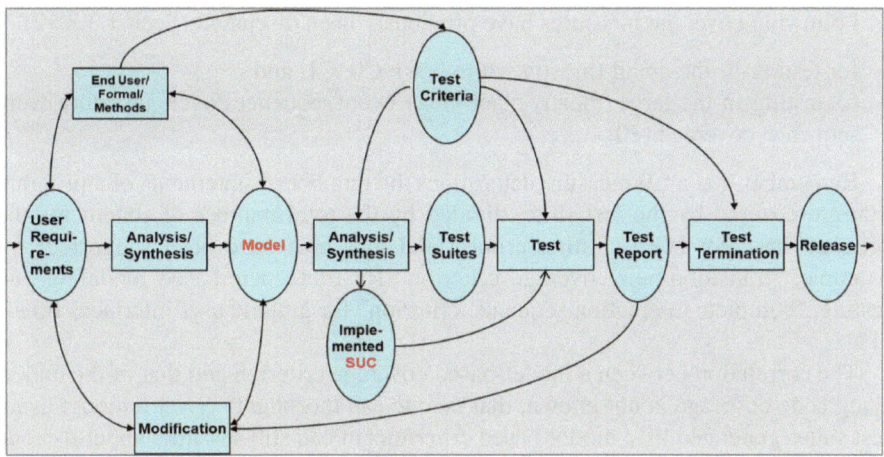

Fig. 5.5 Model-based analysis and testing process

Figure 5.5 refines and illustrates the above-mentioned steps of MBT that start with analyzing the user requirements and constructing a model. Sets of well-defined test suites can be constructed using meaningful test criteria. The phrases "well-defined" and "meaningful" remind of the requirement that end-user(s) should be involved in this process.

Provided that the SUC is already implemented and thus executable, it can then be tested. Depending on the test report, which summarizes the outcomes of the test, the test process can be concluded (*terminated*), or continued. To avoid an endless testing loop, the model, together with the test suite, should be carefully reviewed and revised each time, reconsidering the user requirements, which may have been updated in the meantime.

5.2.1.2 Test Adequacy and Test Coverage

After analyzing and validating the model, test cases can be generated and collected in test suites that correspond to well-defined test objectives. This process aims at "covering" the structure or functionality of the model by constructing test cases that will then be run to test the SUC.

A test *coverage*, as a measure of test thoroughness, is used to quantifiable describe the extent to which the model/source code is tested by a particular test suite.

Coverage is not a size, has no dimension, and thus is measured in percentage (%). Neither can it be measured by the number of test cases, but it controls this number. Thus, the coverage is the quotient of the part of the model/program that will be, or has been, tested to the whole size of the model/program. A model/program with high coverage is supposed to have been more thoroughly tested and thus has a lower chance of containing software bugs than a program with low code coverage.

Following coverage measures have previously been discussed (Sect. 1.3.3.12).

- for testing-in-the-small (mostly white-box): C0, C1, and
- for testing-in-the-large (mostly black-box): event sequence coverage, faulty event sequence coverage [20].

Remember that a C0 measure determines the number of statements of a program that are covered by the test suite, divided by the total number of statements the program has. For MBT, further criteria and measures have been suggested; for example, "transition-pair coverage criterion" for finite-state-based modeling and testing, "complete interaction sequence criterion" for graphic user interfaces (GUI) [19].

The correlation between a model-based coverage criterion and that of the underlying code coverage is not known, that is, one can thoroughly cover a model using test suites generated by a model-based criterion but can still say little about its code coverage.

Note 1: A test that achieves a 100% coverage does not necessarily guarantee that the SUC contains no faults.
Note 2: Coverage does not necessarily depend on the number of test cases.

5.2.1.3 Is Model-Based Testing Always Beneficial and Simple?

Model-based testing (MBT) has many advantages, for example, the following:

- MBT elegantly overcomes one of the toughest problems of testing, namely, predicting the test results (*test oracle*), because the model will be constructed exactly in the way that the test process and test results steer the modeling process.
- Test suites can be used all the time, no matter how often the implementation is modified—as long as the requirements are not changed.
- Changes on requirements can be adapted to the model, and test cases can then be updated and generated automatically, provided that the model is technically sound; at best, formal.

Even though modeling techniques—for example, using flow charts, decision tables, FSA—can be learned quickly, MBT is not easy to deploy. Putting the reality in a model, that is, selecting the relevant aspects and ignoring others, is hard and necessitates not only domain insight but also some experience. Thus, model construction requires avoiding unnecessary detail and focusing on the relevant issues. This needs expert-guided training.

Automatic test generation requires formal models based on mathematic methods. Unfortunately, not all testers like math.

So, what is the value of MBT?

- If prepared carefully, it can be guaranteed that no test will be run unnecessarily and unintendedly more than once. Moreover, there are good chances unexpected faults

will be detected by unintentionally exposing side effects. However, no guarantee can be given by MBT that SUC contains no more faults.
- Last but not least, the "richer" the tester, that is, the higher his/her test budget, the finer and the more comprehensive the test suite can become. This consequently increases confidence in the SUC.

5.2.1.4 Some Widely Used Models

The success of MBT depends on the selection of adequate modeling means. Modeling medium should possess as much expressive power as needed to describe the system or process under consideration (both abbreviated as SUC, since we use system and process interchangeably), that is, not less, but also not more. The reason is simple: The more expressive power you have, that is, semantic richness and terminological variety and diversity, the more effort you will need to handle and utilize them. Accordingly, the test suites should be adequate, corresponding to the requirements: not too big, not too small.

As mentioned in the previous section, test criteria are defined by means of the models' coverage measures. These criteria are based on either of the following subjects:

- *structural* issues of the SUC *(implementation-oriented/white-box)*, or
- its *behavioral/functional* description *(specification-oriented/black-box)*, or
- both, if both implementation and specification are available *(hybrid/gray-box; Sect. 3.3.11.5)*.

Further, criteria are needed to control the test process and judge the effectiveness of the test suites and to determine the point in time at which to stop testing (test termination problem).

The model under consideration usually focuses on one or more of the following features of the object under consideration:

- *Notational* expressiveness is rich if a description can be formed comfortably.

  - *Descriptive* notation stresses the description of the behavior/functionality of the SUC, for example, Roman digits (I, II, III, …) or Arabic digits (0, 1, 2, 3, …).
  - *Prescriptive* notation stresses the explanation of how to produce/obtain the SUC, for example, an algorithm to calculate trigonometric functions sine and cosine.

- *Operational* strength enables to apply logic/mathematical operations to the model, that is, calculate on and with the model, for example, arithmetic calculus of addition/subtraction/multiplication/division.
- *Rationalizing* (amplifying) models aim at reducing the effort to operate on and with the model.

- *Generating* models produce solutions/prototypes, for example, production of words of a formal language using its formal grammar [48].
- *Recognizing* models help to decide whether a solutions/prototypes is valid/correct or not, for example, checking whether a word belongs to a formal language using its formal grammar [47].
- *Cognitive* models enable reasoning, for example, to prove a theorem.

Existing models can be categorized into following groups:

- Graph-Based models
- Rule-Based Modeling
- Algebraic Modeling.

Some of them, which are relevant to testing, will be discussed in the following sections. See recommended references at the end of this chapter [10, 15, 25, 47, 79, 86].

5.2.2   Model-Based Testing of Reuse-Oriented Software

The techniques listed in the mid branch of Fig. 1.1 (Sect. 1.3.3.11 and Sect. 5.1.2), that is, *test methods at component/system level*, are generally applicable to testing CBSE-based software and COTS. However, they need some adaptation effort. The discussion in Sects. 6.4 and 6.5 explain why and where this additional effort is necessary: Assuring their compliance with application- and context-dependency. Needless to say, the component has to be "built for reuse" (Sect. 4.3.4 and Fig. 4.4 therein). Accordingly, it has to fulfill the requirements concerning its structure and interface (Sect. 4.1.1, Figs. 4.2 and 4.3). Therefore, the test of a reuse component has the character of an integration test and is to be performed at the customer site. A final acceptance test has to be performed, also at the customer site, after completing the integration with other inter-operating systems and hardware, considering also environmental and energy-consuming aspects (see Sects. 2.2.4, 3.3.2 and 7.2; [24]).

Testing of COTS software corresponds by and large to testing of CBSE-based software. However, COTS software can also be delivered without source code. This situation necessitates specific care [23, 78].

Generally, test cases for reuse can be generated using criteria in accordance with the model used to specify the behavior of the PL. Such models can be *state-oriented*, for example, state machines that deliver state-transition coverage as a test case generation criterion. *Causal* modeling can be realized using formal logic, operationalized by Boolean algebra [12, 15]. In this case, criteria are constructed to cover control instances of the software under consideration, for example, condition/decision coverage, or modified condition/decision coverage (MC/DC) [12]. This book prefers *event-oriented* strategy using event-sequence graphs (ESG; see Appendix A).

The rest of this section is dedicated to model-based testing of software product lines and product variants that can be derived from these product lines.

5.2.3 Criteria for Sampling Configurations

One of the major challenges of testing reuse software is the sampling of product configurations, that is, product variants, for testing. The reason is simple: In most cases in practice the variability is very broad and thus a thorough testing is, due to costs, not feasible. Some of the strategies to meet this challenge are as follows:

- Choose as many different products as possible so that each of the test cases can be executed at least once.
- Consider the interactions with its environment (other systems, user) on the behavioral level.
- Sample product configurations from the product line's test cases, in accordance with the selected features.
- Minimize the number of configurations.
- Minimize the number of tested products that represent configurations and so the test effort.

The challenges mentioned above are partly contradictory. Finding a trade-off represents an optimization problem in constructing a test suite of m-test cases subject to [57]:

- few/many configurations not exceeding m,
- small/large configurations by means of selected features,
- diverse configurations, and
- combinations thereof.

Another challenge is the test adequacy given by criteria for the coverage of the structure or functionality of software under consideration. Various criteria are suggested to decrease the risk of releasing a faulty product. Some of them will be reviewed in the following sub-sections.

5.2.4 Model-Based Testing of Product Lines

Imagine that you order a car with specific capabilities and accessories. When it is time for delivery and you are required to sign to confirm you are getting the car you ordered, you will usually check following:

- Is this car a car at all? In other words, can you get inside it? Does it have the capability to drive, to brake, to be steered, etc.?
- Does it have the specific features and accessories you ordered and paid for, for example, is it red and a convertible? Does it have automatic transmission, plug-in hybrid engine, and everything else you wanted?
- Will it will be delivered at the time and location of your choice, with the promised price reduction for your old car as a trade-in; payment in monthly instalments, free gas and free electric power charge for the first three months?

- Moreover, you would like to be informed in case an ordered accessor is missing and be told the reasons for the problems that caused this inconvenience, such as disruptions in supply chains, a global pandemic.

Accordingly, SPL testing has to validate that any product variant will fulfill the requirements concerning the features it is expected to possess and will comply with the behavior its potential user expects. Moreover, the combined model also has to include the events of the relevant environment that might impact on the SPL and thus its product variants. These objectives require answers to following questions;

- To which extent and to which efficiency are *all* of the potential variants covered?
- To which extent and to which efficiency are the *behavior* of the variants and their *interaction* with the user and environment covered?

The first question leads to covering the feature model, while the second one to covering the behavior model. So, these questions concern the *variability* and *behavior* of the product family.

The variability aspect requires, as a first step, the analysis of the feature model to check redundancies, inconsistencies, flaws in structuring, etc. Apel et al. give a good review about *variability smells* that are undesirable properties. Examples are dead features in a FM that cannot be reached and, thus can never be selected, or unused variability that is not needed and, as a result, also can never be selected [11].

The behavioral aspect can be handled by models discussed in Sect. 5.2.1. The variability aspect will be discussed in the following sub-sections. The combination of both aspects, feature and behavior, has already been discussed in Sect. 5.2.5, and its deployment for testing SPL will be explained in depth and illustrated by an example in Sect. 5.2.5.

5.2.4.1   Binding of the Variability

As frequently discussed in the previous sections, variability is a key factor of reuse. It not only determines the structure and number of producible variants but also the costs of reuse testing. Assigning a variability factor to a specific, predefined value—also called instantiation—entails the *binding* process of the variability. Also, producing a 100% model out of a 150% model forms a *variability binding* (Sect. 4.2.6).Accordingly, model-based testing of reuse-oriented software engineering can be carried out in two alternative ways, depending on how the question is decided.

When to Bind the Variability?

The variability can be bound during domain engineering or application engineering. These two strategies will be reviewed and discussed, based on the approach introduced by Lackner, et al. [55, 58, 60, 61, 62, 65, 84].

5.2.4.2 Domain-Centric Testing

For this strategy, the binding of the variability takes place during domain engineering, *after* the test case design based on the PL model. Test generation uses the
150%, complete model of the product family. Therefore, this strategy will also be
called *product line-centric* testing (PLCT). It is a *bottom-up* strategy (in the sense
of integration testing, see Sects. 3.3.11 and 5.1.1) since no product test model is
necessary before the binding of the variability. Based on the life-cycle concepts of
SPL (Fig. 4.15), Fig. 5.6 outlines this strategy that consists of two subsequent steps:
PLCT1 (product line testing1) for the generation of the PL test cases from the PL
test model, and PLCT2 (product line testing2) for the variability binding and the test
generation for the product variants.

Note that a single domain-centric test case can apply to several product variants
that share one or more features. Accordingly, test cases for products will be selected
from a common test suite, using the corresponding features as criteria. Thus, tests
can be carried out for all products or for a minimal number of variants.

The quality of the bottom-up testing will be evaluated by considering the following
aspects (see also the questions in Sect. 5.2.4):

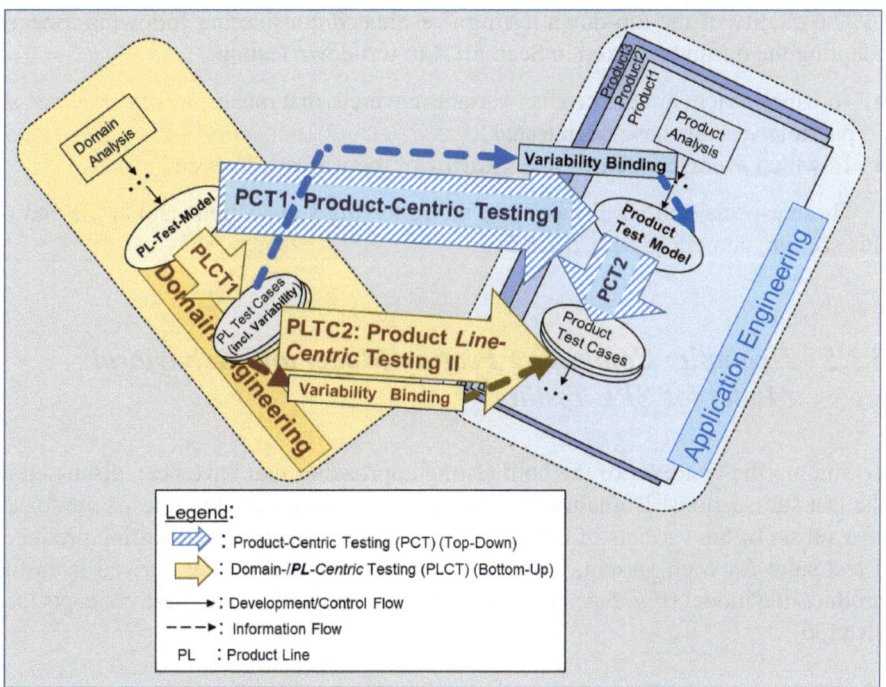

Fig. 5.6 PL/domaincCentric bottom-up Testing versus product-centric, top-down testing (based
on [62])

- To which extent is the *common features* of the product variants are covered?
- To which average extent is the *behavior* of the system that is *common* to all variants covered?

These aspects will be discussed in the following sub-section and at the end of this present section (Sect. 5.2.5.5).

5.2.4.3   Product-Centric Testing

For this strategy, the binding of the variability takes place during application engineering and *before* the test case design, based on the PL model. Test generation uses the 100% models of the product *variants*. Therefore, this strategy will also be called *product-centric* testing. It is a *top-down* strategy as product test models are used by selecting all features or unselecting some features, using selection criteria to group variants or choosing a single, special product to be tested. Selection criteria can also be generated by pairwise/n-wise testing (see Sects. 3.3.11.3, 5.1.3.1 and 5.1.3.2).

Figure 5.6 outlines this strategy that consists of two steps: PCT1 for the variability binding and product test model generation, and PCT2 for the generation of the product test cases from the product test model and the test generation for the product variants.

The quality of the top-down testing is evaluated considering following aspects, adapting the questions asked in Sect. 5.2.4 to top-down testing:

- To which extent are the product variants covered, that means, *what percent* of all possible variants have been tested?
- To which average extent is the *behavior* of the variants covered?

These aspects will be discussed in the following sub-sections and at the end of this section, namely in Sect. 5.2.5.5.

5.2.5   Example: Combining Feature Model and Behavioral Model for SPL Testing

To sum up the concepts of the both testing approaches that have been discussed in the last sub-sections: Domain-centric testing requires generating a test suite for all or a subset of the variants of a product line, while product-centric testing produces a test suite for each variant, based on the model of this variant derived from the product line model (Fig. 5.6). The following examples illustrate these concepts (see also [30]).

5.2.5.1 Test Cases, Test Suites, and Coverage Criteria for SPL Testing

The combined feature/behavior model of a SPL can be covered by product variant generation, that is, by generating all potential product variants ("brute force" approach). This can be, however, too costly, even intractable if the SPL under consideration possesses a great deal of features, leading to a broad variability of unmanageable extent. Worse, some of the variants that can be derived might prove to be useless in case the feature diagram includes some alternative features that mutually excludes each other. Consider the example depicted in Fig. 4.12: a high security level is required only if the payment will be by credit card. Other methods of payment will be carried out following low security regulations. So, a derived variant that is provided by high and low security mechanisms and enables the payment by bank account or e-coins, but not credit card, would be out-of-the market. Some techniques for avoiding such hapless faulty issues for a controlled coverage of the feature model will be discussed in the next sections.

Several techniques are available for covering the behavioral model of a SPL (Sect. 4.2.1). Many authors suggest transition coverage of state machines. This book follows the approach introduced by Weißleder et al., but advocates the coverage methods suggested by the ESG approach and to exploit its advantages summarized in Sect. 4.2.5 and Appendix 1 [21]. The ESG approach enables the conventional *positive* testing of the system under consideration, validating that it correctly does everything as required. ESG concept also realizes *negative* testing that validates that the system under consideration does not do anything that is *not* required (Sect. 1.3.3.5). Moreover, the scalability of the approach enables to adjust the test costs with the increasing length of the test sequences. Last but not least, tool chains are available for generating and optimizing test suites for both positive and negative testing [16, 31] (Appendix A).

Test cases to validate the SPL under consideration form *complete* sequences of events (*CES*) that start at the beginning of the ESG (noted by the bracket "[") and are concluded at its end (noted by the bracket "]"), representing *desirable* (or correct/legal/acceptable) sequences. Coverage criteria are formed by varying the length of the events to be covered, that is, 1, 2, 3, etc. Covering a sequence of the length 1 means *event* coverage, length 2 means coverage of *event pairs*, length 3 means coverage of *event triples*, etc. A CES alone might not be sufficient to fulfill the required coverage criterion, such as covering all event pairs. In this case, more than one CES needs to be constructed, forming a *test suite*. It is also likely that a CES or a test suite covers some of the events (event pairs, event triples, etc.) multiple times (see Appendix A).

Any pair of event sequences that are not included in the ESG forms an *undesirable* (or faulty/incorrect/illegal) pair of events (*FEP, faulty event pair*). Any sequence of events that starts at the beginning of the ESG and includes a FEP is a *complete FEP* (*CFES*). FCESs are used to realize negative testing (Sect. 1.3.3.5).

5.2.5.2 Domain-Centric Testing of the Example

Domain-centric testing of a SPL forms a bottom-up strategy, merging the feature model and 150% behavioral model of the product line under consideration to generate all potential test cases using an appropriate coverage criterion (Sects. 3.3.11 and 3.3.13, [43]). The feature model is used as a control mechanism to adapt the model to the features of an individual variant (Fig. 5.7, based on Fig. 5.6).

We return now to the idea of behavioral augmentation discussed in Sect. 4.2.5 (Figs. 4.10 and 4.11) and link the behavioral model to the feature model (Fig. 4.12). This linked part of the 150% state machine will then be used for automatic test generation of only valid product variants (based on [84, 85]).

The feature search is optional, that is, it can be included or skipped. Accordingly, we need a "guard" that is set (or selected) for deriving variants that include this feature. For variants that do not include the search feature, this guard will be reset (or de-selected). Another guard is needed for the feature security that will be set if the payment will be made by a credit card. The feature low is not set as it would exclude the payment by credit card. The feature high, which is necessary for payment by credit card, is set as it enables the payment by alternative ways as well. Thus, this version to run all paths of the 150% model.

The test suite list summarized in Table 5.2 consists of two parts. The first part reproduces the test suite generated to cover all events, which are the transitions of the state machine in Fig. 4.10 and borrowed from [84].

The second part of the list represents test suites generated by the test tool *TSD* (*Test Suite Designer*, [31] for the ESG given in Fig. 4.11 (as augmented with FM, as in Fig. 5.8) and consists of following:

- A test suite to cover all event sequences of the length 2 (event pairs) for positive testing. All events are covered, as well, some of them multiple times.

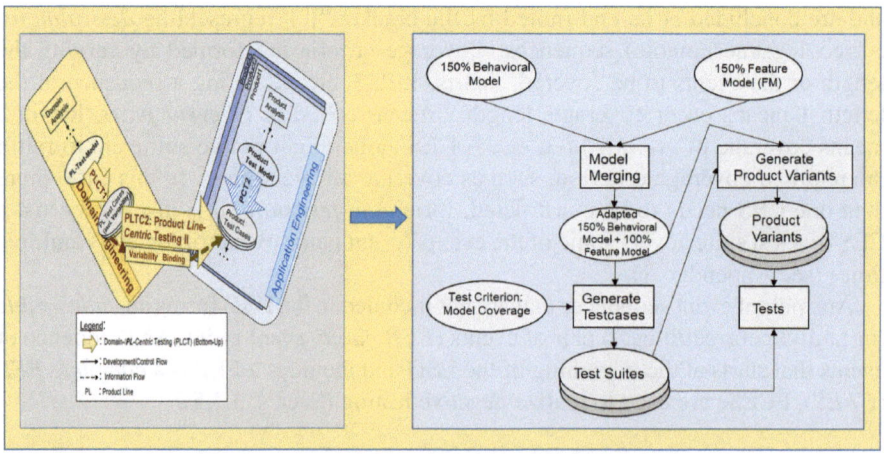

Fig. 5.7 Domain-centric, bottom-up Testing of the example, concept (based on [84])

Table 5.2 Test suites for bottom-up testing of the example (Fig. 5.8)

TEST SUITES for

Domain-Centric, Bottom-Up Testing of the Example

A. TEST SUITE GENERATED WITH SM (TWcall3)

Sequence Length to be Covered: 1 (Transitions as Events):
 # Test Cases: 2; # Test Steps: 27; # Transitions (Events) Covered: 22

```
1-21: OpenProductCatalog, ProductDetailsFor, ReturnToCatalog, SearchFor, Product-DetailsFor, AddProductToCart, ReturnToSearchResults, ReturnTo-Catalog, ToCart,
      RemoveProduct, ProceedToCheckout, Select-BankAccount, ProceedPayment, Invalid, SelectPayment-Method, SelectECoins, ProceedPayment, Invalid, Cancel.Payment,
2-6: OpenProductCatalog, ProceedTo-Checkout, SelectPaymentMethod, SelectCreditCard, Proceed-Payment, Valid.
```

B. TEST SUITES GENERATED WITH ESG

Sequence Length to be Covered: 2 (Event Pair = 2 Transitions);
 # Test Cases (CES): 11: # Test Steps: 85; # Events Covered: 17
 # Event Pairs Covered: 36 (Transition Pairs)

B1. Positive Test Cases (Complete Event Sequences – CES)

85: [,

```
1-04: OpenProductCatalog, SearchFor, OpenProductCatalog, ProductDetailsFor,
2-11: OpenProductCatalog, SearchFor, ProductDetailsFor, SearchFor, ProceedTo-Checkout, SelectPaymentMethod, SelectCreditCard, ProceedPayment, Valid, Cancel.Payment,
3-09: OpenProductCatalog, ProductDetailsFor, ToCart, ProceedToCheckout, SelectPaymen-Method, SelectCreditCard, ProceedPayment, Valid, Cancel.Payment,
4-08: OpenProductCatalog, ProductDetailsFor, ProceedToCheckout, SelectPaymentMethod, SelectCreditCard, ProceedPayment, Valid, Cancel.Payment,
5-10: OpenProductCatalog, ProductDetailsFor, AddProductToCart, SearchFor, ProceedTo-Checkout, SelectPaymentMethod, SelectCreditCard, ProceedPayment, Valid, Cancel.Payment,
6-10: OpenProductCatalog, ProductDetailsFor, AddProductToCart, ToCart, ProceedToCheckout, SelectPaymentMethod, SelectCreditCard, ProceedPayment, Valid, Cancel.Payment,
7-09: OpenProductCatalog, ProductDetailsFor, AddProductToCart, Select-PaymentMethod, SelectCreditCard, ProceedPayment, Valid, Cancel.Payment,
8-04: OpenProductCatalog, ProductDetailsFor, AddProductToCart, AddProductToCart,
9-11: OpenProductCatalog, ProceedToCheckout, SelectCreditCard, ProceedPayment, Invalid, SelectPaymentMethod, SelectE-Coins, ProceedPayment, Valid, Cancel.Payment,
10-09: OpenProductCatalog, ToCart, RemoveProduct, ProceedToCheckout, SelectPaymentMethod, SelectBankAccount, ProceedPayment, Valid.
]
```

B1. Negative Test Cases (Faulty Complete Event Sequences – FCES) (not complete)

B1.1 Length 2

```
[, Open Product Catalog, Open Product Catalog
[, Open Product Catalog, Add Product to Cart,
[, Open Product Catalog, Remove Product,
[, Open Product Catalog, Select Payment Method,
[, Open Product Catalog, Select Credit Card,
[, Open Product Catalog, Proceed Payment,
[, Open Product Catalog, Invalid,
[, Open Product Catalog, Valid,
[, Open Product Catalog, Cancel Payment,
[, Open Product Catalog, Select E-Coins,
[, Open Product Catalog, Select Bank Account,
```

B1.2 Length 3

```
[, Open Product Catalog, Product Details For, Product Details For,
[, Open Product Catalog, Product Details For, Remove Product,
[, Open Product Catalog, Product Details For, Select Payment Method,
[, Open Product Catalog, Product Details For, Select Credit Card,
[, Open Product Catalog, Product Details For, Proceed Payment,
[, Open Product Catalog, Product Details For, Invalid,
[, Open Product Catalog, Product Details For, Valid,
...
```

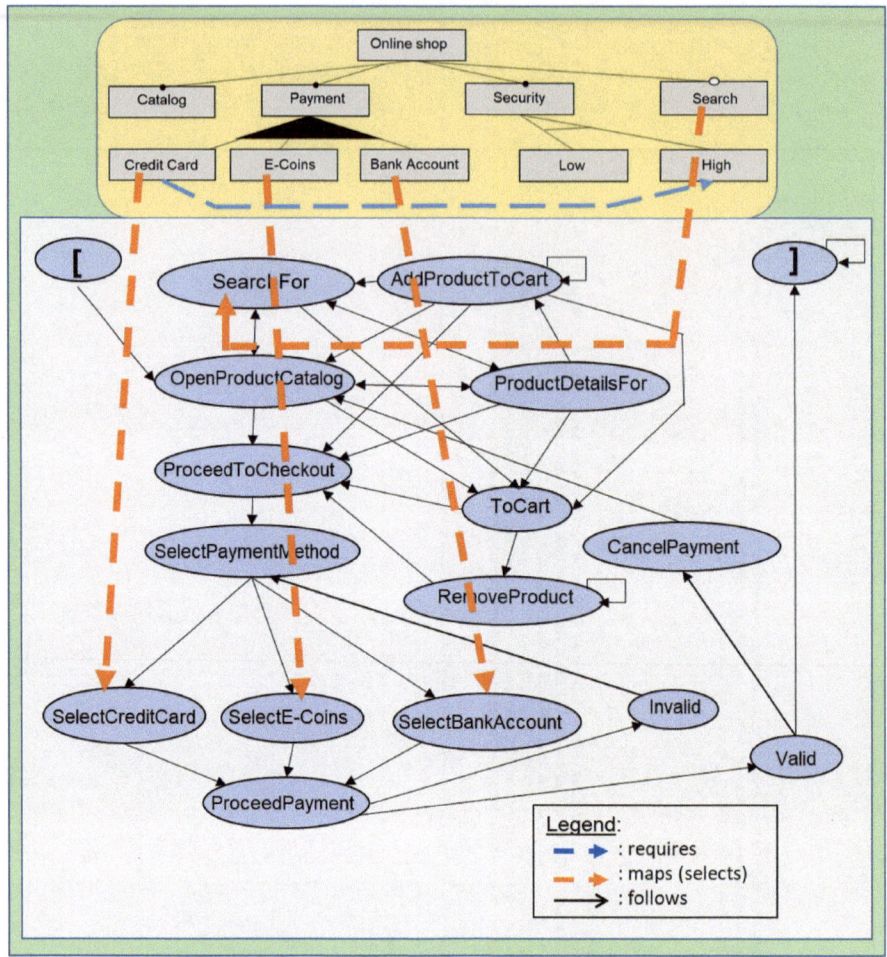

Fig. 5.8 Bottom-up, domain-centric testing of the example, 150% model

- Test suites for negative testing to cover all faulty complete event sequences of the length 2 and 3.

5.2.5.3 Product-Centric Testing of the Example

Product-centric testing of a SPL represents a top-down strategy that starts, analogously to the bottom-up strategy, with the 150% behavioral models (Figs. 4.10 and 4.11) and merges it with the 100% feature models (Figs. 5.10 and 5.11); the latter has already been discussed in Sect. 4.2.4, Fig. 4.10).

The resulting model will then be used to generate product variants and, subsequently, 100% models. Test suites will be generated to test these variants, using an

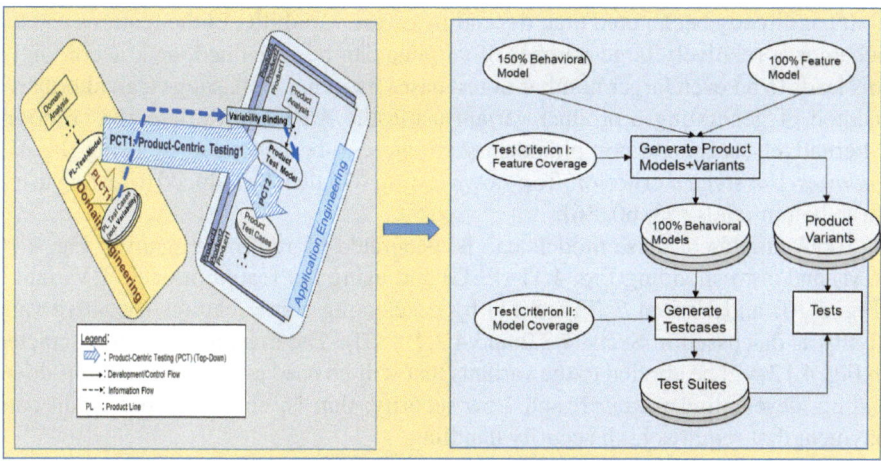

Fig. 5.9 Product-centric, top-down testing of the example (based on [84])

appropriate coverage criterion, for example, event pair coverage (Fig. 5.9, based on Fig. 5.6).

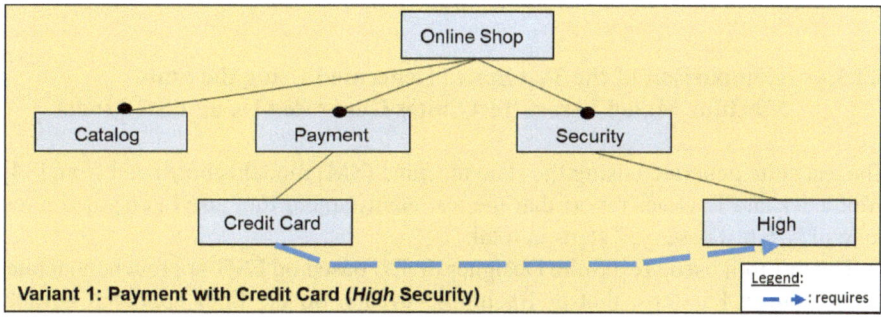

Fig. 5.10 Product variant 1 for top-down testing (based on [84])

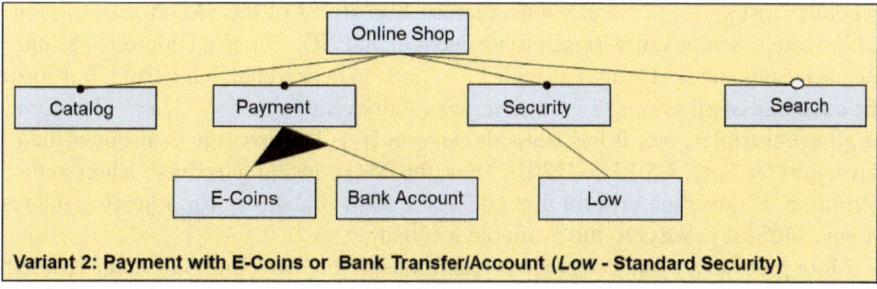

Fig. 5.11 Product variant 2 for top-down testing (based on [84])

It has already been noted that, depending on the variability of the features and the behavior, a relatively large number of variants can be generated, and, accordingly, this leads to an even larger number of test cases. Lackner et al. suggest an alternative method of generating a product variant using the *All-Features-Selected* criterion. Alternatively, a second complementary variant can be generated by using the *All-Features-Unselected* criterion. Top-down testing will then be reduced to testing using these both models [58, 60, 86].

100% models of these models can be generated by returning again to Fig. 4.10 (SM) and corresponding Fig. 4.11 (ESG) and using the feature model of Variant 1 (Fig. 5.10) and Variant 2 (Fig. 5.11) by deselecting some features (negative variability as discussed in Sects. 4.2.2 and 4.2.4; [85]). The mapping function depicted in Fig. 4.12 will be applied to the variants that will be used as examples for top-down testing, deselecting `search` and `low` security, that is, enabling only credit card payment that requires high security handling.

Figure 5.12, which is an augmented version of Fig. 4.12 and has already been used as an example in Sect. 4.2.6 (Fig. 4.14), depicts the 100% model of Variant 1. Test suites covering all transitions of the SM (not included) and event pairs of the ESG (Fig. 5.12) for positive testing and all faulty event pairs are given in Table 5.3).

Accordingly, the complementary part of the 150% model leads to the 100% model of Variant 2 (Fig. 5.13). Applying the transition coverage criteria also to this model delivers the test suites given in Table 5.4.

5.2.5.4   Comparison of the Test Suites Generated Using the State Machine Model Versus Test Suites Generated Using ESG Model

The test suite generated using the state machine (SM) model is borrowed from [84]. Weißleder and Lackner report that the test environment they used generated a test suite of two test cases, 27 steps in total.

The test tool used, Test Suite Designer (TSD) based on ESG-approach, generates a test suite in 85 steps, that is, tracing 85 events. So the ESG approach is more expensive if costs are measured by the number of test cases.

The difference between the costs is caused first of all by the fact that the TSD test suite covers event *pairs*, that is, sequences of events of the length two. SM strategy generates test suites to cover events that are transitions of the SM. A transition in a UML state machine corresponds to an event in an ESG. The ESG approach requires the event sequences of length i, with $i > = 1$, to be covered, whereby $i = 1$ forms the coverage of all events, $i = 2$ coverage of all event pairs, $i = 3$, $i = 3$ coverage of all event triples, etc. It has been shown that $(i + 1)$- coverage is stronger than i-coverage (see Sect. 3.3.13 and [20]). Thus, the ESG concept introduces a hierarchical definition of coverage criteria that enables a good scalability for adjusting the test targets and test process to the available test budget.

More precisely, event coverage is analogous to the Traveling Salesman Problem, while event pair coverage is analogous to the Chinese Postman Problem (CPP) [9, 20, 33]. TSD deploys specific CPP algorithms to optimize the number of test cases

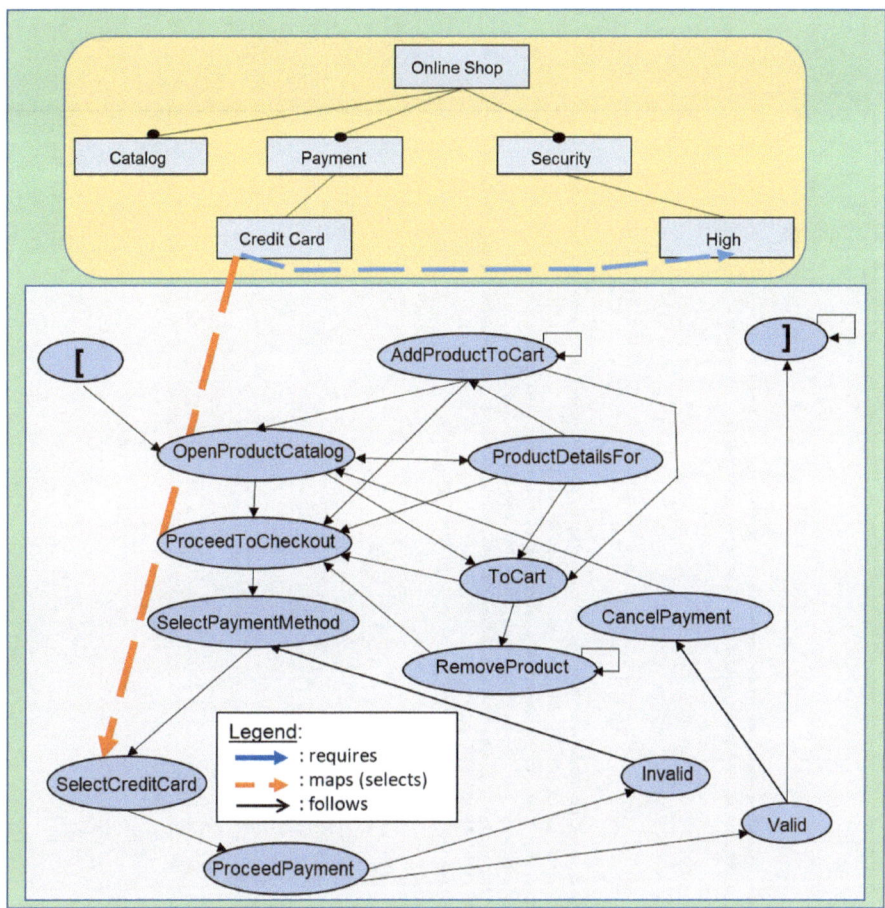

Fig. 5.12 Variant 1—Credit-card-payment-only as ESG model

generated, also for the event sequence of the length greater than 2 [17]. The strategy TSD is based on determines the least number of test cases of minimal total test length. This means, running a test suite with only one test case (in ESG jargon: CES, complete event sequence, see Appendix A) will be favored, because running more than one test case would cause interrupts between the test runs caused by (i) the reset of the system under test each time that it is necessary to return to the initial state, (ii) then to evaluate the test last run, and finally (iii) to prepare the next test. These three steps are avoided as far as possible by TSD. SM strategy prefers running more tests by keeping the total test length as small as possible. Accordingly, the test suite proposed by Weißleder and Lackner includes more than one test run to cover all events (note: not event pairs!).

Table 5.3 Test suites for top-down testing of the variant 1 of example (Fig. 5.12)

TEST SUITES for

Product-Centric, Top-Down Testing of the Example Variant 1

A. TEST SUITE GENERATED WITH SM ([WeLa13])

Sequence Length to be Covered: 1 (Transitions at Event);
 # Test Cases: 2; # Test Steps: 19; # Transitions (Events) Covered: 22

1-09: OpenProductCatalog, ProductDetailsFor, AddProductToCart, Return-ToCatalog, ToCart, RemoveProduct, ProceedToCheckout, CancelPayment;

4-10: OpenProductCatalog, ProceedToCheckout, SelectPaymentMethod, SelectCreditCard, ProceedPayment, Invalid, SelectPaymentMethod, SelectCreditCard, ProceedPayment, Valid.

B. TEST SUITES GENERATED WITH ESG

Sequence Length to be Covered: 2 (Event Pair = 2 Transitions);
 # Test Cases (CES): 5; # Test Steps: 62; # Events Covered: 14
 # Event Pairs Covered: 25 (Transition Pairs)

B1. Positive Test Cases (Complete Event Sequences – CES)

62: [,

1-11: OpenProductCatalog, ProductDetailsFor, OpenProductCatalog, ProductDetailsFor, ToCart, ProceedToCheckout, SelectPaymentMethod, SelectCreditCard, ProceedPayment, Valid, CancelPayment,

2-08: OpenProductCatalog, ProductDetailsFor, ProceedToCheckout, SelectPaymentMethod, SelectCreditCard, ProceedPayment, Valid, CancelPayment,

3-10: OpenProductCatalog, ProductDetailsFor, AddProductToCart, ToCart, ProceedToCheckout, SelectPaymentMethod, SelectCreditCard, ProceedPayment, Valid, CancelPayment,

3-09: OpenProductCatalog, ProductDetailsFor, AddProductToCart, ProceedToCheckout, Select-PaymentMethod, SelectCreditCard, ProceedPayment, Valid, CancelPayment,

4-15: OpenProductCatalog, ProductDetailsFor, AddProductToCart, Open-ProductCatalog, ProceedToCheckout, SelectPaymentMethod, SelectCreditCard, Proceed-Payment, Invalid, SelectPaymentMethod, SelectCreditCard, ProceedPayment, Valid, Cancel-Payment,

5-09: OpenProductCatalog, ToCart, RemoveProduct, RemoveProduct, ProceedToCheckout, Select-PaymentMethod, SelectCreditCard, ProceedPayment, Valid.
]

B2. Negative Test Cases (Faulty Complete Event Sequences – FCES)

B2.1 Length 2

[, Open Product Catalog, Open Product Catalog
[, Open Product Catalog, Add Product to Cart,
[, Open Product Catalog, Remove Product,
[, Open Product Catalog, Select Payment Method,
[, Open Product Catalog, Select Credit Card,
[, Open Product Catalog, Proceed Payment,
[, Open Product Catalog, Invalid,
[, Open Product Catalog, Valid,
[, Open Product Catalog, Cancel Payment,

B2.2 Length 3

[, Open Product Catalog, Product Details For, Product Details For,
[, Open Product Catalog, Product Details For, Remove Product,
[, Open Product Catalog, Product Details For, Select Payment Method,
[, Open Product Catalog, Product Details For, Select Credit Card,
[, Open Product Catalog, Product Details For, Proceed Payment,
[, Open Product Catalog, Product Details For, Invalid,
[, Open Product Catalog, Product Details For, Valid,
[, Open Product Catalog, Product Details For, Cancel Payment,

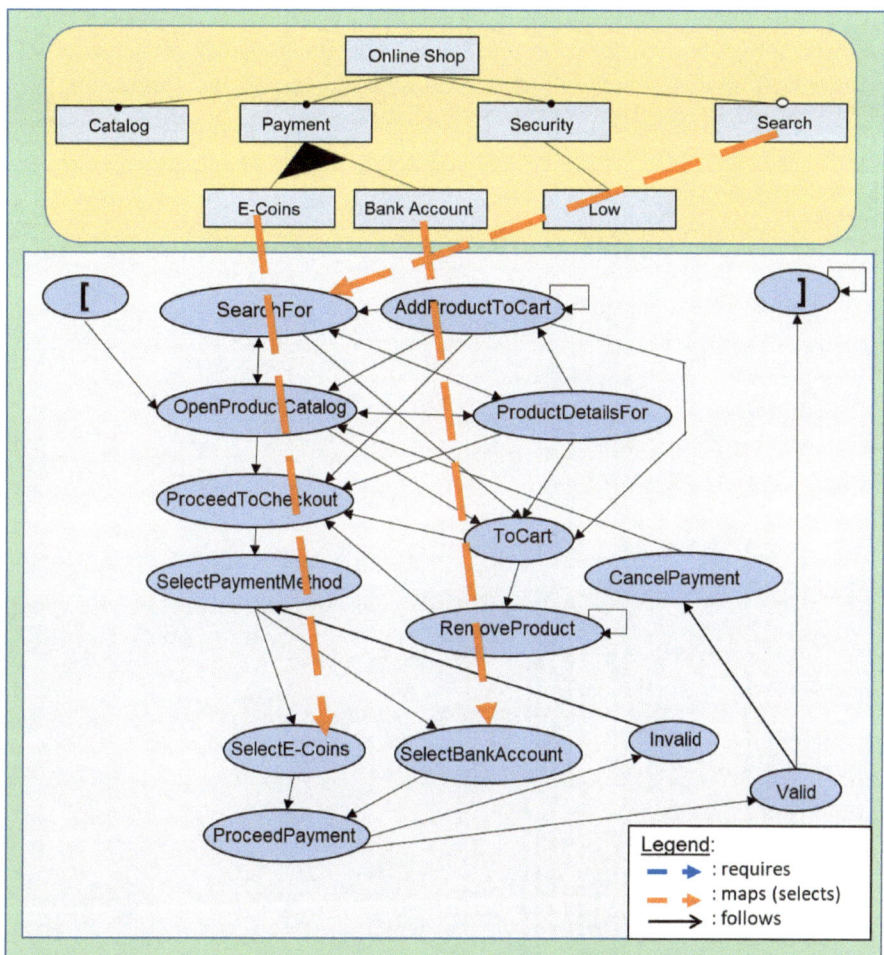

Fig. 5.13 Variant 2 - E-Coins or bank account-payment with search option as ESG model

5.2.5.5   Discussion: Which Modeling Approach, Which Test Coverage Criteria?

As discussed above, event sequence coverage is stronger than event coverage. This means, any test suite to cover event pairs also covers events. Research work found that event pair coverage detects more faults than event coverage [20]. Consequently, covering event triples is supposed to do an even better job. This is true, but the test costs increase excessively with the length of the event sequence to be covered. Experiments show that the coverage of event pairs is the most cost-effective strategy.

Research and development activities in the designing, modeling, and testing of SPLs are continuing, and interesting novel approaches will be proposed in increasing

Table 5.4 Test suites for topdDown Testing of the variant 2 of example (Fig. 5.13)

TEST SUITES for

Product-Centric, Top-Down Testing of the Example Variant 2

A. TEST SUITE GENERATED WITH SM ([WeJa13])

Sequence Length to be Covered: 1 (Transition as Event):
Test Cases: 2; # Test Steps: 24; # Transitions as Events: 22

1–11: *OpenProductCatalog, SearchFor, ProductDetailsFor, AddProductToCart, AddProductToCart, RemoveProduct, ProceedToCheckout, CancelPayment;*

2–13: *OpenProductCatalog, ProductDetailsFor, AddProductToCart, ReturnToCatalog, ProceedToCheckout, SelectPaymentMethod, SelectECoins, ProceedPayment, Valid*

B. TEST SUITES GENERATED WITH ESG

Sequence Length to be Covered: 2 (Event Pair=2 Transitions);
Test Count (CES): 7; # Test Steps: 85; # Events Covered 17
Event Pairs Covered 36 (Transition Pairs)

B1. Positive Test Cases (Complete Event Sequences – CES)

05: *I,*

1–15: *OpenProductCatalog, ProductDetailsFor, ToCart, RemoveProduct, ProceedToCheckout, SelectPaymentMethod, SelectE-Coins, ProceedPayment, Invalid, SelectPaymentMethod, SelectBankAccount, ProceedPayment, Valid, CancelPayment,*

2–21: *OpenProductCatalog, SearchFor, OpenProductCatalog, ToCart, ProceedToCheckout, SelectPaymentMethod, SelectBankAccount, ProceedPayment, Valid, CancelPayment, OpenProductCatalog, ProductDetailsFor, AddProductToCart, SearchFor, ProceedToCheckout, SelectPaymentMethod, SelectBankAccount, ProceedPayment, Valid, CancelPayment,*

3–09 *OpenProductCatalog, SearchFor, ProductDetailsFor, ProceedToCheckout, SelectPaymentMethod, SelectBankAccount, ProceedPayment, Valid, CancelPayment,*

4–12: *OpenProductCatalog, ProductDetailsFor, OpenProductCatalog, ProductDetailsFor, SearchFor, ToCart, ProceedToCheckout, SelectPaymentMethod, SelectBankAccount, ProceedPayment, Valid, CancelPayment,*

5–12: *OpenProductCatalog, ProductDetailsFor, AddProductToCart, OpenProductCatalog, ToCart, ProceedToCheckout, SelectPaymentMethod, SelectBankAccount, ProceedPayment, Valid, CancelPayment,*

6–10: *OpenProductCatalog, ProductDetailsFor, AddProductToCart, ToCart, ProceedToCheckout, SelectPaymentMethod, SelectBankAccount, ProceedPayment, Valid, CancelPayment,*

7–06: *OpenProductCatalog, ProceedToCheckout, SelectPaymentMethod, SelectBankAccount, ProceedPayment, Valid,*
]

B2. Negative Test Cases (Faulty Complete Event Sequences – FCES)

B2.1 Length 2

[, *Open Product Catalog, Open Product Catalog,*
[, *Open Product Catalog, Select Bank Account,*
[, *Open Product Catalog, Select Payment Method,*
[, *Open Product Catalog, Select E-Coins,*
[, *Open Product Catalog, Proceed Payment,*
[, *Open Product Catalog, Valid,*
[, *Open Product Catalog, Invalid,*
[, *Open Product Catalog, Remove Product,*
[, *Open Product Catalog, Add Products to Cart,*
[, *Open Product Catalog, Cancel Payment,*
[, *Open Product Catalog, Select Bank Account,*
...

B2.1 Length 3

[, *Open Product Catalog, Proceed to Checkout, Open Product Catalog,*
[, *Open Product Catalog, Proceed to Checkout, Select Bank Account,*
[, *Open Product Catalog, Proceed to Checkout, Proceed to Checkout,*
[, *Open Product Catalog, Proceed to Checkout, Select E-Coins,*
[, *Open Product Catalog, Proceed to Checkout, Proceed Payment,*
[, *Open Product Catalog, Proceed to Checkout, Valid,*
[, *Open Product Catalog, Proceed to Checkout, Invalid,*
[, *Open Product Catalog, Proceed to Checkout, To Cart,*
[, *Open Product Catalog, Proceed to Checkout, Remove Product,*
[, *Open Product Catalog, Proceed to Checkout, Product Details For,*
[, *Open Product Catalog, Proceed to Checkout, Add Products to Cart,*
[, *Open Product Catalog, Proceed to Checkout, Search For,*
[, *Open Product Catalog, Proceed to Checkout, Cancel Payment,*
[, *Open Product Catalog, Proceed to Checkout, Select Bank Account,*
...

numbers. This book can offer only snapshots of some momentary developments in this fast-moving area of often short-lived techniques. It is the author's hope that the techniques discussed in this section will help the reader gain an idea of the big picture and provide some hints for following the state-of-the art in the years ahead.

5.2.6 Model-Based Mutation Testing of Product Lines

The discussion on a fault-oriented analysis and testing technique, namely model-based mutation testing for PLs, concludes this chapter.

The concept of mutation testing has been briefly explained in Sect. 3.3.11.4. Belli et al. introduced a concept of model-based mutation testing that will be used in the following discussion. This concept is based on elementary mutation operators, "delete" and "insert," and their combinations and concatenations applied to the model of the SUC [18]. For example, "replace" can be built by a "delete" step, followed by an "insert." The elementary operations, "delete" and "insert," can be applied to the leaves of a feature diagram, which forms a tree, or to the nodes of an ESG to mutate the PL model and/or product models that will be used to generate test cases. On the other hand, sample product configurations will be used to develop product variants. The generated test cases will then be applied to these variants. If the test fails, the mutant is "killed," that is, a fault is detected (see Sect. 3.3.11.4). Otherwise, the variant is free of this (and only this) fault (Fig. 5.14). A problem can be caused by *equivalent* mutants exhibiting the same behavior as the original model, at least in some cases, so that they cannot be killed, even though they deliver faulty results in some other cases. Approaches that follow this concept will be briefly reviewed in the following.

Ferreira et al. suggest operators for mutating the feature diagram to generate fault models and test cases [36]. A FM-Analyzer carries out tests using the generated test cases "against" the original, correct FM and mutants. Consequently, a mutant is considered dead if the outcome of a test of this mutant differs from the outcome of the test of the original FM using the same test case. Mutation scores can then be determined to check the efficiency of the approach and to compare it with similar approaches.

Another FM-based approach was introduced by Filho et al. to select products associated with mutation operators for modeling faults in the FM [37]. Multi-objective and evolutionary algorithms are used to minimize the test set generated. The operators and mutation score are used to evaluate the approach.

Henard et al. propose deriving dissimilar test suites by applying heuristic methods to compare variants and detect faults in FM, and, based on the number of those faults, determine the mutation score of the examined test suites [46].

Last but not least, Lackner and Schmidt review existing paradigms for specifying software product lines and discuss likely faults in specification processes [59, 60].

See also Sect. 5.1.4 for other methods of redundancy-based testing.

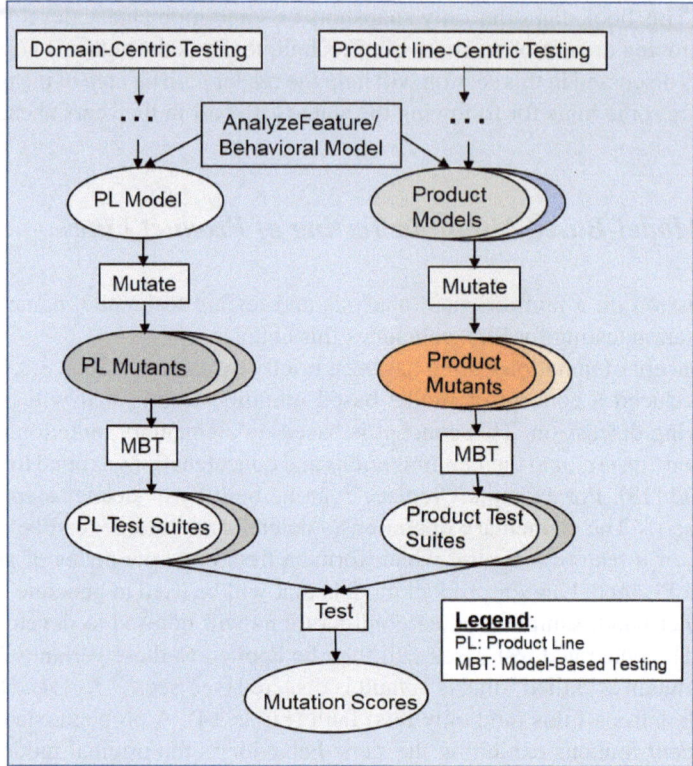

Fig. 5.14 Mutation testing of product lines (based on [59, 60])

5.3   Economical Aspects and Metrics for Reuse

Just to be clear, we are talking about software systems built for around one-tenth the cost. With around one-tenth the faults. Delivered in around one-tenth the time. If you know of another way to achieve such a staggering combination of better-faster-and-cheaper, please let me know.

Paul Clements, Software Engineering Institute, Carnegie Mellon University (in [64]).

The motivation behind reuse stems primarily from the expectation of potential saving through reduced time-to-market and risks, and thus saving costs on the one hand and, on the other hand, increasing quality through ensuring dependability, standardization, and interoperability of the products developed and offered by a company's business area. These benefits and their costs need to be analyzed in the course of reuse economics as a subject of not only managerial interest but also as a technical research area. A brief review on the discussion of these aspects, that is, of the question, "When is it worthwhile to involve reuse technology into the development, and when is a development without reuse favorable?," will conclude the first part of this book.

5.3.1 Benefits of Software Reuse

The benefits of reuse, if successfully realized, can be summarized in following aspects:

- Improved productivity
- Competitive advantages
- Improved quality.

These benefits might be rather qualitative, that is, not directly measurable, but they lead to following economic advantages and opportunities that can be measured:

- *Cost reduction* through savings of development and maintenance costs.
- *Cost avoidance* through excluding likely expenses; for example, for missing deadlines, for compensating the risks of development caused by personnel fluctuation, or over-crowding.
- *Profit increase* realized through shortened time-to-market and customer-oriented, customer-perceived quality increase.

These benefits of reuse, especially profit, can enable a commercial, market-driven organization to meet new market windows and niches that this organization would otherwise miss with no reuse.

5.3.1.1 Costs of Software Reuse

Reuse is not free of costs. Direct costs of reuse are caused by development and maintenance of reusable assets. Hidden costs can be caused by risks of obsolescence due to ignoring technical progress or user needs. Also a performance degradation can arise since reusable components often include more functionality than a specialized application requires, that is, reuse assets usually include functional redundancy to meet different market needs (see Sect. 3.3.16.5). And last but not least, a reusable component can pose a security or safety risk because information included may be critical for a potential user who does not want this information made publicly available; for example, the nature of the function of the reused component.

Initial and recurring costs of reuse can be summarized as follows:

- Investigation costs; for example, expenses of cost-benefit analysis.
- Development and maintenance costs.
- Personnel to run reuse; for example, to produce, to select, to administrate.
- Tools to develop, maintain, store, retrieve the assets.
- Continuous education of the personnel.

5.3.1.2 Return on Investment of Reuse, Economic Models

It is evident that the introduction of reuse technology in a company causes initial costs (Sect. 1.2.3). Even before starting, it is necessary to establish domain engineering to identify and develop reusable components. The more reusable assets are available, which, hopefully, will be successfully marked and deployed in new products, the more attractive will be the cost/benefit ratio [11, 25, 64, 72].

5.3.1.3 Return on Investment of Reuse, Economic Models

It is evident that the introduction of reuse technology in a company causes initial costs (Sect. 1.2.3). Even before starting, it is necessary to establish domain engineering to identify and develop reusable components. On the other hand, the more reusable assets are available, which, hopefully, will be successfully marked and deployed in new products, the more attractive will be the cost/benefit ratio [11, 26, 64].

A company that plans to adopt reuse techniques will, first and foremost, ask the question: How long must I wait for a return of my investment? This is a difficult question to answer which, on the one hand, depends on the degree, extent, efficiency, and adequacy of the reuse techniques deployed. The success depends on, as always, the marketing situation and the efficacy of the sale management and its marketing strategy.

An organization needs patience and stamina for the long haul to survive the costs of installing systematic reuse by one of the technologies recommended in this book, namely CBSE, COTS development or SPLE. Figure 5.15 illustrates this situation; the saving effect does not arise before deploying or selling a reusable asset a "certain number" of times. Fortunately, several economic models help to determine this "certain number" question, and further questions that can be categorized as follows [26, 63, 79]:

- Models for cost-benefit analysis
- Models to determine ROI (return on investment)
- Models to determine cost avoidance.

Figure 5.16 tries to shed some light on this situation. A relatively large amount of investment is necessary at the beginning to launch an adequate reuse technology [50, 69]. It will then take a while until the first sign of benefits are visible. The good news is that the volume of investment necessary to continue the process will decrease.

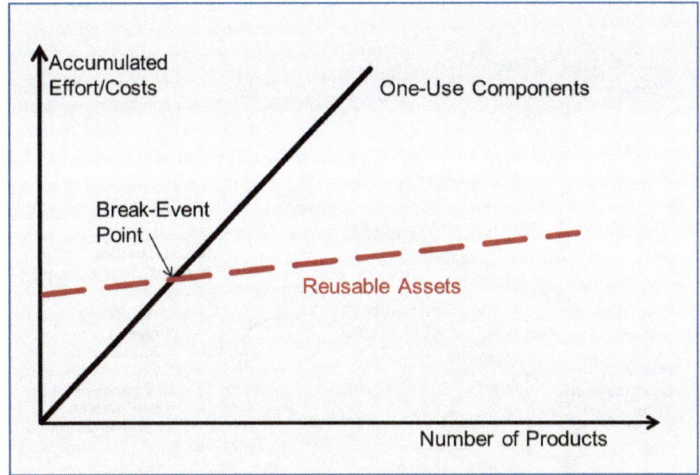

Fig. 5.15 Comparison—One-use versus reusable assets (based on [11] and [64])

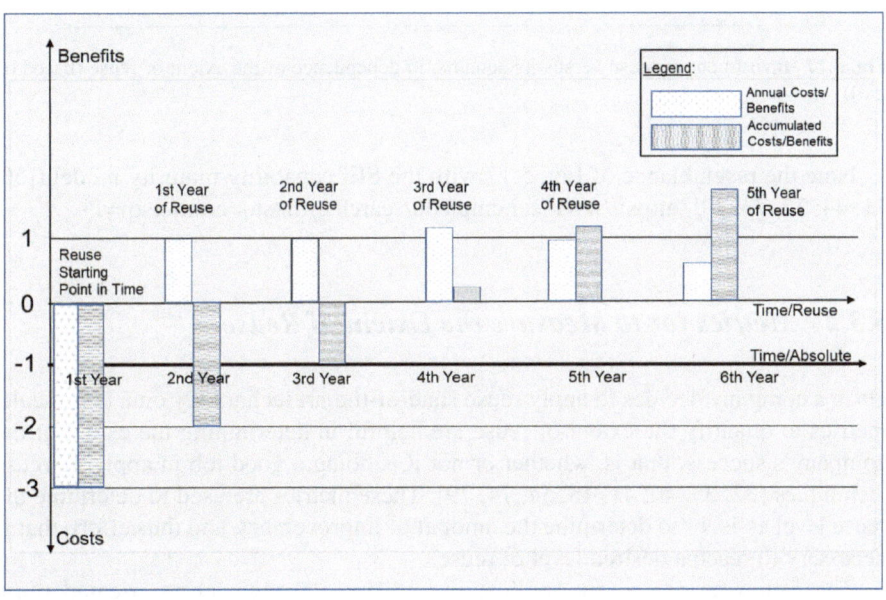

Fig. 5.16 A running chart of investment—costs (below the Line) and benefits (based on [50, 69])

Last but not least, Fig. 5.17 illustrates the benefits of reuse in terms of improved quality, time-to-market, etc. vs. costs. Even informal code reuse can reduce development time. The next level of reuse, black-box reuse, can reduce maintenance costs, etc.

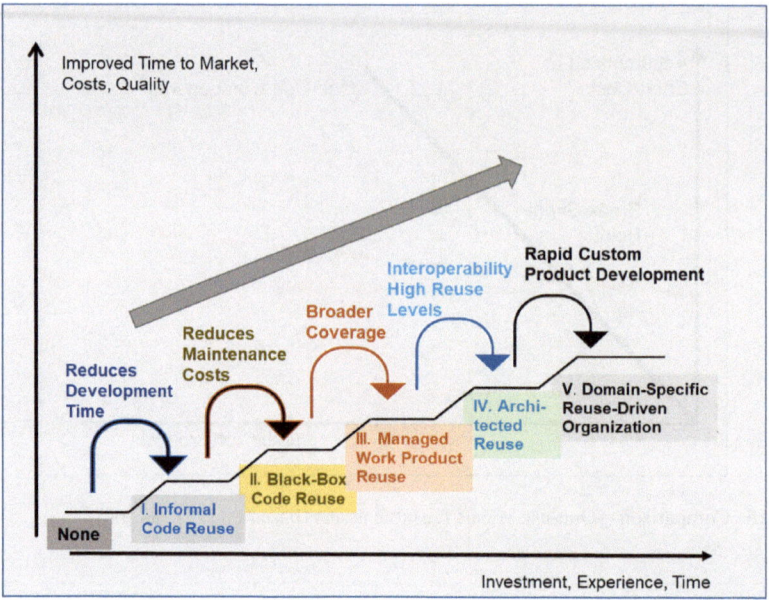

Fig. 5.17 Investment of reuse versus its benefits, in dependence of the extent of reuse (based on [50])

Note the resemblance of Fig. 5.17 with the SEI capability maturity model ([50, 71, 44, 76, 73, 79], https://www.sei.cmu.edu/search.cfm#stq=cmm&stp=1).

5.3.2 Metrics for to Measure the Extent of Reuse

Once a company decides to apply reuse state-of-the-art technology on a large scale, metrics to quantify the extent of reuse are helpful in determining the extent of the company's success, that is, whether or not it is doing a good job in applying reuse techniques [32, 35, 40, 41, 48, 54, 74, 79]. These metrics are used to determine the reuse level as-is, or to determine the amount of improvement, and thus effort, that is necessary to reach a desired level of reuse.

The following are a few application and domain engineering-oriented reuse metrics:

- Reuse Distribution—Percentage of

  - new code
  - reused code (internal, external)
  - adapted code
  - project-level return on investment.

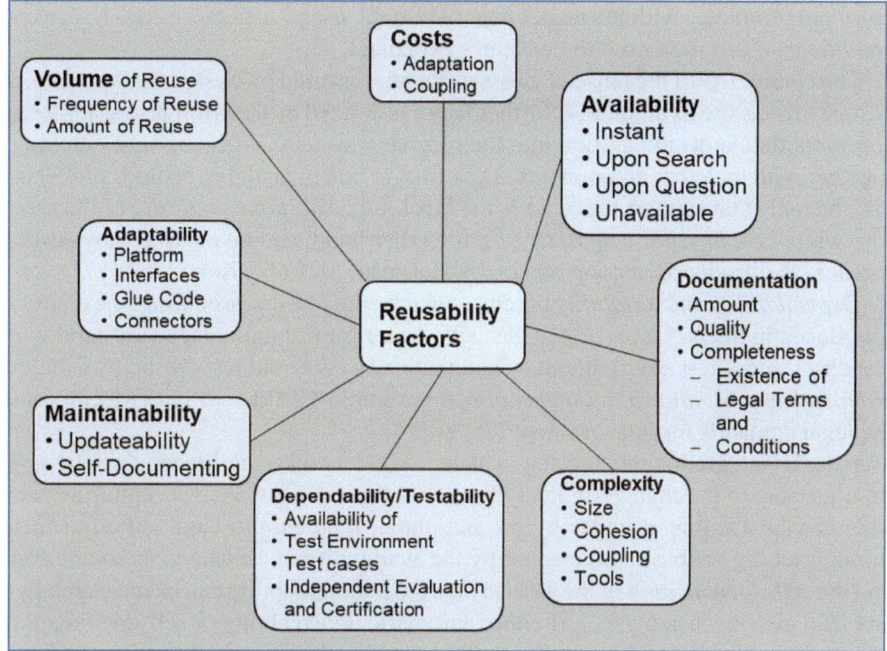

Fig. 5.18 Reusability metrics (based on [49])

- Percentage of reusable code the company uses.

Figure 5.18 illustrates general aspects to measure the extent of reuse that will briefly be discussed below (see also Sects. 1.2.2 and 2.1.2).

The *amount* of reuse can be measured by the size of the reused code, for example its lines of code. Equally important is the frequency with which this code will be reused. These two factors define the *volume* of reuse a company performs.

The functional *adaptation* effort of a reusable asset to be deployed widely defines the *costs* of its reuse. However, the adaptation costs are increased by the additional effort to efficiently *couple* this asset in the receiving, composed system while allowing for its dynamic constraints, for example concerning the limitation of its execution time.

The most important benefit of the reuse of an asset is measured by its obtainable *price* in the market. However, its cost *saving* factor is also important since, among other things, it enables the company to seize new opportunities.

A reusable asset can be *instantly available*, that is, at any time and without any time loss, or with some effort, upon *search* or *question*. However, time constraints can make the asset in latter cases *unavailable*.

The *amount* and the *quality* of a reusable asset's *documentation* must be balanced, that is, brief but easy to understand and to the point. Furthermore, the documentation must be complete, without neglecting important usage aspects. *Legal terms* and *conditions of business* need to be clearly explained.

The *complexity* of the reuse of an asset can be measured by its *size* and *coupling*, as is done to measure its amount. A further factor is defined by the effort necessary for its *cohesion*; that is, its integration into the receiving system and cohesively "sticking" together with the other components. The effort to understand the methods of reusing and the tools that support the reuse is the black-box *external complexity* of the asset. The white-box *internal complexity*, on the other hand, can be measured by several metrics, as introduced, among others, by Halstead, McCabe, Kemmerer [87].

Dependability and *testability* of reuse can generally be assessed using the methods introduced in Sects. 1.2 and 1.3. A decisive factor for testability is the availability of test cases and a test environment to automatically carry out tests. It helps using an *independent evaluation and certification*, for example, CMM or SPICE, for enforcing the legal demands for improvement [75, 80].

After its integration into the new system, a good *maintainability* requires the asset be adjustable to the changes in likely future versions of a potential accepting system. This can be facilitated either by an automatic *updateability* and self-adjustment through setting visible parameters or by the availability of a changeable source code and the *self-documenting* of the asset. Thus, the maintainability can be measured by a Boolean metric ("true/false"): If either automatic updateability or self-documenting of a changeable source code is given, then the asset is maintainable, otherwise it is not maintainable.

An asset's adaptability is closely related to its maintainability. Adaptability, however, primarily concerns the asset's *interfaces*, presuming, as with maintainability, the availability of a changeable source. Secondary factors are the programming language and the asset's implementation platform. Glue code and connectors are further requirements of a good adaptability. Thus, adaptability, like maintainability, can also be measured by a Boolean metric ("true/false").

Key Points, Exercises, Recommended Further Reading, References

Key Points

- Reusable software is also software. Thus, existing methods of software testing is, in principal, also applicable to software reuse (3.3.8).
- However, the combinatorial explosion of the variability and features of products in product families requires specific care when selecting test methods.
- Pairwise testing and model-based testing are appropriate techniques for testing reuse-oriented software. They take specific characteristics of software reuse into account, and thus ignore some others that are not very relevant.

- While domain-based testing aims at validating families of products, application/product-centric testing aims at validating a specific product of the family.
- Combining feature model and behavioral model is an effective test technique to validate product lines that can be utilized in both domain-based and application/product-centric testing.
- Selecting a technique to validate user-oriented software is also a matter of economic aspects.

Exercises

1. Why is combinatorial explosion likely when testing reuse-oriented software?
2. How can this combinatorial explosion be excluded?
3. Select appropriate methods for variants' testing from the test survey in Fig. 5.1.
4. Which test methods are appropriate for testing reuse-oriented software?
5. What are the differences between domain-based testing and application/product-centric testing?
6. How many test inputs does pairwise test deliver?
7. How many two-way (pairwise) test cases can be generated for the FD in Fig. 5.4?
8. Event-oriented behavioral modeling is primarily done using state machines that are combined with feature models to generate test cases. A simplification of the syntax and semantics of state machines is given by Event Sequence Graphs (ESG, [20]). Would this lead to a more efficient test generation process?
9. Discuss how assets other than programs can be reused, for example, tests, test suites, test specs.
10. To what extent can a test technique affect the economics of reuse? Consider the characteristics of reusability as depicted in Fig. 5.18.

Recommended Further Reading

1. Ammann, P., Offutt, J.: Introduction to Software Testing. Cambridge University Press, UK (2008)
2. Bach, J., Schroeder, P.J.: Pairwise testing: a best practice that isn't. Available at http://www.testingeducation.org/wtst5/PairwisePNSQC2004.pdf
3. Beizer, B.: Software Testing Techniques, 2e. International Thomson Computer Press (1990), and Dreamtech (2003)
4. Binder, R.V.: Testing Object-Oriented Systems : Models, Patterns, and Tools. Addison-Wesley (2006)
5. Kreowski, H.-J., Montanari, U., Orejas, F., Rozenberg, G., Taentzer, G.: Formal Methods in Software and Systems Modeling. Springer-Verlag (2005)
6. Mathur, A.P.: Foundations of Software Testing, 2/e. Pearson (2013)
7. Sommerville, I.: Software Engineering, 10e. Pearson (2015)

8. Zander, J., Schieferdecker, I., Pieter, J. (eds.): Model-Based Testing for Embedded Systems. CRC Press (2012)

References

9. Aho, A.V., Dahbura, A.T., Lee, D., Uyar, M.Ü.: An optimization technique for protocol conformance test generation based on UIO sequences and rural Chinese postman tours. IEEE Trans. Commun. **39**, 1604–1615 (1991)
10. Ammann, P., Offutt, J.: Introduction to Software Testing. Cambridge University Press, UK (2008)
11. Apel, S., Batory, D., Kästner, Ch., Saake, G.: Feature-Oriented Software Product Lines. Springer, Berlin Heidelberg (2013)
12. Ayav, T., Belli, F.: Boolean differentiation for formalizing Myers' cause-effect graph testing technique. In: 2015 IEEE International Conference on Software Quality, Reliability and Security—Companion, pp. 138–143 (2015)
13. Bach, J., Schroeder, P.J.: Pairwise testing: a best practice that isn't. Available at http://www.tes tingeducation.org/wtst5/PairwisePNSQC2004.pdf
14. Bayer, J., Flege, O., Knauber, P., Laqua, R., Muthig, D., Schmid, K., Widen, T., DeBaud', J.-M.: PuLSE: a methodology to develop software product lines. In: Proceedings of the ACM Symposium on Software Reusability, pp. 122–131 (1999)
15. Beizer, B.: Software Testing Techniques, 2e. International Thomson Computer Press (1990), and Dreamtech (2003)
16. Belli, F., Beyazıt, M., Budnik, C.H.J., Tuglular, T.: Advances in model-based testing of graphical user interfaces. In: Memon, A.M. (ed.) Advances in Computers, vol. 107, pp. 219–280. Academic Press, Burlington (2017)
17. Belli, F., Budnik, Ch.J.: Test minimization for human-computer interaction. Appl. Intell. **26**, 161–174 (2007)
18. Belli, F., Budnik, Ch.J., Hollmann, A., Tuglular, T., Wong, W.E.: Model-based mutation testing—approach and case studies. Sci. Comput. Program. **120**, 25–48 (2016)
19. Belli, F.: Finite state testing and analysis of graphical user interfaces. In: Proceedings 12th International Symposium on Software Reliability Engineering, pp. 34–43 (2001)
20. Belli, F., Budnik, Ch.J., White, L.: Event-based modelling, analysis and testing of user interactions: approach and case study. Softw. Test. Verif. Reliab., 3–32 (2006)
21. Belli, F., Beyazit, M., Memon, A.: Testing is an event-centric activity. In: Proceedings of the International IEEE Conference on Software Security and Reliability, pp. 198–206 (2012)
22. Bertolino, A., Fantechi, A., Gnesi, S., Lami, G.: Product line use cases: scenario based specification and testing of requirements. In: Käkölä, T., Duenas, J.C. (eds.) Proceedings of Software Product Lines Research Issues in Engineering and Management, Chap. 11, pp. 425–445. Springer, Berlin (2006)
23. Beydeda, S., Gruhn, V. (eds.): Testing Commercial-off-the-Shelf Components and Systems. Springer (2005)
24. Beydeda, S., Gruhn, S., Mayer, J., Reussner, R., Schweiggert, F., (eds.): *Testing of Component-Based Systems and Software Quality*, GI Edition – Lecture Notes in Informatics, Series of the Gesellschaft für Informatik (GI), Vol. P-58, 2004
25. Binder, R.V.: Testing Object-Oriented Systems : Models, Patterns, and Tools. Addison-Wesley (2006)
26. Boehm, B., Brown, A., Madachy, R., Yang, Y.: A software product line life cycle cost estimation model. In: Proceedings of the International Symposium on Empirical Software Engineering (ISESE), pp. 156–164 (2004). https://doi.org/10.1109/ISESE.2004.1334903
27. Bosch, J., Krueger, C. (eds.), Proceedings of the 8th International Conference—Software Reuse: Methods, Techniques, and Tools (ICSR), LNCS 3107 (2004)

28. Clarke, E.M., Orna Grumberg, O., Peled, D.A.: Model Checking. MIT Press (1999)
29. Cohen, D.M., Dalal, S.R., Fredman, M.L., Patton, G.C.: The AETG system: an approach to testing based on combinatorial design. IEEE Trans. Software Eng. **23**(7), 437–444 (1997)
30. Devroey, X.: Behavioural model-based testing of software product lines. Ph.D. Thesis, University of Namur, PReCISE Research Center (2017)
31. ESG Test Suite Designer (ESG-TSD) homepage. URL: http://download.ivknet.de/
32. Edwards, S.H., Kulczycki, G.: Formal foundation of reuse and domain engineering. In: Proceedings of the 11th International Conference on Software Reuse (ICSR). Springer Science & Business Media (2009)
33. El-Fakih, K., Hierons, R.M., Turker, U.C.: K-branching UIO sequences for partially specified observable non-deterministic FSMs. IEEE Trans. Software Eng. (2019). https://doi.org/10.1109/TSE.2019.2911076
34. Engström, E., Runeson, P.: Software product line testing—a systematic mapping study. Inf. Softw. Technol. **53**, 2–13 (2011)
35. Ezran, M., Morisio, M., Tully, C.: Practical Software Reuse. Springer Practitioner Series (2002)
36. Ferreira, J.M., Vergilio, S.R., Quinaia, M.: Software product line testing based on feature model mutation. Int. J. Software Eng. Knowl. Eng. **27**(05), 817–839 (2017). https://doi.org/10.1142/S0218194017500309
37. Filho, R.A.M., Vergilio, S.R.: A mutation and multi-objective test data generation approach for feature testing of software product lines. In: Proceedings of the 29th Brazilian Symposium on Software Engineering, pp. 21–30 (2015). https://doi.org/10.1109/SBES.2015.17
38. Fragal, V.H., Simao, A., Mousavi, M.R., Turker, U.C.: Extending HSI test generation method for software product lines. Comput. J. **62**(1), 109–129 (2019). https://doi.org/10.1093/comjnl/bxy046
39. Fragal, V.H., Simao, A., Mousavi, M.R.: Validated test models for software product lines: featured finite state machines. In: Proceedings of the 13th International Conference on Formal Aspects of Component Software (FACS), Revised Selected Papers, pp. 210–227 (2016). https://doi.org/10.1007/978-3-319-57666-4_13
40. Frakes, W.B., Terry, C.: Software reuse: metrics and models. ACM Comput. Surv. **28**(2), 415–435 (1996). https://doi.org/10.1145/234528.234531
41. Frakes, W.B., Isoda, S.: Success factors of systematic reuse. IEEE Software, V **11**(5), 14–19 (1994). Available at http://ieeexplore.ieee.org/stamp/stamp.jsp?tp=&arnumber=311045
42. Geppert, B., Li, J., Rößler, F., Weiss, D.M.: Towards generating acceptance tests for product lines. In: Proceedings of the 8th Internationakll Conf. – Software Reuse: Methods, Techniques, and Tools, pp. 35–48 (2004)
43. Grönniger, H., Krahn, H., Pinkernell, C., Rumpe, B.: Modeling variants of automotive systems using views. In: Proceedings of the Modellierungs-Workshop MBEFF, Berlin, Informatik-Bericht 2008-01 (2008). Available at www.se-rwth.de/publications
44. Hallsteinsen, S., Paci, M.: Experiences in Software Evolution and Reuse—Twelve Real World Projects. Springer Science and Business Media, 18.09.1997
45. Henard, C., Papadakis, M., Perrouiny, G., Klein, J., Le Traon, Y.: Assessing software product line testing via model-based mutation: an application to similarity testing. In: IEEE Sixth International Conference on Software Testing, Verification and Validation Workshops, Luxembourg, pp. 188–197 (2013).: https://doi.org/10.1109/ICSTW.2013.30
46. Heuer, A., Stricker, V., Budnik, C.J., Konrad, S., Lauenroth, K., Pohl, K.: Defining variability in activity diagrams and petri nets. Sci. Comput. Program. **78**, 2414–2432 (2013)
47. Hopcroft, J.E., Motwani, R., Ullman, J.D.: Introduction to Automata Theory, Languages, and Computation (2e). Pearson Education (2000)
48. Hossain, Sh.: Rework and reuse effects in software economy. Global J. Comput. Sci. Technol. C Software Data Eng. **1**(4) Version 1.0 (2018)
49. Hristov, D., Hummel, O., Huq, M., Janjic, W.: Structuring software reusability metrics for component-based software development. In: Proceedings of the Seventh International Conference on Software Engineering Advances, pp. 421–429 (2012)

50. Jacobson, I., Griss, M., Jonsson, P.: Software Reuse—Architecture, Process and Organization for Business Success. ACM Press–Addison Wesley Longman (1997)
51. Kaindl, H., Mannion, M.: A Feature-Similarity Model for Product Line Engineering. In: 14th International Conference on Software Reuse (ICSR), Lecture Notes in Computer Science, vol. 8919, pp 34–41. Springer, Berlin (2015). https://doi.org/10.1007/978-3-319-14130-5_3
52. Kamischke, J., Lochau, M., Baller, H.: Conditioned model slicing of feature-annotated state machines. In: Proceedings of the 4th International Workshop on Feature-Oriented Software Development (FOSD), pp. 9–16 (2012). https://doi.org/10.1145/2377816.2377818
53. Kienzle, J., et al.: Toward model-driven sustainability evaluation. Commun. Assoc. Comput. Mach. **63**, 80–91 (2020)
54. Kim, Y., Stohr: Software reuse: survey and research directions. J. Manage. Inf. Syst. **14**, 113–147 (1998)
55. Konrad, S., Lauenroth, K., Pohl, K.: Formal definition of syntax and semantics for documenting variability in activity diagrams. In: Bosch, J., Lee, J. (eds), Software Product Lines: Going Beyond, SPLC 2010, Lecture Notes in Computer Science, vol. 6287. Springer, Berlin, Heidelberg. https://doi.org/10.1007/978-3-642-15579-6_5
56. Kreowski, H.-J., Montanari, U., Orejas, F., Rozenberg, G., Taentzer, G.: Formal methods in software and systems modeling. Springer-Verlag (2005)
57. Lackner, H.: Model-based product line testing: sampling configurations for optimal fault detection. In: Proceedings of the 17th International SDL Forum on SDL 2015: Model-Driven Engineering for Smart Cities, vol. 9369, pp. 238–251 (2015). https://doi.org/10.1007/978-3-319-24912-4_17
58. Lackner, H.: Domain-centered product line testing. Ph.D. Thesis, Humboldt University, Berlin (2016)
59. Lackner, H., Schmidt, M.: Towards the assessment of software product line tests: a mutation system for variable systems. Software Product Analysis Tools (ACM SPLC), vol. 1, pp. 62–69 (2014)
60. Lackner, H., Schmidt, M.: Potential errors and test assessment in software product line engineering. In: Proceedings of the Tenth Workshop on Model-Based Testing, EPTCS 180, pp. 57–72. Springer (2015). https://doi.org/10.4204/EPTCS.180.4
61. Lackner, H., Schlingloff, B.-H.: Advances in testing software product lines. In: Memon A.M. (ed.), Advances in Computers, vol. 107, pp. 157–217. Elsevier (2017). https://doi.org/10.1016/bs.adcom.2017.07.001
62. Lackner, H., Thomas, M., Wartenberg, F., Weißleder, S.: Model-based test design of product lines: raising test design to the product line level. In: Proceedings of 7th International Conference on Software Testing, Verification, and Validation (ICST), pp. 51–60 (2014)
63. Leach, R.J.: Software Reuse, Second Edition: Methods, Models, Costs, 2nd edn. (2012)
64. van der Linden, E., Schmid, K., Rommes, F.J.: Software Product Lines in Action. Springer (2007)
65. Lisboa, L.B., Li, J.J., Morreale, P., Heer, D., Weiss, D.M.: An evaluation to compare software product line decision model and feature model. In: Proceedings of the 9th International Conference on Evaluation of Novel Approaches to Software Engineering (ENASE), pp. 1–8 (2014)
66. Lity, S., Nahrendorf, S., Thüm, T., Seidl, C., Schaefer, I.: 175% modeling for product-line evolution of domain artifacts. In: Proceedings of the 12th International Workshop on Variability Modelling of Software-Intensive Systems (VAMOS), pp. 27–34 (2018). https://doi.org/10.1145/3168365.3168369
67. Lochau, M., Oster, S., Goltz, U., Schürr, A.: Model-based pairwise testing for feature interaction coverage in software product line engineering. Software Qual. J. **20**, 567–604. https://doi.org/10.1007/s11219-011-9165-4
68. Lochau, M., Schaefer, I., Kamischke, J., Lity, S.: Incremental model-based testing of delta-oriented software product lines. In: Brucker, A.D., Julliand J. (eds) Tests and Proofs. TAP 2012, Lecture Notes in Computer Science, vol. 7305, pp. 67–82. Springer (2012). https://doi.org/10.1007/978-3-642-30473-6_7

69. Martínez-Fernández, S., Ayala, C.P, Franch, X., Marques, H.M.: REARM: a reuse-based economic model for software reference architectures. In: Proceedings of the 13th International Conference on Software Reuse—ICSR, pp. 97–112 (2013)

70. Mathur, A.P., Foundations of Software Testing, 2/e. Pearson (2013)

71. Mili, A., Chmiel, S.F., Gottumukkala, R., Zhang, L.: Managing software reuse economics: an integrated ROI-based model. Ann. Software Eng. **11**, 175–218 (2001)

72. Oracle Practitioner Guide, Determining ROI of SOA through Reuse (2012)

73. Paulk, M.C.: How ISO 9001 compares with the CMM. IEEE Softw. **12**(1), 74–83 (1995). https://doi.org/10.1109/52.363163, see also https://resources.sei.cmu.edu/asset_files/Techni calReport/1994_005_001_435267.pdf

74. Pfleeger, S.L.: Measuring reuse: a cautionary tale. IEEE Softw. **13**(4), 118–127 (1996). https://doi.org/10.1109/52.526839

75. Poulin, J.S., Caruso, J.M., Hancock, D.R.: Business case for software reuse. IBM Syst. J. **32**(4), 567–594 (1993)

76. Reifer, D.J.: Practical Software Reuse—Strategies for Introducing Reuse Concepts in Your Organization. Wiley (1997)

77. Reuys, A., Kamsties, E., Pohl, K., Reis, S.: Model-based system testing of software product families. In: Pastor, O., Cunha, J.F. (eds.) Proceedings of the International Conference on Advanced Information Systems Engineering (CAiSE). LNCS 3520, pp. 519–534 (2005)

78. Rocha, C.R., Martins, E.: A strategy to improve component testability without source code. In: Testing Commercial-off-the-Shelf Components and Systems, pp. 47–62

79. Schmid, K., John, I., Kolb, R., Meier, G.: Introducing the PuLSE approach to an embedded system population at Testo AG. In: Proceedings of the 27th International Conference on Software Engineering (ICSE), pp. 544–552 (2005). https://doi.org/10.1109/ICSE.2005.155 3600

80. Schmietendorf, A., Dimitrov, E., Dumke, R., Foltin, E., Wipprecht, M.: Conception and experience of metrics-based software reuse. In: Proceedings of the International Workshop on Software Measurement (IWSM'99), pp. 178–189 (1999)

81. Seidl, C., Wille, D., Schaefer, I.: Software reuse: from cloned variants to managed software product lines. In: Automotive Systems and Software Engineering: State of the Art and Future Trends, pp. 77–108. Springer International Publishing (2019)

82. Sommerville, I.: Software Engineering, 10e. Pearson (2015)

83. Vouk, M.A.: Back-to-back testing. Inf. Softw. Technol. **32**(1), 34–45 (1990)

84. Weißleder, S., Lackner, H.: Top-down and bottom-up approach for model-based testing of product lines. In: Proceedings MBT 2013, EPTCS 111, pp. 82–94. Springer (2013). https://doi.org/10.4204/EPTCS.111.7

85. Weißleder, S., Wartenberg, F., Lackner, H.: Automated test design for boundaries of product line variants. In: Proceedings of the ICTSS 2015, LNCS 9447, pp. 86–101 (2015). https://doi.org/10.1007/978-3-319-25945-16

86. Zander, J., Schieferdecker, I., Pieter, J. (eds.): Model-Based Testing for Embedded Systems. CRC Press (2012)

87. Zuse, H.: Software Complexity—Measures and Methods, Programming Complex Systems, vol. 4. De Gruyter (1991). https://doi.org/10.1515/9783110866087

Part II
Dependable Reuse of E&E Components and Products

Chapter 6
Background

Part B of the book starts with a summary of the state of Ecodesign knowledge and connects it with know-how about refurbishment/recycling. For a reader this information should be helpful to understand the major principles of Ecodesign and why it is a win–win situation for both the reuser and the refurbishing company, and also for a producer of a new product if the manufacturer combines new and as-new components. As the development of Ecodesign rules for a HW product is now more than 20 years old the focus wasn't directed to the still evolving aspect of SW all factors were concentrated on the reuse of parts and materials. Now that SW techniques are more sophisticated, it is the SW which becomes a more and more important part, but always in combination with HW.

Therefore in part A the reuse of SW is discussed, as a subject both alone and in combination with HW.

This basic knowledge is important for all refurbishment processes of HW because the producer requires either the latest SW for an update, the compatibility of old SW with the new one or new SW programs to harmonize all. In this case the customer can think about complex plants, planes or other systems or environments where older and newer components always have to be combined. Consumer products might be easier to update because updated SW might be available and only one piece of code is often required.

In part B the focus directs to the combination of HW **and** SW!

A new product with an old piece of SW might consume too much energy and can no longer be sold. The answer will be Green IT. This means the SW is optimized for low energy consumption and more environmental considerations.

The SW should be married with the HW and it is important that all subjects have to be tested. That is another chapter in this book. Now not only dependability of HW has to be guaranteed, but the tests also have to cover the quality and dependability of the latest HW and SW combination. The environmental aspects also have to be tested. An old-fashioned product with too much energy consumption should be removed and discarded.

© The Author(s), under exclusive license to Springer Nature Switzerland AG 2021   183
F. Belli and F. Quella, *A Holistic View of Software and Hardware Reuse*,
Studies in Systems, Decision and Control 315,
https://doi.org/10.1007/978-3-030-72261-6_6

Refurbishment is another subject of this book. A state 'as-new' is the easiest to be tested. Several also standardized models have less strict requirements. So in comparison, readers get information about different states of 'new' they can buy. Standardization is a mandatory pre-requisite, therefore an overview of the important standards is also presented.

Nevertheless, all legal requirements are discussed inside the Ecodesign Directive and the requirements which re-users have to expect. SW is now a legal requirement for reuse of several types of household equipment.

The book finishes with several appendices covering the rules for ecodesign, low energy consumption or requirements from the ecodesign directive.

6.1   Preliminaries

Ecodesign is meanwhile widely accepted in industry and more or less integrated in product development schemes as a means to minimize environmental impacts and to reduce costs. In legislation worldwide ecodesign is established covering design, substance restrictions, or recycling/refurbishment. A combination with software (SW) reuse is usually not established in law. When reusing or refurbishing a hardware (HW) component or a complete product or system SW reuse will be an essential part of restoring. So, SW reuse will be part of all new and old HW components, from the smallest up to those used in big plants:

- Software (SW) reuse might be necessary during the development phase of a new HW product, causing potential environmental impact and influencing later reuse. SW development works usually in parallel to HW development (treated in part A).
- Dependability and environmental impacts of the existing SW of a component will in addition be part of the refurbishment process (treated especially in part B).

The *Ecodesign Directive* [1] requires manufacturers for every energy related product to install more or less a complete system starting with take back over reuse, till disposal. According to this directive "it must be assessed whether possibilities for reuse, recycling and recovery of materials and/or energy can be taken into account" (*Ecodesign Directive*: Annex 1, Sect. 1.2.c). In addition, manufacturers should provide proof of whether reuse might or might not be possible. A declaration about the fulfilling of these requirements is part of the *conformity declaration* of this directive. In so-called "Implementing Measures" the Commission has formulated special requirements for many E&E products, the number of which will be extended in the future. The latest state of these measures should be checked on the homepage of the European Commission [1]. It seems the legislation has largely concentrated on HW alone. But a combined view on the cooperation of HW and SW would be required. Nevertheless, many products and components do no longer work without a SW application. In the new EU directives about repair of household equipment (see

Sect. 11.1.2), it is required that the manufacturer offer all necessary SW updates free of charge to keep the product on the highest level of efficiency.

The Main subject of this book is to give more information about reuse of HW components and products in combination with SW to give some hints about how to achieve a combined application. SW is not only important for refurbishment, but also for remanufacturing and SW up-grade of a component or a complete plant. SW might also be relevant for too high energy consumption [2]. This subject is in detail discussed in Sect. 7.2.

The *Ecodesign Directive* also contains many rules covering how a company should manage the whole design process as an ecodesign process. Helpful assistance is given in the book *Ecodesign –the competitive Advantage* [64]. In 7.1 a short summary about the state of ecodesign is presented. If SW reuse is applied during the development process, then the rules of Section A are valid. But as there are no hints for SW reuse in the *Ecodesign Directive* itself, and there were no cases describing the combination of SW with HW reuse, Section B deals with several potential cases which could occur, such as:

- Old HW and old SW to be reused after refurbishment,
- Combination of old HW and old SW with new components,
- Upgrade of components (HW, SW) and their combination with an old or a new system and tries to assist an applicant in these cases.

The other aspects of HW components, for example their recycling for materials recovery are treated in this book briefly and only for completeness. Detailed information about this subject is published in the booklet "Utilization of Used Components" [3].

By only reusing old equipment or performing a little bit of restoration a manufacturer could hope to avoid the duty to apply this directive contrary to when supplying a new product with reused components. But with the later-on discussed subject of a new product containing as-new parts the producer will put a new product containing as-new parts on the market and will have to comply with the directive anyway.

Decisive for a manufacturer to adopt the desired practices should not be only to comply with the law, but also to experience the many commercial and environmental benefits combined with reuse and the state of as-new. In the following the description of many opportunities should persuade applicants that reuse is really a benefit.

6.2 Requirements for the Reuse of Hardware Components and Systems

A financial benefit is generally the driving force for reuse of products and components. In Fig. 6.1 most of the economic benefits of reuse are summarized. Reuse is usually applied if the total costs from 'take back' over 'refurbishment to' sales are lower than the earnings for the resold product or component. Decision making

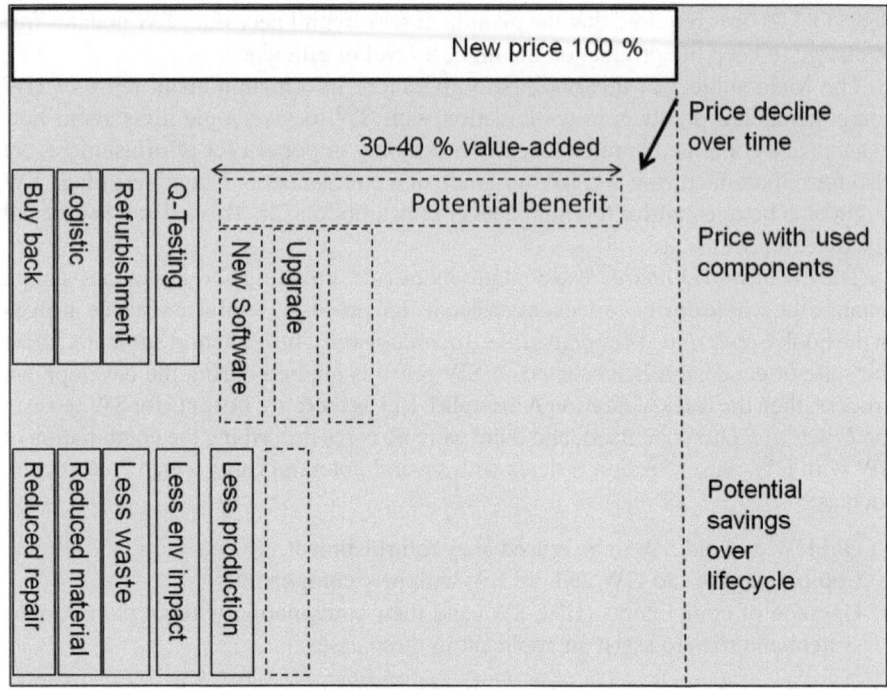

Fig. 6.1 Benefits for the reuse of components in new products. Potential savings over lifecycle are usually not direct input into price calculation

factors of this figure are discussed in detail in Sect. 6.2.3. Furthermore pro's and con's are discussed in the booklet "Utilization of Used Components" [3].

It is a question of legal product liability to guarantee to a customer a certain quality of the refurbished component or product for a defined guarantee time. Therefore customers should distinguish between the states of the reused or refurbished product they get.

Remark/Footnot As-new, like new, equivalent-to-new are identical in meaning for the purposes of this document: The component of this state cannot be distinguished from an absolutely new component or has only minor deviations, like optical ones which have no influence on the product properties. Later-on the acronym Qagan (*Qualified-as-Good-as-New*) is used as a technical terminus which assumes that the used, as-new product has been qualified by tests equivalent to the ones the manufacturer carries out to release new products.

The comparison of the properties of a used component with the properties of new components is a rather simple possibility to check whether a used component is still equivalent-to-new (ETN) or nearly like new. All tests can be directly made within the incoming inspection of all new components, which does not distinguish between used and new. So, a used component which does not show differences to the new one or has only defined slight deviations such as color, will be ETN; sometimes the used component cannot be distinguished from an absolutely new one at all, as might happen in the case of a stainless steel screw (however, take care: It must be known whether a potential over-straining of the steel component could have happened, as then a safety problem could occur). As no additional tests to the incoming inspection are usually required, only one set of criteria must be compared, and the same production plant for new products is applied to produce absolutely new products or those with ETN components. This concept seems to be the one with the lowest costs for reuse of components. The concept is described in more detail in chapter VII, together with the state of corresponding software.

Apart from the already mentioned *Ecodesign Directive*, some other legal requirements, for example released by the European Commission (WEEE directive [4]), make it increasingly important to collect certain volumes of E&E products, and in addition to refurbish and reuse these components and products, whenever possible. The trend is enforced by the continuously increasing demand for a higher recycling quota by the European Commission. The basis of these requirements is a hierarchical system starting with avoidance of

- waste,
- reuse, and
- at last disposal of old products to create a circular economy.

In the end a circular economy is created for material from cradle to cradle. Such concepts are meanwhile laid down in the legislation worldwide. So, the pressure from the public to reuse waste products is increasing. Some authors add remanufacturing to reuse focusing on the special requirements of remanufacturing as easy disassembly or assembly. The word reuse here is applied for all kinds of reuse: as it is or prepared for remanufacturing. The word refurbishment includes also a restoration if necessary and includes also remanufacturing steps. Often its meaning is the same.

It is obvious that not only countries but also companies which practice reuse or remanufacturing need a strategy to implement their own system covering all product aspects over the whole life-cycle, and a management system to achieve and manage practical solutions. In the EU national governments have not only sharpened their legislation to achieve a higher collection quota, but have also set a special focus on some problems like recovering precious metals [5]. Many companies have already organized in parallel take back and refurbish product systems, such as for aviation, copying, defense, medical equipment or telecommunication plants. So, the current situation of reuse is no longer in the beginning, but concentrates today more on capital goods than on consumer products. Huge reuse markets are available, for instance for the reuse of medical equipment. Many examples can be found in the internet. Opportunities for reuse of consumer products like household equipment are now

more in the focus of new research (http://susproc.jrc.ec.europa.eu/Washing_mach
ines_and_washer_dryers/index.html). All basic information from theory to testing
or marketing of HW components is explained in [3]. It will not be repeated here in
detail.

6.2.1 Technical Requirements

The standard *IEC 62309*, "Dependability of products containing reused parts –
requirements for functionality and tests." [6] defines the quality criteria for the
treatment of new products that contain used parts. It explains the rationale for the
dependability and functionality of re-used components, and the methods necessary to
use them in new products. Additionally, it presents which kinds of test and selection
criteria are necessary to fulfill the criteria of Qualified-As-Good-As-New (QAGAN).
If the properties of the used components are largely identical to those of new components or in the best case they cannot be distinguished, then these QAGAN components can be reused to manufacture new products after having fulfilled the incoming
inspection criteria.

The purpose of this standard, therefore, is to guarantee that after testing the
dependability and functionality of a new product with reused components, it will
be comparable to a product containing only new components. The application of
this standard enables a manufacturer to give customers a product with "qualified-
as-good-as-new" components along with a full guarantee, just like they would get
for a new product. This standard is the reference standard for our following discussions. But the focus will also be extended to less qualified components or products
if required.

One essential part of the standard shows the curve of the failure rate versus time
for electronic components in general. In reality, there would be a graph like this
for each type of component (see Fig. 6.2 for electronic components with or without
"burn-in"). This curve is evidently valid—however it is frequently different for some
components. By means of failure rate curves and practical tests, the manufacturer
has to prove that the designed life of the component doesn't fall into the steeply
increasing right side of the curve.

During the process of "Burn-in", new components of a product, assembly or
system, are pre-stressed as a part of the manufacturing cycle. The intention of the
pre-stressing (burn-in process) is to take the components past the point on the bathtub
curve where the initial failures occur ([3], there in Sect. 1.5, Fig. 1.2) and get them
to the point on the graph where failures are very rare, making the product inherently
more reliable. It is essential that the bathtub curve used for these components reflects
the true nature of this component in this application [45].

Besides the failure curve, technical criteria have to be fulfilled for a successful
reuse of a component. There are many influencing variables for a decision about a
potential reuse and it is advisable to use a checklist to analyze them. Whether or not
the criteria mentioned are exclusion criteria for the components very much depends

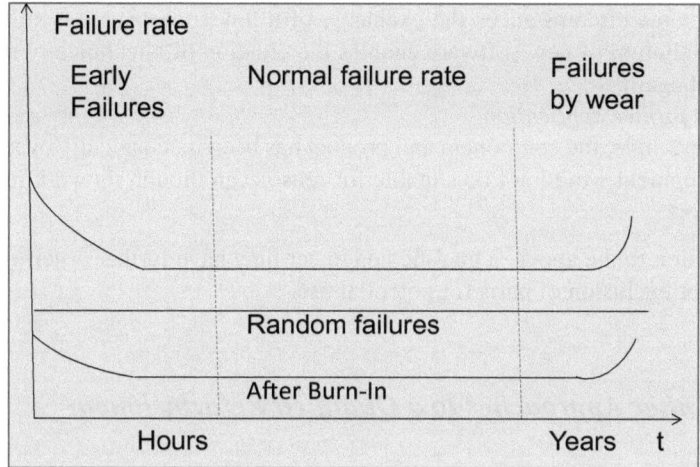

Fig. 6.2 Dependability of electronic components with or without "burn-in": Components as supplied without "burn-in" (upper curve) and components as supplied with "burn-in" (cf. IEC 62309)

on the kind of application they will be used in. The following aspects have to be checked preferentially:

- *Technological state of the product or of the component*
  In this case the software might cause problems if it could not be updated to the current state of the art. Also, if a technology change will be planned, a component could be removed and discarded.
- *Suitability for disassembly*
  The main aspects here are the suitability for disassembly of a connection, the number of elements, work of disassembly and to disconnect a joint, as well as the accessibility.
- *Product structure*
  How high is the ratio of standardized components? These might be suitable for different applications. Platforms allow better reuse. How many different product levels have to be dismantled before access to the inner components is given? Is the product structure simple?
- *Testability*
  As high safety is a basic condition for a new use, the properties of the components have to be evaluated. If evaluation is not sufficiently possible this should be a 'go/no go' criterion. If the test software is not available the component could be sent to the OEM for testing.
- *Ratio of technically sound components or boards*
  This property determines the value of the components and of the whole product.
- *Possibility for modernization or technical upgradeability*

Under some circumstances the exchange of a few components is sufficient, or the installation of new software enables the creation of a technically high value product again.

- *Type of former application*
  If, for example, the component and product has been mechanically over-stressed, the equipment would not be suitable for reuse even though its working lifespan was short.

In addition to the above, a qualified engineer may have further criteria to aid the selection or exclusion of parts for potential use.

6.2.2   Other Approaches to a Qualified Refurbishment

It should be mentioned that several other standardization organizations want to assist in the reuse of products with components as-new. In Sect. 11.2, the standards IEC 62309, IEC 24700 and ANSI RIC 001.1206 are compared. The application of IEC 62309 is focused on electrical and electronic products, IEC 24700 concentrates on office equipment, in the aforementioned ANSI RIC the target group might be broader. In all standards the as-new state for the components or equipment should be achieved after processes of selecting, testing, cleaning, sorting, remanufacturing and documentation. In IEC 62309 a new product with as-new components is manufactured. Reliability tests and safety checks have to be applied. Certification of the factory according to ISO 9001 and ISO 14001 are not sufficient: energy consumption, hazardous materials and material recycling have to be checked. Requirements of the *Ecodesign Directive* (there Annex 1) have to be fulfilled.

Those who put new products on the market have to comply with the CE mark directive. This is the case for products remanufactured according to IEC 62309. It is not clear whether this is required from the other standards. IEC 62309 was not written for certification by a third party. A declaration by the company itself is sufficient, which fits with the philosophy of the CE mark. For the standard BS 8887-220 companies like Ricoh have already got certificates stating that they comply with this standard.

A discussion is also driven to distinguish between reuse and remanufacturing and other definitions. But such a distinguishing is only a marketing catchphrase. European legislation only requires reuse, without defining in which way. In some standards remanufacturing includes having to achieve the state 'as-new' (see Sect. 11.2). So someone who wants to search for 'reuse as-new' in the internet will also find many examples under remanufacturing! In this book the word reuse is usually applied instead of remanufacturing because the meaning could differ according to the definition in other standards.

If a company is not sure whether and which law will have to be applied, for example concerning substance restrictions for an E&E product to be put on the market somewhere in the world, they could ask either their national association, for example ZVEI in Germany or one of the European industrial associations in Brussels. Coordinating for these organizations is *Business Europe* [https://www.businessEuro pe.eu], which can also be asked. Best informed for really product specific questions might be the sector specific associations, in the same way as *COCIR* is for medical products.

6.2.3   Economical Criteria for Decision Making

After the technical criteria, the economic ones also have to be evaluated. The following criteria must be considered:

- Take back volume, organization and management of take back,
- Degree of distribution of the equipment in the market,
- Value of the equipment (primary and residual value),
- Branding policy,
- Third party or own product,
- Substitution: That is, the question has to be answered as to whether the function required could be substituted during the planned reuse phase. Example: The functions of a fax machine are integrated into other products and the fax machine no longer stands alone, as it did before.)
- Costs of procurement of new components,
- Sales markets (countries),
- Additional benefits such as the usability of the component in several products,
- Environmental impact of the products and environmental consciousness of the customers,
- Upgradeability as in ungraded software.

This list can also be extended by each decision maker according to their own experiences.

The take back volume is decisive for an economical reuse. After the beginning of sales of a new product usually it requires some years until a product becomes redundant and will be given back. For capital goods this time can be longer, for trains up to 30 years. Some consumer products often live on average only 1 year. So the volume of desired components differs from product to product. As a continuous flow of components should be guaranteed for a longer period, reuse should be planned for several product generations.

If the market share for a certain product will be very low it could be difficult or cost intensive to get a product back. In this case a reuse might not be successful.

It is known from cars that their values dramatically shrink after the first day of sales. For medical equipment for example, its value stays much more stable for a longer time. In both cases components could fulfill the criteria for a qualification as 'as-new' and can guarantee financial benefit.

Branding policy could be an obstacle for reuse. For a high-priced product the expectations of customers could prevent the CEO from organizing reuse. A company known as environmentally friendly will have customers who are much more interested in reuse.

A company using mainly their own components for production has the advantage of having more knowledge about failures and could reduce the work for qualification of the component to be reused. Cooperation with third parties is also possible.

Changes in technologies could finish a reuse concept, especially if the new technology will be cheaper. This is a risk not only for every reuse planning but also for the actual product.

Procurement costs can also influence a decision. If there will be a continuous and strong decrease of the costs of new components it might be easier to buy new ones.

The sales markets also play an important role. Especially in some countries it may not be allowed to export used products because of fear of misuse. Then the market of such a country might not be big enough for effective remanufacturing.

Opportunities for reuse are increased if components are standardized and identical platforms or structures are applied. Then the volume of reusable components is bigger, and the component could be used for different products.

The energy consumption of reused components or of the whole product shouldn't be higher than for other components or products. Those components should be de-listed.

As the planning for reuse will cover several product generations upgradeability will become important. Especially SW upgrades must be possible. Now this is also required by the EU for household equipment for repair. On the other hand SW reuse of legacy SW should also be an option.

It is obvious that such a concept requires a financial calculation of the overall benefit and a management system to achieve the planned targets.

In Fig. 6.3 some cost savings for reuse and reduction of costs together with lower environmental impacts are summarized. The benefits can be rather high but are usually shared with customers.

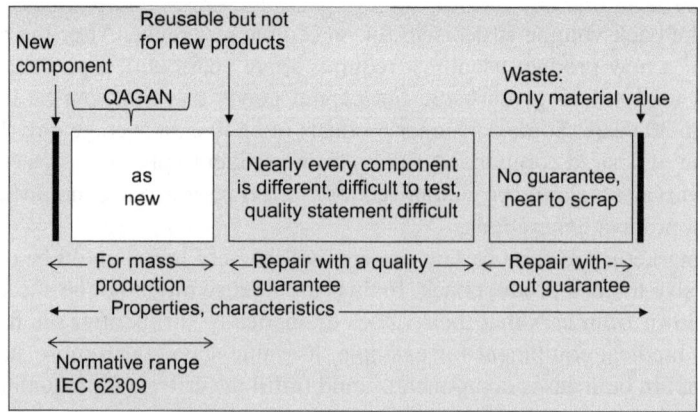

Fig. 6.3 QAGAN definition in IEC 62309 visualized

6.2.4 Environmental Preconditions

In *Waste Framework Directive* of the European Commission *waste* is defined as "an object the holder discards, intends to discard or is required to discard". A second life of the product or a component thereof or its reuse usually requires an inspection of its quality and an assessment of the possibility of its reuse. Knowledge about the materials record of a component, low energy consumption, and absence of prohibited or other risky substances are environmental preconditions for reuse of components. If such information is not available reuse should not be accepted.

The role of software on environmental impact is often underestimated. Computer programs can lead to high energy consumption or waste generation. If this is a problem this situation must not urge customers to substitute the product. There are many examples in which a change of a computer program can become the solution. Or for a defined HW a control unit enables energy reduction. As an example, a motor will become environmentally better compatible by adding a control unit. A dramatic energy saving will be achieved by this action.

A collection and recovery program should be established by the manufacturer for all kinds of E&E products. In many countries such an initiative of a producer is expected by the public. The value of materials is often high enough so that recycling should already be done voluntarily or recycling is required by law like by WEEE [4] or *Ecodesign Directive* [1]. A target of 5% reuse is discussed by the European Commission as part of the recycling quota (www.ewwr.eu/docs/ewwr/reuse_RRE USE.pdf).

6.2.5 Reuse in the Combined View from SW Reuse and HW Reuse

One of the often-forgotten subjects with the reuse of HW components gathered out of E&E products is the fact that some components require SW like controllers and nearly all complex products require many SW blocks. During the refurbishment process a mix of the SW state of the components of different ages could occur which meets different SW states in the complete products and these might not be compatible or reliable. The subject is now to integrate updates into HW and to exchange SW elements in a variety of different states of components and products. Where necessary the information about the HW treatment in the following is borrowed from the book [3] in combination with the SW part. As IEC 62309 [6] is focused only on HW this subject could not be dealt with in this standard. On the other hand in IEC/PAS 62814 [26] only the SW reuse is widely treated. But, also the combination of HW and SW requires some additional solutions or it may cause problems.

While HW and SW for a new product are developed together, SW reuse in combination with HW reuse could cause more problems, for example because of different technical states which have to become over-bridged. A repository of reusable qualified SW building blocks might help to offer a selection of useful SW elements for legacy SW and for upgrades.

The special focus on SW reuse in combination with HW reuse is not only on the complete product but also on components that contain SW. Some of these components might be peripheral, like large printers, which are often also stand-alone products, and others might be for example controllers in the product with self-regulating SW. At the beginning in the new product HW and SW were adjusted with each other. But during repair or refurbishment, new and reused components will be combined with the product to be reused. The problem will be that it might become difficult to combine these SW elements because the SW state in a component might be unknown. For the overall product or system, the situation could be that up-to-date SW is required. But for communication of older components with the whole system, legacy SW elements have to be checked establish whether they/interact with all other components of the system correctly. Also, communication with an outside network could become necessary. The outside network might contain non-compatible elements. For bigger plants or defense equipment such problems could occur.

After some tests, like incoming inspection, a component might be classified as-new or QAGAN. Also if the component carries certain SW and it can be definitely confirmed that this SW is original, incompatibility with other components and their SW could occur.

Smith [46] mentioned from practical experience of combined HW and SW reuse in robots that the SW of the product should not be too complex. In addition:

- SW reuse should be handled in checked and standardized units.
- Programming should enable cross programming language.
- Also HW upgrades have to fit to program commands.
- HW components may vary in a product depending on the configuration, a SW might encounter problems with that.
- HW device and low level SW must be standardized to support portability,.
- It is helpful that a HW abstraction layer of the SW is applied instead of a detailed HW level. By using such a platform for example, noisy signals of a sensor would not directly be transferred to the highest programming level and can be "normalized" before.
- Also a virtual machine could be helpful to test all applications of the program before going into practice.

A much stronger challenge will be the integration of an already developed part like an engine, a product like an inventory database, or a legacy system like a telemetry processing system, into another context or component. This might be the case for Defense Systems [57]. The authors have found that documentation of the legacy system, availability of subject matter expertise, and complexity/feasibility of integration are key factors that must be analyzed prior to reuse. After starting it is uncertain whether or not the project managers will efficiently and effectively reuse HW and SW legacy systems based on cost, schedule, risk, operations, maintenance, and performance. Very basic questions about documentation and feasibility of Integration have to be put about such systems.

Documentation: Type and quantity of available documentation. If documentation does not exist or is deemed insufficient for current needs, what re-engineering efforts must be done to understand and document the system moving forward?
Feasibility of Integration: Are there technological gaps that exist? How broad are the technological gaps? If there are technological gaps, is there a path forward that can enable integration?

Also the *expertise of special experts* is required. Costs may be estimated in comparison to similar projects. Especially in very expensive projects HW and SW reuse might be an interesting project.

One target there must be to establish a failure-free communication between all these new, old and exchanged elements. It would be ideal if the state of the SW of all these elements was known and identical or sometimes still better, it would be possible to get the source code for SW reuse or adaptation, if required.

Because of SW complexity it might be difficult to check whether a SW containing component can really work safely and dependably. If this cannot be guaranteed completely a reuse cannot be recommended.

6.3 Equivalent-to-New (ETN) and Qualified-as-Good-as-New (QAGAN)—More Than Buzzwords Introduced by IEC 62309

In IEC 62309 [6] especially the concept of using equivalent-to-new (ETN) components for the reuse of products, especially in new products, is described as a good precondition for HW reuse. As the ETN state might be misunderstood by a customer so that the customer could believe an ETN component was really new instead of only ETN, a new word QAGAN is used for that state, which means the component is qualified-as-good-as-new (QAGAN).

The reason for selecting this QAGAN state of qualification of a component is that for mass or serial production, except for some slight changes like in color or similar, the same properties, that is, the properties of a new component must be valid for every component. Incoming inspection is identical for both QAGAN and new components, saving a lot of costs. In Fig. 6.3 the possible aging states for a component are shown, so the definition area for QAGAN is small.

So, a new and a comparable QAGAN component can be inserted into mass or serial production of a new product. It should be mentioned again that often the difference between new and QAGAN components cannot be distinguished. In this Figure the state of QAGAN is defined explicitly only for HW components without looking at the SW. The volume of QAGAN components can be rather big, other than it looks like from the definition area. As the area of repair will also be of interest for reuse, this subject will also be discussed together with SW reuse.

For the QAGAN state the following features were defined in [3, p. 36]:

- Quality features of QAGAN components according to IEC 62309 are the same as for
  corresponding new components. Therefore, the same guarantees are granted as are for new.
  components/products. Minor restrictions are possible without affecting the quality and.
  usage.
- Guideline conditions are passed quality, dependability requirements, and test conditions in association with the internal and customer-related documentation, and all test procedures.
  The conditions, the characteristic values, etc., are precisely described, stored and also handed over to the relevant departments of a manufacturer for archiving.

The volume of potential QAGAN components can be low or high. It is estimated that about 25% of components or more can be such components. QAGAN can be stainless screws of a value of €1 each or an X-ray tube with a value of thousands of €. Often the reuse also depends on the required components for production and service.

The idea for QAGAN is also transferred in IEC 62309 to a new definition of a "new" product. So according to IEC 62309 a new product is:

> **A product as a whole, including all of its constituent parts, that has not yet been put into normal use** with the remark: A *new product may contain one or more QAGAN parts.*

The world of reuse could be fine with this definition if no problem with the SW could occur. If the producer knows the SW and the state of its components exactly the definition of the QAGAN state will be easy. But if there is a component containing an unknown SW it has to fit into the system of the product SW system and must be free of failures. So, the QAGAN state should be extended if not already done.

For the extension of the definition to HW components by SW the following add-on is proposed by the authors to be applied:

> **A QAGAN component checked for its HW as-new will be accepted for a new product only if the SW and/or the SW controlling state is known, and it could be integrated in the new product or is updated to the state of the new product.**

Also, the SW can be reused. SW reuse is often already applied in SW development but is usually not mentioned by manufacturers. It promotes better SW quality and quicker development. In combination with QAGAN the SW state of a QAGAN component must be that of the new component. This could also mean an upgrade if the state of the SW for the new product has been changed or upgraded.

If doubts in quality or safety will remain, the QAGAN qualification cannot be granted.

For components containing SW a new problem will occur; The SW state might be different from component to component depending on the age of the component. The state might be known and differ from component to component, the state might be unknown and also the complete SW might be unknown because it came from another company. The next question will then be whether the latest state of the SW is required on this component to qualify the whole product or system, or whether the refurbished system with the latest SW could work with the SW version found on the component. These questions have to be answered too before a QAGAN state will be granted.

Components with a doubtful SW state require a test which will guarantee a failure free reuse.

The QAGAN definition was selected from the view of components. A completely different standpoint occurs after the refurbishment of a taken back product. It is obvious that for repair or renewal or upgrade QAGAN components with a definite SW could be applied in the refurbishment process. But for the complete product the state of the SW might not be really clear if not all recommended updates were applied during the product life and the whole system does not contain the latest SW version. Whatever the recycler is doing, for example if the complete SW is exchanged or updated, a compatibility check of the SW will be required. After a SW exchange the compatibility with some components might be disturbed or the communication to some peripheral equipment might experience problems.

For a complete product or system final test SW might be required to evaluate the state of the product or system. In computer production for example 96 h cyclic stress testing is applied before the computer is sold. These SW tests are proprietary to the producer.

After the test it could be found out that the system doesn't work reliably. If the manufacturer of the components and product is identical this problem should normally not occur but nevertheless it might become possible, especially if required updates had not been installed in time.

In the meantime, a PC was designed by iameco which got the first European Ecolabel in 2011 for a PC. The principle behind that is slightly different from the aforementioned system with QAGAN components. It follows IEC 24700 [55]. A consequent Design for Recycling (DfR) is applied, including the selection of resource-efficient components and materials, the design of components and the product architecture and the selection and use of joints, connectors and fasteners [58]. A system was installed which allows simple upgrading by the user, repair or take back by the manufacturer. The advantages of this DfR are: Accessibility, replace-ability and ability for extension or disassembly, and the impression of stability of value. Especially the customer is able to do the work of keeping a long life and always up-to-date product by managing their own exchanges. The planned lifetime is three times longer than usual for a PC; 10 years. About SW quality nothing is mentioned, but it seems that changes in SW might be done from outside, such as upgrades, as they are usually done by the PC program or by the user alone. Interfaces assist the customer. Only two requirements are defined according to IEC 24 700 [55]:

- The equipment must be tested using the equivalent procedures as defined by the manufacturer, regardless of whether or not the product contains all new components or if it contains reused components.
- If the equipment is originally manufactured in an ISO 9001 certified factory, then this condition must be maintained. Factory certification to the ISO 9001 standard may ensure that the product conforms to the product's established design.

These requirements are less strict than those of IEC 62309. They enable for a manufacturer more flexibility. Nevertheless, both methods could be applied.

Matsumoto and Ijomah [92] describe the following hierarchy of secondary market processes:

- Remanufacturing
  The process of returning a used product to at least OEM original performance specification from the customer's perspective and giving the resultant product warranty that is at least equal to that of a newly manufactured equivalent.
- Reconditioning
  The process of returning a used product to a satisfactory working condition that may be inferior to the original specification. Generally the resultant product has a warranty that is less than that of a newly manufactured equivalent. The warranty applies to all major wearing parts.
- Repair
  Repairing is simply the correction of specified faults in a product. Generally, the quality of a repaired product is inferior to that of the remanufactured and reconditioned alternative. When repaired products have warranties, they are less than those of newly manufactured equivalents. Also, the warranty may not cover the whole product but only the component that has been repaired.

The view at these processes are the production processes leading to different qualified products. By remanufacturing a state similar to Qagan state is achieved. Qagan components are in addition combined with new components to produce a new product. So a new product is manufactured by using Quagan or absolutely new components. Furthermore another extension described I this book is the role of software. It will be not sufficient to have it new or like new, it might be necessary to get it into a state working failure free.

6.4 Integrating Concept from Take Back to Disposal

An optimized reuse system requires knowledge about the product market and secondary markets, also about the potential internal and external market of spare parts and co-operation with experienced recyclers to get a high value from the recovery of valuable materials and low costs for disposal of a minimum of waste. No company should only concentrate on reuse of components and products because a higher cost reduction will be possible if all aspects of recycling are considered.

In Fig. 6.4 the different possibilities for reuse are shown. All processes are important and should be managed in a company to get the best result. In combination with SW the focus in this book is especially on the reuse of parts and products. All aspects of these cyclic processes are explained in more detail in [3]. In the next figure some advantages are demonstrated, like forward/backward logistics, or the reuse of components and materials in the same production line. This allows a close connection and also cooperation between customers and suppliers.

Several processes have to be installed, including those with customers, to achieve an overall and optimized process for reuse (see Fig. 6.6) which has to be included in the management system. Forward–backward logistics can be applied to get the used product back (1). Especially the original equipment manufacturer can organize refurbishment and testing and could reuse the tested components (2). The third circulation process addresses the material which today is usually not yet well organized. The reasons for this are especially: bad disassembly of components and materials, poor collection rates and often poor recycling processes (details see [3]). Really interesting for manufacturers might be the reuse of those materials which are like new. For those metals and plastics simple reuse processes can be installed. On the contrary, publicly organized take back or collection systems lead to a broad mixture and may better allow the achievement of the legally required quota but the quality state of the returned products and combined components is often very poor. The medical industry has defined a *circular index* (C.I.) to measure the improvement per year of recycling. An index of 100% should be the target. The index is defined as C.I. = tons reused/tons put on the market (www.cocir.org).

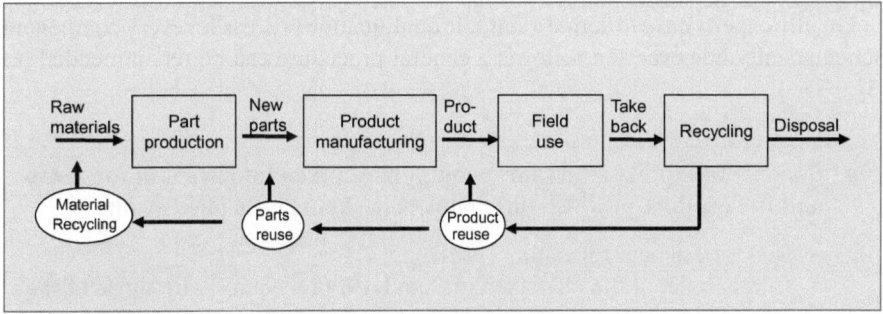

Fig. 6.4 Possibilities of reuse for components, products and material (cf. IEC 62309)

As consumer products have often very short working lives for some products like cellular phones, below 1 year of use this cycle cannot be applied, although many components are rather new. An enlargement of the useful life of the product through the assistance of the manufacturer might here also be a solution. Anyway, this product will differ from a new product like by flexibility for updates and upgrades including potential take back to the manufacturer. If the products would not be damaged there could also be a reuse opportunity for components. The circle would look different from one for capital goods.

If valuable components and materials are identified and can easily be disassembled from a product, a longer useful life can be expected, and with every use a high value is generated again, often equal to the value of a new component.

As shown in Fig. 6.6, QAGAN parts might be reused several times. This will require the planning of reuse of these components. The working lives of the products, for instance in the consumer area, are today one year old or less. In the field of capital goods this time could be ten years and longer. Anyway, worn parts could be exchanged to prolong the working life of a product or component. Also, parts of the as-new designed life which are not sufficient for QAGAN could be reused for some repair purpose.

6.5 Selection and Qualification of Components and Products for Reuse

Qualification of components usually means qualification as QAGAN (qualified-as-good-as-new) components. Depending on the application this means a detailed process from the take back till the application again as spare part or in a new product.

Besides the known state of QAGAN including SW qualification, it is possible to distinguish between new products containing QAGAN components and those products which cannot be defined as new but only as refurbished with a quality statement. The last state is interesting if for example a hazardous substance might be contained in a component or product, which should no longer be put on the market but can be legally reused for some time.

Quality experts have to define a suitable qualification process for every component. Schematically however, the following general procedure can be recommended [cf. [3]:

- Conduct **feasibility study for a component taken into account for reuse and the product** in which this component should be applied again:

  - Risk assessment for failure (FMEA),
  - Comparison of the characteristics and test procedures with those of the new components,

- Calculation of the financial benefit.

- Describe **process and qualification of disassembly**, for example

  - Determine the best disassembly steps and define them.
  - Apply special protection measures such as ESD protection with electronic components; follow hygiene rules; if in doubt contact experts in disinfection.

- **Cleaning** of the components, for example, by carbon dioxide, pressurized air; potentially by disinfection.
- **Qualification of components**: Classify potential risks according to the effort required: Passive components with cleaning; passive components with extra work; dynamic components with wear (refurbishment).
- Test critical properties like product safety at first.
- Perform pre-testing

  + Visual test: Color, state, cracks,
  + Simple technical procedures: Passage of current, voltage,

    - Components with wear: Determination of residual lifetime; potential exchange of worn components.
    - A new step not described in [3] has to be the SW evaluation if necessary:

      - Check state of the SW of the component,
      - Compare with the planned state of product SW and check compatibility,
      - Check potential incompatibilities of peripheral equipment also supplied by different suppliers.
      - Finally implement a test system for the end product with all new and QAGAN components.

- Insert components into incoming inspection,

  - Compare with the specification of the new component including SW,
  - Investigate process capability,
  - Consider failure curves, similarity curves, and
  - Signal-to-noise ratio.

- Insert components statistically into the production line,

  - Investigate failure rates before and after sales,
  - Establish documentation for customers including SW states, give recommendations for compatibility with peripheral SW systems,
  - Compare with quality of devices consisting only of new components.

Table 6.1 Criteria in the selection process for products/systems (Source: COCIR [8]; Industrial standard about "Good Refurbishment Practice," concept of the medical associations of the manufacturers of medical devices from EU, Japan and USA)

System selection according to	Analysis	Result of analysis
– Performance	– On-site or after supply	– Available components
– State	back	– Visual impression
– Age	– According to offer	– State of the wear-bearing
– Availability of spare parts:	– By customer order	components
typically 5 years	– Disposition	– Available accessories
– Service history		– Documents
– Model change in preparation		– Regular maintenance
– Upgradeability of HW		– State of the SW
and SW		

Beyond this general plan there are many detailed steps which can be handled in the form of a checklist, updated by practitioners and individual for every company.

The second important step is the selection of reused products or systems. Before final selection some general information has to be gathered about the product or system. In Table 6.1 an example from the medical industry is shown. In this checklist much information is contained about the SW of the product, components or intersections. After the analysis a decision could be taken about the value and the suitability of a product for reuse or whether only components should be extracted.

In addition, decisions about many more criteria have to be made. In the following steps of the corresponding treatment phase, criteria are listed and necessary preconditions for reuse are identified. The points in Table 6.1 can be used as a 'to do' checklist for the required steps. Which checklists should be applied depends on the kind of product. For decision making, the results of the answers to questions arising from the checklists have to be analyzed.

According to [8] the procedure happens in the following steps:

- System selection for refurbishment,
- De-installation, packaging and transport,
- Refurbishment,
- Re-installation,
- Professional service.

Manufacturer information belongs to each of the working steps described in the following (see [8]; as well as rules on workmanship and measurements, which are not discussed here, in further detail). Much information about the SW state is gathered by system tests and the available technical documentation (Device master record-DMR). Afterwards further tests can be initiated and further SW upgrades have to be planned. In this phase the potential SW states of components also have to be taken into account and corresponding compatibility tests have to be planned. The potential new options from, for example, peripheral equipment, and the final state of the SW of the complete product, must be defined. A summarizing checklist is shown in Table 6.2.

Table 6.2 Checklist for work procedures and information starting from de-installation until refurbishment according to [9]

1. De-installation/packaging	2. Planning of refurbishment	3. Refurbishment
and transport, supply to manufacturer	Technical documentation (Device Master Record) as a basis for refurbishment	Cosmetic works/coating
Test of the system before de-installation	Test for the latest mandatory updates of hard- and SW	Substitution of damaged components
First evaluation before de-installation	Secure removal of all data	Build-in of options and new display equipment
First cleaning/disinfection	Planning of the necessary upgrades	Detailed testing of components and subsystems
Disassembly by qualified service providers	Planning of all refurbishment works	SW updates to the latest issue
Packaging into original transport packages	Planning of system tests	Update of the technical documentation
Transport to site of refurbishment	Preparation of a GRP[a] complying declaration	Complete system test by original test equipment and according to original test procedures
Incoming analysis	Planning of packaging and transport	Update of the Device History Records in comparison to the Device Master Records
Complete cleaning/disinfection	Planning of re-installation and start test during re-installation with the customer	Hand out of a quality certificate, marked by quality seal (GRP[a] declaration)

4. Re-installation with customer[b]	5. Professional service Guarantee as for a new system	
Packaging according to service instruction,	Availability of the original spare parts	
Transport by qualified service provider for logistics	Contracts for maintenance	
Professional installation	Update management by manufacturer	
Putting into operation according to service instructions	Application training for employees	
Application training for customer	Financing, service, contracts	
Update and hand-out of certificates and customer documentation	Qualified contract partners worldwide	
Information about materials defect liability like with new products		
Spare parts availability of typically 5 years		
Full availability of service contracts and financing solutions like with new products		

[a]GRP = Good Refurbishment Practice according to the guidelines of COCIR [8]
[b]Info, for example: www.healthcare.siemens.com/refurbished-systems-medical-imaging-and-therapy/refurbishing-process

For re-installation in step 4 instructions for the new SW have to be provided, and besides the new documentation, the service engineer has to get necessary information about SW service for this refurbished product. It would be ideal if there will be no difference when compared to a new product.

At this point the important HW recommendations are not described in detail. Many examples were already given in [3]. But for SW reuse a checklist from practical experience might be helpful.

There will also be the question of what to do if professional SW would no longer be upgraded and might therefore become unsafe. This risk might occur with older components.

As described in IEC 62309 the following requirements are also valid, which are largely hidden inside of the manufacturers' instructions. They are inserted here from [3] for completeness.

Refurbishment of components/product

- ESD compatibility/instructions of the manufacturer must be considered.
- Housings must be visually intact, a second coating is acceptable.
- Reconditioning—exchange of a worn element is allowed if afterwards all of the quality criteria are passed.

Evaluation of dependability

- Expected working life of a QUAGAN component must be at least equal to the NDL of a new product.
- Verify by failure distribution curve of new components and of the remaining working life of the reused components; or with random samples from the quantity of the components being QAGAN.
- Characteristic curves enable statements about the residual life of wear dependent components.
- Final test inclusive of functional testing of the products with reused components required. Identical for products with only new components.

Performance and results of the qualification check have to be documented as explained in section X. Beyond this, design requirements for the products of the next generation have to be defined (Sect. 7.1) and within the management system, planning for components to be reused for several product generations has to be implemented.

6.6 Definition of New Product, Refurbished and Used Product

As a customer should not be supplied with defective products but instead will be provided with new highly valuable ones containing QAGAN components, there is no reason to hide anything. Therefore, unmistakable information about the potential availability of reused components and their state will be necessary. Offers, data sheets, product descriptions, sales literature and contracts are suitable for supplying this information.

It is important to direct the customer to the information about the performance of all important and suitable tests and about the product safety including the CE mark, because the product has been newly put on the market.

of a worn element is allowed A reference inside the customer documentation according to *IEC 62309* makes sense. For example it could read"This new product contains used but "qualified-as-good-as-new" components according to IEC 62309".

Liability period and liability conditions have to be the same as products containing only new components. At the end the message will read: *The customer will not be exposed to extended risk in addition will often get an economic benefit.*

> **Principle:** Detailed and clear information for the customer about the availability of the QUAGAN components will be absolutely necessary!

All relevant deviations, properties, characteristic values, liabilities and other important information have to be written into the contracts. Also, an alignment with the contracts of the new products is required.

Used product with QAGAN components

Many refurbished still very valuable products might not be new products even though QAGAN components have been applied. One reason could be that the state of SW could not be maintained to the latest SW state. In this case a used product will be sold with a quality certificate, but it has to be explained why this state is granted by the selling company.

Also according to other philosophies with continuous updates like that for PC from iameco [58] for consumer products, or according to IEC 24700 [55], other standards for reuse could apply company specific requirements. Anyway the information for customers about the state of a product should be given as is mentioned above. If there is not sufficient information for customers the sales offer should be taken with care.

Special recommendations for service or exchange of parts have to be provided. This state could be required if for example military equipment has to be reused, or if for example older plants have to be controlled by equipment where compatibility problems could occur with a very modern SW.

Key Points, Exercises, Recommended Further Reading, References/Websites

(1. Preconditions) Key Points

- The validity of the bathtub curve of failures is a **technical** precondition for reuse. This curve determines the residual life of a component. Whether a component might be reused depends on the technological state of the product or of the component, the structure of the product including easy disassembly, testability, upgradeability, value and potential benefit and many other factors affect the decision about the reuse of a component or product.
- **Economic** reasons for reuse consist of the whole market situation in competition, legal acceptance, enough volume of take back and production. Value-add for components on average can be estimated to be in the range of 30–40%. Costs occur for testing, buy back, cleaning or refurbishment.
- **Environmental** requirements are laid down in laws or standards and demand high take back and recovery quotas, especially in Europe, including for reuse. Also over the life cycle reuse reduces the environmental impact, for example by the use of less resources such as materials or energy.
- The **role of SW** becomes more and more important the more components use and reuse SW. SW reuse might keep legacy SW alive in the form of reusable building blocks. But the SW compatibility with the overall SW of the product must be guaranteed before reuse. Also different HW types might not fit with the same SW.
- HW components require more and more SW. For the purpose of reuse in a used or a new product, maybe as part of a production line or for repair, HW and SW have to be qualified both. As not only new SW might be used a repository of **qualified SW** might be built up. Especially in complex HW systems SW could meet programming language challenges, different SW levels or different product configurations.
- For qualification of a component as-new **not only the HW but also the SW state has to be checked** and may be upgraded. Not only for a state as-new but for reuse in general, for example in expensive capital goods like in defense projects, also legacy systems could be combined during a new development project. This will be a challenge for HW and SW experts.

Exercises

(1) Which components could be reused without further testing only after visual inspection? Use own experience. Please name some components and explain why you have selected them.

(2)   Which environmental reasons could occur such that components, for example of a mobile phone, should not be reused? Why is material recovery interesting for mobile phones? Mention some reasons.

(3)   How could an average of 5% reuse of components or products be achieved if the useful life of a product is 5 years and the sales of the same kind of new products ends after 6 years. Make some assumptions and discuss some possibilities.

(4)   Search in the internet about remanufacturing, reuse and read some case studies about successful reuse of products by companies. How many different industry branches do you find which already apply reuse/remanufacturing? Which ones concentrate on as-new components?

(5)   Mention some SW problems that could occur with components after take back and before the components are newly integrated into a new product or system?

(6)   Will it be always sufficient to exchange the old SW for the SW of the latest state if a product or component is remanufactured? Explain some problems.

Recommended Further Reading

In [3] find case study about X ray tube, testing.

Case studies of reuse or remanufacturing, for example on the homepage of the European Recycling Network (ERN) https://www.remanufacturing.eu.

Study about potential remanufacturing of household equipment in the US:

A. Bustani, S. Sahni, T. Gutowski, S. Graves, Appliance remanu-facturing and energy savings, Environmentally Benign Manufacturing Laboratory, 28 Jan 2010, MITI-1-a-2010, Sloan School of Management: web.unit.edu/ebm/www/Publications/Mitei-1-a-2010.pdf.

References/Websites

Referred literature: [1, 3, 5, 26, 45, 46, 57].

Webpages of the European union about reuse: for example.

- www.ReBorn-en-project.org (about plants equipment to be reused).
- www.ewwr.eu/en/project/mainfeatures.
- www.ewwr.eu/docs/ewwr/reuse_RREUSE.pdf (more general about reuse).
- The European Remanufacturing Network (https://www.remanufacturing.eu) includes business models, examples and case studies.
- Study: Ecodesign and Energy label revision: Household Washing machines and washer-dryers: http://susproc.jrc.ec.europa.eu/Washing_machines_and_washer_dryers/index.html.

(2. Concepts) Key Points

- The QAGAN state of a component is an artificial word for the state 'as-new' after repair or refurbishment of a component. It is defined in IEC 62309. Such components should be used in new products together with the new components.
- "New" means now the product is newly manufactured and all components are new or like new. They do not differ in useful lifetime or quality but are cheaper than products with absolutely new components.
- Potential customers must be informed about the state of the product which they want to buy.
- In addition to IEC 62309 the SW quality also has to be checked and has to match the state of a new component if not differently required. Also the SW might be reused.
- The state of the component as-new includes now the up-to-date SW.
- Other approaches for reuse or remanufacturing have less strict requirements or work differently. For example, a system for the life time extension of a PC was built up, which is designed for upgrading to keep its state always at the latest version. Customers are much more involved in keeping in touch with the manufacturer.

Exercises

(1)   What could be a real difference between absolutely new and the QAGAN state? Give some examples.
(2)   Which problems could occur to name a complete reused product, not only components, to become QAGAN?
(3)   Try to create a similar (and not really existing) system as in [5] for a cellular phone. Take the examples from [58] (to be downloaded from www.iameco. com) and explain what you would do to get a similar system. Some ideas will be sufficient for the beginning.

Recommended Further Reading [3]

Examples from Siemens Healthineers Refurbished Systems for medical equipment: https://www.healthcare.siemens.de/refurbished-systems.
    Other strategies for an upgradeable PC see examples from: www.iameco.com.
    Environmentally better compatible phone: fairphone. https://www.fairphone. com/.

References/Websites [6, 55, 58]

New PC with extended lifetime:www.iameco.com www.micropro.ie.

(3. Integration) Key Points

- A take back system for capital goods requires planning of the logistics of products and components to get them completely back into the factory. After refurbishment they could get new designed lives in new products.
- For consumer products this way seems to be difficult because they are often damaged and real waste. In this case a much longer working life could be designed with upgradeability by the customer. Also take back could be organized which is different to the actual situation.

Exercises

(1) The iameco PC has an NDL three times longer than for a conventional PC. Why could it not be dealt with like other products in Fig. 6.5?
(2) What could a process look like if mainly wood is used and incineration of the material should not be the next step after take back?
(3) How a manufacturer can save money if a cycle like that in Fig. 6.6 is installed?

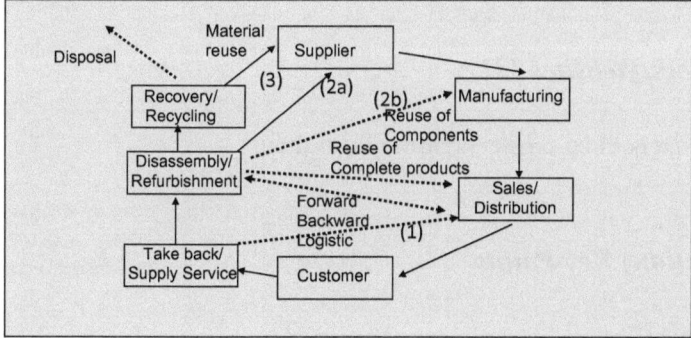

Fig. 6.5 Material flows from supplier to customer until recovery (disposal) according to [3]

Fig. 6.6 Repeated life cycles of used QAGAN components according to IEC 62309. Legend: NDL: New designed life; ANDL: As-new designed life

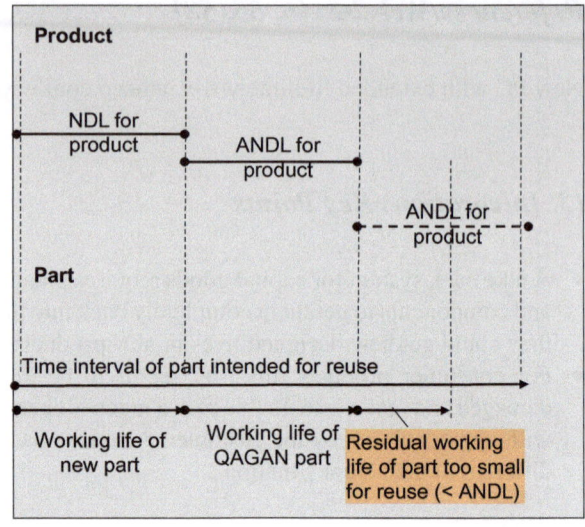

Recommended Further Reading [3]

Examples for different company rules for ecodesign, take back systems in: Goodship, V., A. Stevels, WEEE Handbook, Woodhead Publishing, Philadephia 2012 ISBN 978–0-85,709–089-8.

Recommendations for refurbishment, industrial standard with international reputation: Good refurbishment practice for Medical equipment: http://www.cocir.org.

References/Websites [3]

http://www.reborn-eu-project.org/downloads.html.

(4. Selection) Key Points

Components to be qualified run through several processes.

- System selection for refurbishment,
- De-installation, packaging and transport,
- Refurbishment, remanufacturing,
- Re-installation,
- Professional service.

In the beginning the state of the system will be checked, including the SW. Decisions are taken about what should happen with the system. If accepted for reuse (in total or parts of it) it is sent back to the factory, where refurbishment is done (disassembly, cleaning, restoration, repair, qualification or testing). Afterwards, resale as a certified product or re-installation with a new customer are the next steps. Service for a new life starts. In other cases when it is pure waste, it might directly be sent to a material recycler for recovery. All processes are in detail special for the kind of products involved.

- Depending on the requirements of the product, like hygienic cleaning for medical equipment, a special program must be developed to achieve the state as-new or a similar qualified one. An extra focus has to be set to the SW state and its update.
- Checks of dependability and testing are required, they differ also for different components and products corresponding to the test of new components.

Exercises

(1) Why are often only capital goods treated like the examples given and usually not consumer products, even though they are often sold in a much higher degree? Give some reasons.
(2) In which phases of take back till resale must the SW state be checked, and what has to be documented for SW, after a product was sold to enable better take back at the end-of-life?

Recommended Further Reading

See [3] for the recommended refurbishment processes according to IEC 62309.

A study from the European Commission for remanufacturing of household equipment ([70] http://susproc.jrc.ec.europa.eu/Washing_machines_and_washer_dryers/docs/Prepstudy_WASH_20150601_FINAL_v2.pdf) shows chances for future potential refurbishment of consumer products.

The potential for reuse of household equipment is presented for the US market in the study from.

Boustani, A., Sahni, S., Gutowski, T., Graves,S., Appliance Remanufacturing and Energy Savings, January 28, 2010, Environmentally Benign Manufacturing Laboratory, Sloan School of Management, MITI-1-a-2010, web.unit.edu/ebm/www/Publications/Mitei-1-a-2010.pdf.

For benefits, strategies for reuse processes see: Wimmer, W., Lee, K.M., Polak, J., Quella, F., Ecodesign - The competitive advantage, Dordrecht/Netherlands et al., Springer 2010, ISBN 978–90-481–9126-0.

References/Websites [8, 9]

COCIR industry standard: Good refurbishment practice (both see: www.cocir.org).

(5. New vs. Refurbished) Key Points

- Declarations about the state of a reused product or a new product with reused QAGAN components are required in combination with the sales of a product.
- CE marking is only required if the product is newly put on the market. This label guarantees to comply with all relevant EU legislation.
- Anyway a good set of documentation for the customer and a warranty create confidence with the customer.

Exercises

(1) What does it mean about the legislation to be fulfilled if a product is as-new refurbished and should be resold? Mention some differences between a new product and a repaired used product, for example for legal requirements.
(2) Define those requirements and standards which have to be applied to put the CE-mark on the product. When it will not be required to fix a CE-mark on the product?

Recommended Further Reading [3]

See section XII in this book about standards. A comparison of IEC 62309 with other standards gives insight into different reuse philosophies.

References/Websites [3, 55, 58, 65–67]

CE mark: https://ec.europa.eu/growth/single-market/ce-marking_en.

ANSI/RIC 2015 (001.12016): Specifications for the process of remanufacturing http://www.remancouncil.org/files/5fdeCD/RIC2015-Specifications-for-the-Process-of-Remanufacturing-Draft-11-25-151.pdf, Remanufacturing Industries Council, 1335 Jefferson Road #20,157, Rochester, NY 14,602–0157.

Chapter 7
Ecodesign, Design for Reuse

7.1 Elements of Ecodesign

Ecodesign requires rules, strategies and especially a design strategy. The target will be to create a product with a minimized environmental impact. In addition, legal requirements have to be considered. Last but not least reuse will only be successful in bigger volumes if the product planner also covers future product generation.

7.1.1 Rules

Since the first *Ecodesign Directive* [1] of the European Commission, ecodesign is required for the development of nearly all E&E products and also energy related products as a precondition for putting them on the European market. In Annex 1 of this directive many generic requirements for Ecodesign are listed, although they do not contain SW and its influence. As this Annex 1 is important for reuse it is also included in this book, also in Appendix A.

> The requirements of Ecodesign Directive include design for reuse for a new product. In case a new product also consists of used components the ecodesign requirements of Annex 1 of *Ecodesign Directive* will be valid as for a completely new product.

An overview about subjects which all potential rules should address is presented in Table 7.1 (taken from VDI 2243 with extension by the authors).

Detailed ecodesign rules are published for example in IEC 62430 [11] (cf. [3]) or in [64].

© The Author(s), under exclusive license to Springer Nature Switzerland AG 2021
F. Belli and F. Quella, *A Holistic View of Software and Hardware Reuse*,
Studies in Systems, Decision and Control 315,
https://doi.org/10.1007/978-3-030-72261-6_7

Table 7.1 Simplified scheme as a basis for the development of detailed ecodesign rules

Level of detail	Building structure	Connection technology	Material	Functional units
General	Recycling concept	Non-destructive disassembly	Ability to recycle	Ability to reuse
Product	Modularity of the total construction	Kinds of connection and their selection	Interacting compatibility; suitability for NDL	Division into functional units
Board level	Exchangeability	for example sticking, clipping on	for example with printed circuit board halogen-free flame retardant	Simple exchange
Component level	Accessibility	Depth of disassembly and disassembly time	Low material diversity	Ability to reuse or repair
Material specific	Separability	Dismantling time	Material selection	Material diversity within functional unit

Ecodesign must not be a complete sustainable design. The newly designed eco-version of a product should show a lower environmental impact than the former version. But the rules usually don't contain social and cost aspects of the product, although application of ecodesign rules usually leads indirectly to lower costs, and lower environmental impact might cause better social conditions. So, sustainability is improved. To achieve better sustainability the product should be, for example, manufactured in fair conditions and the price should be affordable by most people.

Conditions for sustainability might be different from company to company and from country to country due to special circumstances, like the political intentions. Therefore, this subject is not laid out here. Social rules should be agreed before worldwide adoption.

Another important point to be discussed is which strategies are to be applied. In ISO/TR 14062 the following strategies are mentioned. The design strategy shared there is not explained further:

- Company strategy,
- Product strategy, and
- Design strategy.

In the case of reuse a company strategy could be to become a company which is environmentally friendly, offers take back of their products and has organized a recycling concept from reuse to deposition.

A product strategy might cover subjects like sales of environmentally sound products. Also the environmental image of the company can be improved by creating new products including QAGAN components.

The design strategy described in detail below could cover the total structure of the product and the way that it could easily be dismantled into all components for reuse or recycling of the required elements.

Design strategies are not easily derived [21]. Much has to be taken into consideration. It is not only the product, it is also the lifecycle which has to be considered (lifecycle concept). The kind of manufacturing, the situation in which the producer acts: is it own production or mainly assembly of 3rd party components. To explain the difficulties one example might be given:

With a so-called 1-plastic strategy (mono-material target) many components can be combined, recycling of the material can be guaranteed, value can be kept constant, enough volume for recycling can be generated, application of the material is possible for several product generations, etc. [3, 65]. This target means to reduce the number of different kinds of plastics to very few ones. Afterwards many components or their housing can be combined into one, for example by injection molding. The overall number of components is then strongly reduced.

But a choice of the best material can anyway be difficult. A renewable material like wood can today be applied for the housing of a PC, which might have lower impact on the environment. But the recycling of wood is not far developed. Usually after end-of-life it will be burned. Although a whole cycle of longer life could follow: for example second life of the housing, cut in to pieces and use as laminates, use as raw material for the chemical industry and at last perhaps incineration. Such a longer life might be given with a plastic, like that from ABS, which allows several cycles of remanufacturing. Both ways could be interesting but for a development engineer the practical availability today and proximity of the company location might be important for the decision. In Table 7.1 the different choices for strategic design options can be seen. They also prove the complexity and necessity of design strategies.

Additionally, some conditions for Green IT are mentioned. Green by IT and combined SW and HW reuse are explained in Sect. 7.2. They must become part of a product and a design strategy.

7.1.2   Design for Reuse, Requirements for a New Product

In the following points some important design requirements that are necessary for
reusability or easy disassembly are summarized:
   Follow ecodesign standard/instructions for developers like.

- Easy disassembly/modular design,
- No screws/no connections by adhesives, no non-solvable connections,
- Planning for several product generations, or enlarge the useful life of a product so
  that it covers several product generations,
- Determination of product age and wear of components (counters, measurement
  of wear etc.),
- Suitable, aging resistant metals and plastics; alternatively, shorter living, often,
  cheaper but recyclable materials,
- Decision for material strategies. For example selection of very valuable materials,
  realization of the long-term value after take back,
- International rules about ecodesign, published in ISO/TR 14 062"Integration of
  Environmental Aspects into Product Design and Development" [25] and in IEC
  62 430 "Environmentally conscious design for electrical and electronic products
  and systems [11]." [64].

Many companies already have developed their own standards for an environmentally compatible product design, which a newcomer in this field should also
have a look at (for example from BMW, Philips, Siemens). Every company dealing
with ecodesign of its own products should develop their own guidelines for ecodesign based on the international state of the art. These guidelines will represent the
experience of their own engineers and continuously further development through
experience could be spread throughout the company which will be very beneficial.
A combination with best practice examples will be helpful. Some companies have
many rules, others like Philips have concentrated on the so-called "quick 5" (weight,
energy consumption, recycling and disposal, packaging, hazardous substances) most
important rules.

Ecodesign and Design for Reuse in many companies have become the same
because nearly everybody wants all requirements of Ecodesign to be included in
one system of rules. A design for a circular economy could also be a design for reuse
in a wider sense (covering the whole life cycle). From production site the Design
for remanufacturing is introduced. This concept seems similar to a DfR (Design
for recycling) approach, especially if the target is the same; as-new. Nevertheless
some strategic differences between the approaches occur. Upgrading could be a
central factor, like that mentioned for the iameco PC, or it could be reuse for several
generations with components as-new. The focus could also be durability, standardization and compatibility, ease of maintenance and repair, disassembly and reassembly,
modularity and others see [72].

Rules for reuse of HW components and reusable products in combination with
SW are rare. Some points for the impact reduction of service should therefore be

mentioned, according to Vautier and Philippot [71] who started as the first team with an Ecodesign of SW:

- Optimize marketing needs requirement (everything that doesn't have to be supplied to the customer saves CPU space)
- Select the best architecture based on optimized SW library
- Write the SW coding scheme efficiently to minimize energy consumption.

7.1.3 General Strategies

Many Design-for-Recycling (DfR) rules simply added together do not result in a design strategy at product level! Also, in *ISO/TR 14 062* [25], the entire sequence starting from the required design strategy, and ending in marketing and product strategy wasn't dealt with systematically. Systematic and detailed approaches to achieve an environmentally compatible product are missing. To close the gap, an approach was made in [24], see Fig. 1.9. In VDI 2243 the clarification of the following recycling-oriented targets is seen as unavoidable for strategy development:

- Consideration of actual market and customer requirements,
- Investigation of actual and future regulations (legislation, guidelines, standards…),
- Inclusion and consideration of the actual recovery situation in important countries,
- Analysis of forerunner products/competitor products.

These steps were discussed previously in more detail in [3]. But only the requirements for the development were defined there! There are other issues that concern the development engineer, such as the question of 'when does the product need updating?'. They shouldn't just look at that update; it is better to do it at the point when the whole design should be looked at, for increased efficiencies.

> The problem culminates in the end in the question of how to translate different, often contradictory requirements into one solution.

Strategies will also have to cover innovation [3, 24]. If reuse of components or products is planned, sometimes for several product generations the development engineer has to care for the requirements of a new product. So, as a minimum, the frame conditions for this product must be known for the ability to reuse its components. It is inevitable to standardize components and to use platforms.

The same must be valid for the SW reuse. It might not be known which SW programs might be required, but a plan must be made for what HW conditions might be necessary and how SW components could be reused without higher energy consumption. Therefore, tested components should be available in a library.

7.1.4 Design Strategies

Reduced part numbers, easy disassembly or reduced volumes of hazardous substances require design strategies. As explained in [21] such strategies are usually not systematically applied for Ecodesign but should be. They were also forgotten during the development of ISO/TR 14062. Instead, often unstructured and too many rules are applied. But for an optimized product structure, design strategies must be recommended. Otherwise the optimization effect will not be sufficient.

At the end of product life take back and refurbishment can be organized more easily for capital goods, if the products are designed for the remanufacturing process, which means they could be sent back several times to a factory. For consumer products strategies for reuse might be different. Today these products are often damaged or cannot be upgraded. A much longer useful life with upgrades can be a different solution. But it could also be a process of remanufacturing if the products given back were not damaged as much as they are in the public collection systems.

In Fig. 1.8 a practical way is presented for different strategic approaches:
The creation of

(1)   functional units,
(2)   of integrated parts or components, and
(3)   of a structure which is created by commercial components.

Combined with that is the reduction of manifold materials and easy assembly/disassembly. Requirements for product optimizations for better refurbishment, remanufacturing or upgrading have to be fulfilled. Additional could be to prepare platforms, standardize parts and components. Also the depth of disassembly can be reduced (Fig. 7.1).

It should be mentioned here that all the changes in material, components and structure influence severely the production process. In a recurring activity, the changes should be used to improve the production processes, too, and reduce their negative environmental impact.

In ([3], pp. 109–114) several such strategies are described:

- *1-plastic strategy*: 1 plastic material for (nearly) all applications,
- *Reduction of types and parts*: Simplification of product structure,
- *Strategy of functional units*: Easy exchange of the functional unit, like the motor or power supply, such as for repair or reuse.
- In addition to [3] also a special design for upgrades could be a solution: The core product could stay the same but upgrades enable the product to be kept up-to-date.
- Other models can be developed, at the moment from every design rule an own design strategy is derived, which doesn't seem to make sense.

If one of the reduction targets combined with such a strategy cannot be achieved with the conventional product and material structure, it is worth looking for different materials or technologies. It is impressive how products and their structure can be changed, for example with a new stainless steel HSX^R compared with the former

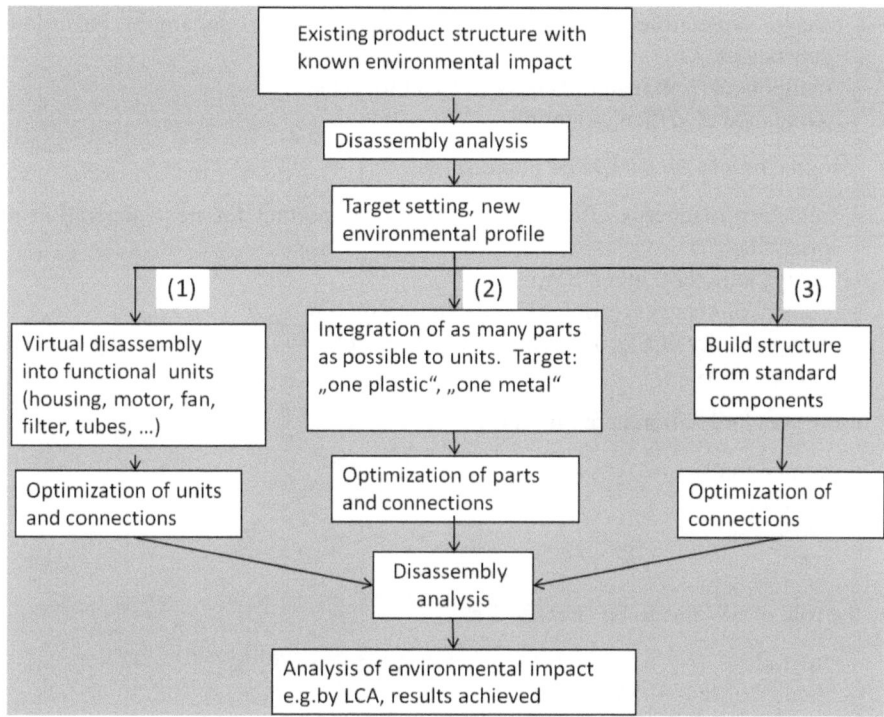

Fig. 7.1 Examples of strategic paths to get optimized product structures. Additional strategic elements such as those for better reuse, easy disassembly or upgradability have to be considered [21]

version. Plastics might now be substituted by light-weight steel, production processes are simplified, and in their number reduced; reuse becomes then easier, the weight of the product can be strongly reduced (see [22]). Ecodesign rules for the design of a product only are not enough to optimize resource efficiency. Using rules, the kind of materials might be changed following a strategy, for example production processes and the whole lifecycle will also contribute to the improvement.

For promotion of reuse

- the product structure should be simplified, for example by

    - formation of standard units, platforms,
    - low level of product hierarchies,
    - solvable joints, easy disassembly,
    - easy accessibility.

- material structure should be simplified by

    - grouping of materials within special units for recyclability,

- easily collectible rare substances, and by easily separable hazardous substances,
- components and materials easy to clean for reuse,
- avoidance of critical materials.

- resource efficiency should be planned by

  - selection of energy-efficient components (important for next generation of product),
  - energy-efficient, green SW,
  - control of energy (resource) consumption by SW,
  - compatibility of SW of components and product by library of tested green SW
  - Components.

- more standardization is required for

  - components,
  - materials,
  - processes,
  - SW.

- the role of SW has to be checked for

  - upgrades,
  - quality of reused SW,
  - networking,
  - energy reduction via SW.

In Appendix B of this book practical rules are repeated from [3] as a more detailed assistance for a recycling oriented product design. They are widely based on VDI 2243 [23] but improved by one's own experience. Such requirements have also to be fixed in the product development scheme. It is obvious that only a single strategy is not enough to get an improved product. The strategy should also be visible during the different phases of the development scheme.

A Case Study
Eco-Design or Design for reuse does not end if only one hazardous substance in a component is substituted by a less hazardous one. But it is often not easy to achieve a stronger success, by for example combining all plastics or metals to mono-materials, getting them to be easily recyclable, with all properties required, and in parallel achieving easy disassembly, simple cleaning, remanufacturing and as-new properties, combined with a long life. Often the proposed material might be more expensive and the development engineer doesn't see a chance for stronger changes, and so the improvement for the environment stays low.

If a stronger improvement is not possible with the materials just applied, the engineer has to look for materials with different properties. Usually they also have some disadvantages. This will mean not only a newly designed product or component, but also a new system of production processes. But as the material costs are about

45% of the total costs, change of materials will be a starting point for a product improvement.

Such new materials are the HSX^R stainless steels [24]. They enable.

- Light weight building and reduce material consumption (ca. 25% less weight),
- Easier cutting, longer tool application times,
- Production to be 100% from metal waste,
- Higher resistance (hardness) according to loads,
- Easy recycling,
- Much lower production costs balancing the higher price,
- Fewer production process steps (no further heat treatment etc.), quicker throughput, and
- Also, in many other processes, for example together with suppliers simplified processes.

It is obvious that many components made from such a material can easily be refurbished and reused, often directly after cleaning. The product should be optimized in the use of such a material, also plastics might be substituted. As is often the case, nothing damaging happens to such a steel component, a reuse can happen for several product generations. Under the name *Ecodesign 3.0* the study [24] investigates the further potential of such a material in many other processes over the product lifecycle.

A strategy is also needed for the upgrade of SW, which includes documentation, development state of the SW, or its compatibility. This is often already applied during the refurbishment process. But especially in legacy systems the end-of-life/obsolescence of such a component could be reached if the energy consumption of the component has been too high.

Green IT is not yet common in Ecodesign, and the same is valid for Green by IT (definitions see chapter 28). Both subjects offer additional chances for reuse. Green IT means the assistance of HW by SW, for example to reduce energy consumption. Green IT requires a different paradigm of programming, with respect to the environment, consuming less resources for its own running compared with the forerunner version. *Green by IT* can make a motor, with the addition of control by SW, into an energy saving one with up to 50% less energy consumption. For the reuse of components, a refurbishing company might be happier to have knowledge about the version number of the SW edition of the component. As failures could also occur during the remanufacturing process, a deposit of tested reusable SW elements should be available.

In this field it might be a solution to have tested reused SW to reduce the energy consumption of the older product, and/or to have SW to reduce the energy consumption of an older HW by adding a control unit. Both could also be verified by a connection to a SW network.

7.1.5   Planning for Several Product Generations

In the first step of such a plan, a vision of the future market development should be created as well as the consideration of which components could be applied in the next two (or more) generations [20].

- Xerox, for example, has noted the price decline of their small appliances and the tendency for more functions to be integrated into large appliances, leading to higher values of new products. This was possible because of the long-period price decline of electronic components which enabled the inexpensive large-scale integration of more and more functions.
- Technical leaps like miniaturization limit the possibility of reuse. Also, of products containing new functions, such as scanner integrated in printers.
- From the view of SW, limitations occur if the new SW will not fit into the old storage equipment. So, the HW reuse might fulfill all criteria, but the SW update might be too large. As mentioned in Sect. 7.2 about Green IT, this problem could be overcome by more efficient SW.
- Classifications of the components or products according to their state of age or SW can help to enable reuse as QAGAN either in new products or according to the planned refurbishment concepts. Here one could distinguish SW states from 'direct reusable, 'update possible' or 'only reusable for certain repair purposes' of equipment in the market.
- Uncertain or non-predictable failure rates caused by higher digitalization can disturb future planning because a planned rate of reuse might not happen.

Problems with SW could be avoided especially if tested reused SW would be applied. If a SW deposit with tested green and efficient SW components would exist, this could open the door for efficient reuse of HW products and a new programming of, for example, updates using the components from the data library. It would guarantee that only tested SW components would be applied and, as a result of using efficient components, overflow of data would be avoided. So, planning for several product generations will also pay off by SW reuse.

Planning for several product generations is only one option for the extension of product life. If it would be possible to have the same core for the end-of-life of a product then only upgrades would be enough. Other application models, like product sharing, could also be an alternative. The interesting situation at the moment in the market is that many different options could be taken.

7.2   Green IT

When talking to people involved in ecodesign about the impact of SW on the environment many of them would say there will be nearly no influence! The contrary is true! Especially power consumption of a product is usually controlled via SW. It is remarkable that many "Implementing Measures" following the *Ecodesign Directive* of the European Commission do not deal with the role of SW and the need to change SW for lower energy consumption by these products! This doesn't mean that in one of the last studies of European Commission (on lot 9) on "Enterprise servers" the role of power management, load balancing or interoperability and the connection of these with HW wasn't seen [74]. Improvement targets are set but a general solution for all products is not recommended.

Even looking only at computer devices, the energy consumption amounts to more than 20% of the total energy consumption of the western world [31]. Schall et al. [30] inform that energy costs for the use of servers could overcome their purchase costs after some years. So, they recommend a benchmark to estimate energy consumption caused by SW. It should distinguish between the aspects of *Green IT* and the aspects of *Green by IT*. The latter may contain both elements.

Murugesan [29] developed a definition of Green IT:

Green IT is the study and practice of designing, manufacturing, using, and disposing of computers, servers, and associated subsystems (such as monitors, printers, storage devices, and networking and communications systems) efficiently and effectively with minimal or no impact on the environment.

Naumann et al. [28] added a definition for *Green SW.*

Green and sustainable SW is SW, whose direct and indirect negative impacts on economy, society, human beings, and environment that result from development, deployment, and usage of SW are minimal and/or which has a positive effect on sustainable development.

For the authors of this book, *Green by IT* might read as follows:

Use of SW programs to reduce environmental impacts caused by E&E products. Applying networks, control units, algorithms etc. The SW should by itself be Green SW.

In Table 7.2 some examples are presented for *Green IT* and *Green by IT*. In cases (2)–(4) also Green IT should not be forgotten although Green by IT might be more important.

Some additional examples according to the subjects of Table 7.2 are explained in the next cases:

(1)   Metri et al. [2] stated: "Traditionally power consumption is treated as a HW problem, whereas SW focuses on features and flexibility. However, power

Table 7.2 Influence of SW on energy consumption or waste generation in products and systems (SoC, system on chip)

Component	Examples of Measures	Examples of Applications
Chips like SoC	New energy saving design	Reduced battery loading
SW of a product/system	Organizing stand-by if required	Many E&E products like TV sets
(1) SW of a component	Load-depending motor/pump	Control of drives/pumps to reduce energy consumption
(2) SW acting from a product	Network with other product	Industry 4.0 allows energy-saving network
Other environmental effects caused by SW	Update inefficient size of capacity of old product, Too many consumables wasted due to inefficient or not consumer-oriented SW	Product or component thrown away, Too much ink or paper wasted by wrong commands like with PC

consumption is highly SW dependent and does not necessarily correlate well to thermal design power specifications". By co-design of SoC (system-on-chip) devices they could optimize battery life. This is, simply spoken, achieved by switching off those functions including CPU which are not used at the same time.

(2) Other examples for too high energy consumption are located in commands, which users do not know. For example, computers might have functions for excellent printing and for draft printing. Usually the draft quality is very acceptable for private use. But if the user is not informed about this possibility, and thus only the best (and the most expensive) function is applied, requiring more time, more energy, more ink, and sometimes more paper. Also the speed of finishing a task requires more or less energy. As long as a computer is running all the day this will not be a problem but it costs a lot of energy.

(3) Drives containing a control unit can control the speed of a motor load depending. So about 50% of energy can be saved compared with a normal motor.

(4) Networking allows saving energy because one product can tell another one if a task is done, and for example, a machine can be sent into stand-by. Often all machinery equipment in a production plant is in the highest energy consuming mode although only one machine is really required or the plant is waiting for the next production volume.

(5) Many commands are not optimized to the requirements of the customer. So, the consumer cannot stop unnecessary printing, etc. Water consumption could also be better controlled.

The role of legacy SW is described as a cost-intensive problem if new business requirements have to be achieved [27]. This fact should not hinder a SW reuse. Especially those SW components which were quality checked and green could overcome some of these problems in the future. With such reliable SW, taken from a data

Table 7.3 Examples of tasks for energy saving and potential technique to realize the saving potential

Improvement of energy-saving	Tasks to save energy
Computional efficiency	Efficient algorithms, multi-threading, vectorization
Efficient algorithms	Select appropriate data structure and algorithms for less power consumption
Multi-threading	Use of many threads reduces energy consumption (-25% less)
Vectorization data efficiency	Vectorize code instead scalar code
Data efficiency	Minimize data movement by SW algorithms, memory hierarchies to keep data close to processing element, efficient cache memories
Managing disc I/O	Optimize power consumption by disc I/O
Prefetching and caching	Buffering, reduce drive spin-up, let operating system manage the CPU frequency
Context awareness	Let system react on changes, for example by sensors, to optimize (context awareness)
AC or DC?	Adapt application's behavior to better AC or DC use
Platform power policies	Application SW can use optimized power policies for lower energy consumption
Compilers, Performance, Library sets	Already available instructions can be used to reduce energy consumption [25]

library, the reuse of HW and SW could become quicker, greener, more dependable and cheaper.

Development engineers can select among many energy saving techniques. In Table 7.3 some are summarized according to Sabharwal et.al. [12–15].

Especially in [15] Chabukswar informs about tools of different SW companies for power management functions or considerations, such as handling sleep and resume transitions, preventing system timeouts, designing for extended battery life, responding to common power events, designing for entertainment and media PC scenarios, using power management application-programming-interface (APIs) from managed code, designing for earlier versions of windows, responding to power events within a Windows service, and testing applications for power management. Also assistance for other tools and technologies is given, such as for threading tools or context awareness tools, combined with possibilities to measure energy consumption caused by SW applications.

In all these cases development engineers, manufacturers and customers could achieve rather high energy savings by SW better adapted to special tasks or by more effort to get an optimized structure. A consumer can hardly do this alone; customers may need assistance. Product manufacturers require skilled purchasing people, either to procure more efficient components or to initiate corresponding developments. The overall expected lower energy consumption also has to become a development task for the next SW and product generation. The article [12] shows that more information is required for the applicant to get the maximum of energy saving.

Kern et al. [16] described more fundamental aspects (see table below) of Green SW but they will not be explained here in detail.

With "SW ecodesign" Vautier and Philippot [71] describe the opportunities for a telecom company: Target will be to reduce energy consumption by 50% till 2020 by optimizing

- Network operator context,
- Technical environment such as routers,
- Network infrastructure,
- Service platform,
- Mobile phone,
- Gateway and Decoders.

For this purpose they use the tool "Greenspector", which can analyze the static and the dynamic energy consumption of a SW program, perform correction of violations and measurement after correction. Optimizations happen SW layer by SW layer. The authors claim that the combination of HW and SW has still to be improved. A better tool and methodology is required to identify the correct impacts.

7.2.1 Measurement of Energy Consumption Caused by SW

It is not easy to measure the real energy consumption of SW. For many methods [16] it is claimed to measure energy consumption of SW. Two different approaches are mentioned here:

- After the first approach the energy consumption levels of two products with different SW are compared using a reference product. This step a skilled consumer could also do.
- By the second kind of approach the development engineer can find out which relation between programming code, processors or programming levels and certain energy consumption exists for which to develop an energy saving SW.
- Standardized is the method of ISO/IEC 14756 [17] for computer systems performance and SW (runtime) efficiency, tested in [18].

According to [16] some metrics are recommended for measuring efficiency in Table 7.4.

In detail, existing methodological assistance for SW developing engineers is not always sufficient. Also, in many cases criteria for Green IT and their metrics must be determined from case to case. More standardization will be required.

Chen et al. [20] described a more comprehensive approach using the example of cloud computing. Besides fixed energy consumption they distinguish Energy consumption of storage resources $E_{Storage}$, Energy consumption of computation

Table 7.4 Metrics for different aspects to measure their energy efficiency

Aspect	Metric
Energy efficiency	Energy/Unit of work
CPU intensity	CPU cycle count
Memory usage	Memory consumption
Peripheral intensity	Peripheral Usage time
Idleness	Idle time

resources E_{comp}, and Energy consumption of communication resources E_{Comm}. So total energy consumption will be expressed by.

$$E_{tot} = E_{fix} + E_{storage} + E_{comp} + E_{comm}$$

On this basis they built up an analytical tool with which they measure energies consumed according to 3 classes summarized in Table 7.5: Data intensive, computation intensive and communication intensive.

An allocation to special SW programs seems to be possible by this model.

In the case of HW and SW reuse these points are really important. If a component or a product doesn't seem to have enough storage space it might be reused in older products for repair. But it has become outdated and can no longer be inserted as a QAGAN component into a production process.

While the discussion about energy consumption moves more and more into the public sphere, the discussion about waste caused by insufficient SW programs is not yet intensive. But meanwhile it is known that the computer-based paperless office was just a dream, in reality this leads to more paper consumption! One example might be the not optimized print-out of text or pictures in the wrong resolution. If a computer has these functions, it often requires too many skills to find them, or it is not easy to quickly stop an undesired print-out. Other consumables such as ink

Table 7.5 Classification of energy consumption by SW

Task type	System configuration	Performance parameters
Data intensive	Description of HW and SW resources allocated	CPU utilization Memory Utilization Bandwidth of Disk I/O Task Execution time
Computation intensive	Description of HW and SW resources allocated	CPU utilization Memory utilization Task execution time
Communication intensive	Description of HW and SW resources allocated	CPU utilization Memory utilization Network bandwidth task execution time

or water have to be added to the environmental impact caused by SW. In the end a component or product might also become waste if the SW cannot be updated.

7.2.2 Quality of Green Software

Kern et al. [16] have proven that one of the reasons for the short lifetime of a HW product is that SW engineers designing an update often don't write efficient SW. As HW is cheap and new products usually contain more storage capacity the inefficient SW will be overcompensated. The useful life of the HW product could become much longer if the new SW would become more efficient and would still work in the old product. But this would require more programming effort. The example of Kern et al. shows that non-appropriate SW could cause another environmental impact: Waste products!

According to [16] a set of criteria was developed for a green and sustainable SW and presented in Fig. 7.2.

The model should be seen as an example. Which criteria should be valid will depend on customer requirements. In addition, further criteria could be integrated, such as for the subject of Feasibility in Fig, 7.2: Quantity of consumables, water consumption or whatever for a certain product will be required, others could be left out.

The common quality aspects belong to ISO 25,000 [19], others are more special (see [16]).

In the sense of SW reuse relevant aspects will be discussed in section A.

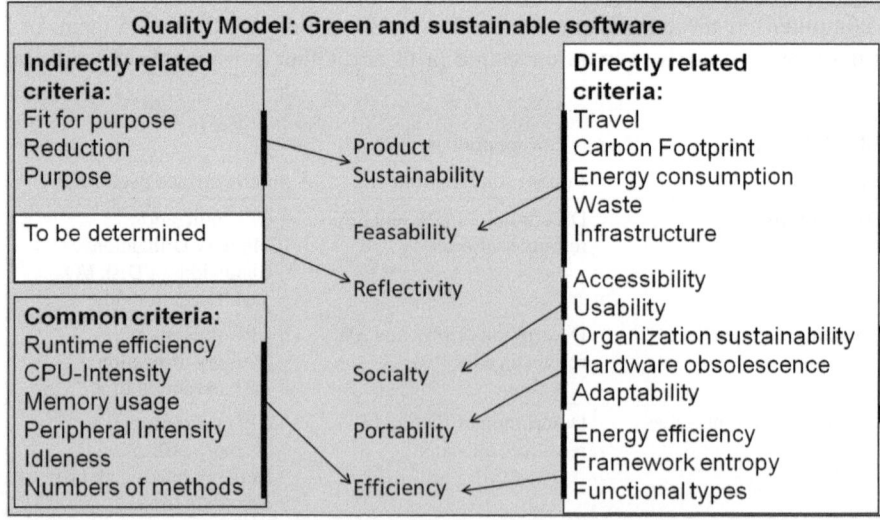

Fig. 7.2 Quality criteria for green and sustainable SW (acc. to [16])

Criteria, directly measurable are, for example, energy-efficiency of a system (related to a reference system) or framework entropy, the usage of external libraries, or different levels of energy efficiency.

Indirectly related criteria are connected with sustainability. New smart technologies can be more sustainable and allow better social compatibility. Product sustainability covers the influence of one product on another one. Fitness for purpose or reflectivity are criteria consisting of different aspects which were not completely defined by [16].

Key Points, Exercises, Recommended Further Reading, References/Websites

(Rules, 1) Key Points

- Ecodesign rules especially for E&E products are meanwhile standardized in different standards and European directives.
- For capital goods many companies apply reuse. Reuse is not so common for consumer products although many reuse projects are promoted now, for instance from the European Commission.
- Rules for reuse are part of ecodesign rules! Nevertheless special rules have to be applied for better reusability, such as for better disassembly if a company will concentrate on this special subject.
- Also a design for remanufacturing will require sometimes different rules than are generally recommended for ecodesign.

Strategies, company, product and design specifics are required to convince the public to reuse, to structure the product and place it in the market, and for the special design of the product.

- The influences of Corporate Social Responsibility (CSR) are additionally part of every environmentally compatible design. Also here reuse gets advantages because the products might be cheaper and more accessible for many people and applications.

(Design for Reuse, 1.2)

- For reuse, simplification of the product structure is necessary, material structure must be simplified to produce as few materials as possible, resource efficiency must be planned, more standardization has to be achieved areas such as parts, materials or platforms for a less complex development and easy reuse.
- The role of SW is important for HW components which use SW. For refurbishment it is required to achieve the quality level of a new component, or to combine new

and legacy SW systems or to reduce energy consumption of the total product, as well as using greener IT or by controlling the energy consumption during stand-by or off-mode.
- A repository of tested, also reused SW, might be helpful. Also rules in development of products for the SW together with HW reuse can be applied, like simplified structures or few unnecessary functionalities.

(Strategies, 1.3)

- Strategies are required for company, product and design.
- The company wants to sell a product in a special market with a special environmental advertisement. The product can be modular and/or long living.
- The design strategy has to cover the whole life cycle. If one new material is introduced, for example a light weight stainless steel, the whole design can be changed including all processes. Also the target to achieve mono-materials leads to fewer parts and strong economic savings.

(Design Strategies, 1.4)

- Planning for several product generations can be managed by planning to apply those components in the next product which may stay identical, such as stainless steel parts. They may also remain very expensive at the end of life and can easily be restored, those like from X-ray tubes. Afterwards they can be resold for the same or another purpose. These requirements are better to be applied for capital goods where an exchange for repair with a customer owning an undestroyed product is easier. Also for consumer products a refurbishment is possible if the product is not too damaged.
- An alternative will be a design for upgrades where the product mainly stays the same but components or SW can be exchanged, always within the same basic product. Such a strategy is applied with the iameco PC at the highest technological level. For consumer products such a strategy should be adopted with more success as the core stays identical and upgrades or repairs could be ordered by a customer.

Exercises

(1)   Why is a design strategy important?
(2)   Develop 5 very important rules for ecodesign from Table 7.1 and compare them with the "Quick 5" from Philips.

(1)   Search the internet for companies with ecodesign rules? How many of them have rules for reuse or remanufacturing?

(2)   Search the internet for design strategies? How many different ones did you find? Can they be combined into one strategy or are there trade-offs?
(3)   Which trade-offs do you see if certain materials should be avoided? What could happen if only renewable materials should be used?

Recommended Further Reading [3, 60, 61, 71, 73]

- About ecodesign strategies: Quella, F.: Ecodesign strategies: A missing link in ecodesign. In: Lee, K.-M.; Kauffmann, J. (eds.): Handbook of sustainable engineering, Dordrecht/Netherlands, Springer, 2013, pp. 269–284. ISBN 978-1-402-08940-4
- About different rules and their application in industry: Luttrop, C, J. Lagerstedt, Ecodesign and the ten golden rules: Generic advice for merging environmental aspects into product development, J. of Cleaner Production 14 (2006) 1396–1408
- Interesting design example and sustainability aspects: www.fairphone.com..
- Examples of circular economies: Ellen MacArthur Foundation https://www.ell enmacarthurfoundation.org/circular-economy.

References/Websites [11, 21–23, 25]

Reuse of factory equipment ReBorn: http://www.reborn-eu-project.org/
Search internet for many new examples!

(2.) Key Points

> **Green IT** should not cause a strong impact on the environment.

- By bad programming especially too high energy consumption could result and prevent further utilization. Waste of many consumables might also be generated, for example by some printers.
- In case of HW reuse of a component, this problem cannot be continued, because it might then be better to throw the whole used product away.
- SW reuse can help to avoid problems with bad programs, if the SW reused was already checked for compatibility and quality.
- Usually a potential improvement will not be introduced for relatively cheap products. In this case the product will be developed anew and also the SW will not be reused.

- In expensive products, like those for the defense industry, aviation and others, old and new components have to cooperate because of their high value. It might be worth developing an energy saving concept.

Measurement

- Different methods are recommended to measure the energy consumption of SW, such as run-time measurement or, depending on the task, it might be data intensive, computation intensive and communication intensive.
- The following metrics are proposed to measure the quality of green SW: Energy/Unit of work, CPU cycle count, memory consumption, peripheral usage time or idle time.
- SW ecodesign becomes a new subject of research. The different SW levels of a whole system are analyzed, such as telecom equipment and its network infrastructure. So SW design strategies cover the programming levels to optimize them. Also, the role of HW has to be investigated. Tools for SW ecodesign are available but require further optimization. In addition, **Green by IT** is understood as assistance for HW to reduce energy consumption.
- **Green by IT** should be discussed together with reuse of older HW. For example by integration of a control unit in a machine. Then energy consumption of a motor or pump can be severely reduced. SW, even a reused program, can be applied to minimize the energy consumption of reusable products.
- A repository of tested green SW might be a good idea for the reuse of green SW elements.

> Green and sustainable IT considerations might influence the whole product: Environmental impact, such as carbon footprint, infrastructure, usability, HW obsolescence, adaptability, functionality and many more.

Exercises

(1) Find out where your PC and your printer at home or in the office require too much time (and consume too much energy) or produce too much waste. Identify possibilities for improvement. What can you do? Which assistance may be offered by the manufacturer of a Green IT product? Which aspects may lead to a sustainable product?

(2) Why is the carbon footprint a quality criterion for SW?

(3)   How could a product working in a network, like a production plant, be made greener by IT? Give an example.

Recommended Further Reading

- Definitions, Measurements, and Quality Aspects: [16]
- Collection of links about green IT: www.greencomputingportal.de, www.eco logee.de/links.
- The article [71] deals with SW ecodesign and gives some recommendations for computing.
- Description of tools for identification of too high energy consumption caused by servers: http://www.spec.org/sert/.

References/Websites [12–16, 20, 27–31, 71]

Server efficiency rating tool: (SERT) by Standard Performance Evaluation Corporation, http://www.spec.org/sert/.

Chapter 8
HW/SW Interferences

8.1 Special Tests for HW and SW Containing HW Components

One problem of reuse of components will be, according to Smith [46], that not only the reused component must be tested in isolation, but the test has to cover its functionality in the whole product or system. With the QAGAN approach such a statement is granted after all tests for new are passed. But the component has also to be thoroughly tested in combination with the rest of the system, just as it does for a standard monolithic system. See Sects. 2.2.1 and 5.1.1, *integration testing* and *re-integration testing*.

It is of decisive importance that any change in HW, for example, word length of the processor or EMC features, is clearly documented, and appropriate conformity tests will be performed. These tests are to be performed in accordance with the tests of the producer IEC 62309 [6].

The same is true for SW changes: Any change in SW might have impacts on HW caused by side effects or inferences, and v.v.

The necessary tests in such situations are regression tests as discussed in Sect. 1.3.3.5.

Eventually, the whole system, that is, HW, SW, are to be integrated and tested. See Sects. 2.2.1 and 5.1.1, *integration testing* and *re-integration testing*.

In Appendix C checklists from IEC/PAS 62814:2012-12 [26] are available concerning the quality of a HW/SW component and the refurbishment of such components.

© The Author(s), under exclusive license to Springer Nature Switzerland AG 2021
F. Belli and F. Quella, *A Holistic View of Software and Hardware Reuse*,
Studies in Systems, Decision and Control 315,
https://doi.org/10.1007/978-3-030-72261-6_8

8.2   Recommendations for a Harmonic Interaction of HW and SW

As mentioned under Sect. 7.2, SW applications can directly be responsible for too high energy consumption, or by the application of SW, the energy consumption of a HW product can be influenced. Additionally, "Green by IT"—SW can help reducing energy consumption. Good examples are drives and pumps. The product without control works nearly continuously. consuming much energy. while the controlled motor only works when required: with up to 50% reduced energy consumption.

For reuse of components and products it is important from the view of SW that

(1)   SW which causes too much energy consumption is substituted,
(2)   SW is installed to control peripheral components to reduce energy consumption, for example, by managing stand-by of a printer in a peripheral application,
(3)   SW is applied that avoids a storage overflow, that is, the capacity of the existing product must be consistent with potential upgrades in the future or with the next generation (otherwise, the component or product will become obsolete),
(4)   As recommended by Hummel [52] a data repository should be built-up or an existing one used, from which reused and quality-tested SW components can be downloaded,
(5)   SW can avoid the consumption of consumables, like paper by a printer, printer ink or unnecessary maintenance,
(6)   For data safety it is very important that before reuse all data are removed from the product.
       From the view of HW it can be important that
(7)   Unreliable HW is controlled by a very reliable SW program Hoffmann [53] to overcome failures,
(8)   HW causing too much energy consumption should usually be addressed; by application of more efficient SW control, this problem can be solved.

From the view of HW and SW costs might be acceptable if

• The volume of components to be reused is high enough,
• The value of a project allows investment in a high quality reuse.

Some cases should be explained in more detail:
    In case (4) above, the authors in Hummel [52] write that a deposit for reliable SW reuse causes some extra work. But they explain also that already today much SW is already reused but not really by qualified persons. Usually the effort is low because it is not difficult to find reused or reusable SW components for a computer program.
    The subject of paper consumption in case (5) seems to have nearly no environmental impact as long as paperless work is propagated worldwide. But instead paper consumption has grown more and more! Every PC applicant knows how difficult it

is to get some print-outs stopped or get them in the required format. Also, remote control of E&E products can help in reducing maintenance or wear or waste.

Case (6) is developed from the astronautic area. As HW can be disturbed or partially destroyed by X-rays a very reliable SW program can detect potential HW failures and prevent the machine from malfunctioning. The researchers see applications for self-driving cars.

But for HW reuse in a much more dedicated way such a SW code can be a chance for extension of HW life, too. So, an idea could be in the case of QAGAN such SW could really prolong the product life. For forecasting dependability, a sufficient amount of data is often not available, and thus the component cannot be qualified as QAGAN. This could be overcome by such SW!

Further influences by an old but reusable HW are described by Smith [46]: Battery stability can influence the signal strength, sensor readings could fluctuate (if wear is measured), or speed signal might not be read correctly. Component-based systems will be easier to evolve and to be upgraded. A decoupling of the control logic from, for example, measured data, such as from sensors and command instruction. This decoupling will ensure that the control is explicitly kept separate and changes to the SW can be made independently. This might be applicable to all control components.

A cost problem could occur with SW reuse in that the amount of HW elements, in which the SW is planned to be reused, must be sufficiently large. Otherwise the work required for reuse planning and preparation could be too expensive. Also, a standardized SW architecture must be available, because if it is not, compatibility cannot technically be guaranteed.

In Appendix D an extended checklist from IEC/PAS 62814 [25] can assist an engineer to pose some questions concerning energy consumption and environmental impact belonging to SW.

Key Points, Exercises, Recommended Further Reading, References/Websites

1. Key Points

- Change in SW can have influence on HW and vice versa.
- The necessary tests in such situations are regression tests.
- Eventually the whole system should be tested.

2. Key Points

- Cooperation between HW and SW is often not planned in advance. One reason might be that they are developed in parallel but not together.

- SW can influence HW severely to create lower environmental impact and vice versa. Also, combinations for example of a few storage media or many storage media and their final measurable energy consumption are not really understood. Detailed investigations are not available or not evaluated for example by the European Commission, such as for lot 9 in *Ecodesign directive* Schischke et al. [73].
- Many rules are published for better HW/SW cooperation, such as a repository for reliable environmentally tested SW which could overcome also unreliable HW.
- HW often is not really embedded in the SW system.
- HW with too great an energy consumption should be eliminated. Final metrics to measure energy consumption of SW are not fully developed.
- HW could also cause incorrect signals for reading by SW, such as poor battery stability.

Exercises

(1) Find out by means of the examples of the Energy Star (http://http://www. spec.org/power/ and https://www.snia.org) where HW and SW need more cooperation.
(2) Which problems for the authors of the impact study do you see in this lot regarding HW and SW cooperation?

Recommended Further Reading

Checklists from ISO/TR 14062:2002–11 [25] in Appendix 4 [73].

Project website for project of European Commission for ENTR Lot 9 Server study in the frame of Ecodesign Directive, Implementing measures: www.ecodesign-ser vers.eu.

References/Websites

Smith [46], Hoffmann [53] and Schischke et al. [73].

Chapter 9
Environment and Dependability

9.1 Environmental Benefit of Reuse

It is important that the negative environmental impact of products with reused components is decreased or, at least, is not increased. Especially energy consumption contributes severely to the environmental impact of a product.

9.1.1 Environmental and Energy Aspects of HW and SW Reuse

An engineer who is going through all recommendations and legislation for the utilization of components will be able to define many criteria for the selection of suitable components or products for reuse. Selection is cooperation between the interests of recyclers, marketing department, customers and production engineers. On the way to circular economy new models for use and reuse were developed, for example for trains or computers, and also will have to be developed, especially for consumer products: Products with longer useful life, sharing models or upgrades. One example for a fair mobile phone is the fairphone, see https://www.fairphone.co [77].

Generally, the environmental impact of a product can be measured by Life cycle assessment (LCA) covering the whole life cycle. As already pointed out in Fig. 6.1, costs and cost savings are important for the sake of reuse. If a complete LCA will be too expensive, life cycle thinking will show many areas where the volume of refurbishment of a product or the quantity of reuse of components can be improved. As required by *Ecodesign Directive* or ISO 14001, during a new development a reduction of the environmental impact should be achieved. By HW reuse a reduction of the impact is already achieved, as many components or materials do not have to be manufactured again. If components or products lead to too high a consumption of energy, or if hazardous substances are included, these products or components should

© The Author(s), under exclusive license to Springer Nature Switzerland AG 2021   241
F. Belli and F. Quella, *A Holistic View of Software and Hardware Reuse*,
Studies in Systems, Decision and Control 315,
https://doi.org/10.1007/978-3-030-72261-6_9

be no longer considered. In those cases where a control unit of energy consumption by SW could be added, such as within production plants, the problem of too great a level of energy consumption could be overcome. The reduction is achieved by a stronger connection of the different production elements via internet (industry 4.0). But it might be too expensive to build in these possibilities at the end of life of old equipment. It would be better if it would be integrated at the beginning of the new development. A chance for a later integration could be seen if the power control is managed by a cheaper external network, such as power control by a cellular phone.

In Sect. 7.2 the impact caused by IT on the environment was described. All SW applied by HW should be checked for its energy consumption. The remanufacturer should know the amount of energy consumption caused by the SW and by which part of this SW. Similar with the monitoring of energy use is the situation with consumables, such as paper consumption. The amounts should be known. Improvements will usually not be achieved during reuse, but could become the initiative for the next development. Anyway, SW for measurement of energy consumption including metrics should be implemented into the product development scheme. The available tools will not all be sufficient today.

It will be a similar case with the application of IT to reduce energy consumption by control SW, such as via a control unit added to a machine. If it was not planned for from the beginning, it can become difficult to insert it afterwards. However, for example during the refurbishment of a drive, the controller could be updated, or new SW with a measurable lower energy consumption could be installed instead of the former one. These additional environmental effects should be checked during refurbishment.

A third chance to control energy consumption by SW is given by IT networks in a household or production plant. There will be an additional option for energy savings.

9.1.2   Environmental Benefit by Reuse of Products

Environmental benefit from the reuse of components or products is obvious. What does not need to be produced saves environmental resources. It can also be seen today where some of the products are collected at volumes of 30–50% of the amount put on the market per year. These volumes are not sufficient. Already after an average of five years of useful life, 50% of the materials would have disappeared, and the question is: where to? Also 90% collection including 90% recovery rate does not seem enough if one looks to billions of cellular phones sold per year (cf. Belli and Quella [3, pp. 48–52]). They all contain rare substances from which some can only be guaranteed to be mined for roughly a further 20 years [10]. So, reuse will be a chance to have all these resources available for many future generations.

Procedures for recovery of materials become easier if the product stays largely the same in its material structure. So, recyclers have a chance to recover enough material for efficient recycling. Mountains of waste should disappear and risks posed by hazardous substances in soil and drinking water can be avoided.

The environmental benefit of reuse might become much higher if following requirements are fulfilled.

- Spare parts from reuse, especially as-new ones, could be resold for many products if they were standardized; a dream, discussed already 30 years ago.
- Different application models, such as rental systems (for example household equipment for a certain time) or upgrade systems, could expand the product life.
- New marketing models must be developed ("sell new products with used parts").
- Design for reuse, remanufacturing, including modularity, easy (dis-)assembly, and planning for circular economies are available.
- The extension of the product's durability might be a new option, too.

There has been always a gap between the remanufacturing of capital goods and consumer products. The first are often in good shape after take back, the second often are not reusable because of damage before and after take back. In a study from the US market, trade-offs are discussed Bustani et al. [71]. One is the higher energy consumption of the older products. They could withstand a reuse. Nevertheless, in such a big market enough opportunities for remanufacturing are expected.

Reuse is in a simple way already achieved if a household product such as a refrigerator is resold privately. This market is interesting for reuse, but it is a market with an untested quality. The end of life of such a product was also not reached and could be longer. So, this second-hand-market should not really be added to the percentage required legally for potential remanufacturing, although it extends the useful life of a product. Energy consumption of these products is also rather high because they may be old and might have some additional problems.

For those products or components which contain SW, the additional potential for energy saving is given by the SW itself and the greening of the HW by green SW. Potentials of 20% of energy consumption of servers and other computer media show that there is a huge chance for improvement given by reuse and learning for the next product generation.

9.2 Quality Testing—Examples/Case Studies, Evaluation of Components

The first point to be checked with a component suitable for reuse will be; Is the component suitable for reuse? If yes, would this be in a new, similar product, or could this component be utilized for other purposes in other products. If no, the question has to be whether the material will still be recyclable or whether the component will be denied for a new application generally, that is, could it be used for another kind of application or disposal would be the last and only option. To answer those questions Sect. 9.2.1 will deal with these subjects:

- Identification of suitable components for reuse;
- Monitoring of stress put on components and products;

- Functionality check of the components before a new application; and
- A new dependability test as well as overload tests for artificial aging.
  (In combinations of old and new components a stress test might be necessary again!)

A case study explains the recommended procedure.

9.2.1   Some Fundamental Relationships

For a used component, the valid relationships are demonstrated in principle in Fig. 9.1 for the wear curve. Consequently, for the reuse of a component after termination of its useful life, it is important to know how long the remaining meantime between failures (MTBF) in the near future will be. If the MTBF will be essentially longer than the usual average lifetime of the product in the market then this component can be used again in a new product of the same kind.

In the following example, it is assumed that the *failure probability* of the component selected for reuse will be described by the distribution function $F(t)$. F(t) is also related to the time between failures (t) within which the component fails, and the corresponding probability. As an example, it is assumed that the Weibull function will be valid. The probability of surviving or *dependability R(t)* the failure behavior of the component observed can be described by:

$$F(t) = 1 - e^{-(\frac{t}{T})^{b}}$$

with

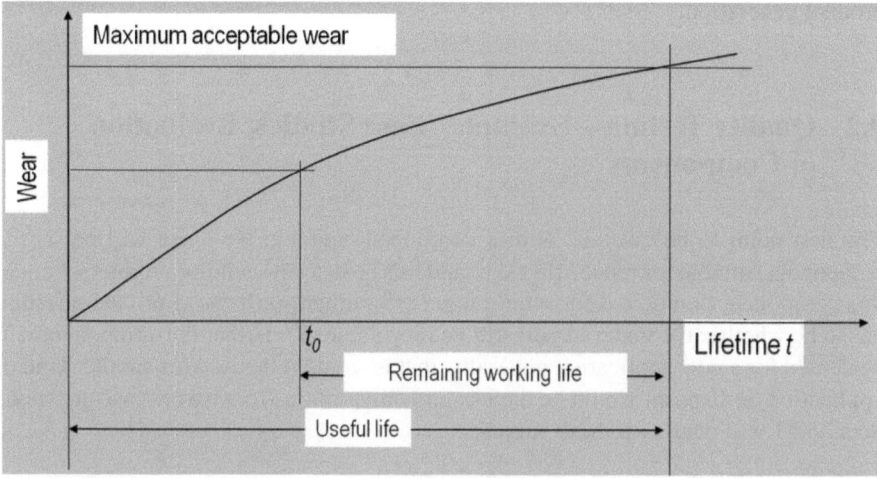

Fig. 9.1   Residual time to failure of a HW component at the end of the useful life of the product

t: time to failure, working life or period of application and
T: characteristic time to failure.

Duration could be mapped not only by time but also by driven km, application cycles etc.

Roughly in practice t = T and b = 1 will be inserted leading to the following special case:

$$F(t) = 1 - e^{-1} = 0.632 \text{ and } R(t) = 1 - F(t)$$

This corresponds to the time until which about 63% of all objects have failed or about 37% of the objects will still be functioning. b is the Weibull slope and usually amounts to between 0.25 and 0.5.

Further examples are contained in the informative part of *IEC 62309*.

9.2.2 Evaluation of the Current State

In general, the following preconditions might be considered for the selection of a HW product or component:

- *Technological state of a product or component*
  An embedded SW program might have no possibility for an upgrade.
- *Suitability for disassembly*
  Some mass products like printers cannot be disassembled.
- *Product structure*
  If standardized components were widely applied it will be easier to exchange them, including from other products.
- *Testability*
  Safety, also under new conditions or applications, has highest priority.
- *Ratio of sound technical components or boards*
  This is important for the value of the refurbished components or product.
- *Possibility of modernization or technical upgradeability*
  For as-new state old fashioned HW or SW is not enough. Maybe for repair of an older state it will be accepted.
- *Type of former applications*
  Humidity or some type of stress could have caused overseen damages.

In more detail, according to *IEC 62309,* procedures have to be applied by which the state of the product with QAGAN components could be evaluated. The evaluation can happen by means of the manufacturer's data sheets, or by lifetime tests of components or modules. Correct application of quality management systems could be helpful in the supervision of the manufacturing and evaluation process (see for example ISO 9001 [54], IEC 60300-1 [7]). For testing, the following procedures may be suitable:

- Visual testing, measurements and functional tests also carried out with products containing only new components.
- Interpretation of run-time counters or consumption counters, which were developed to assist decisions on whether components might potentially be reused. The interpretation is based on the information about the residual life (for example, by characteristic curves such as for wear, Weibull nets).
- Checking whether products with QAGAN housing/parts are visually perfect.
- Appropriate inspections and tests (for example X-ray or ultrasonic tests) have to be included if the mechanical structure will be in some way load bearing.
- Evaluation of the state of SW.

9.2.3 Evaluation of Dependability

By means of analysis and/or tests it has to be validated that the failure rate/failure intensity specified for the product containing QAGAN components is *not* higher than for a product with only new components. In addition, it must be assured that defined limits during the "as-new" NDL (ANDL) of the product (cf. Fig. 10.1) are not exceeded. Preparations for better dependability like burn-in usually do not have to be repeated again. Different is a situation where old and new components are combined. Analysis and tests should be developed by the manufacturer from which it could be ascertained that the product with reused components achieves a degree of performance and a dependability which will be consistent with the ANDL of the product (see, for example, *IEC 60300–1* [7]). These tests should usually already be available for new components. It might happen, however, that certain loads to which the component or product were subjected could not be exactly determined, if, for example, the cycle counter was not working. Then extreme values might be inserted in the calculations, or the component could be rejected.

Final test

Final tests including functional tests of the products with reused components must be the same as for products with only new components. In these cases, cyclic stress tests with the combined testing of electrical characteristics might be used.

If the SW of the component is identical to the new state everything will work without problems. If this is not the case the following tests have to be done:

- Test for failures of the component
- Potentially upgrade and test
- Final test after integration into the new product and its combined system.

9.2.4 Refurbishment of Components

Refurbishment of components can be accepted if all criteria for a successful refurbishment are fulfilled. This especially means: products containing reusable components must be disassembled. These components will be later restored to fulfill their former function and nearly to their former state (as-new) based on suitable instructions, guidelines and procedures. Included will be new complete tests as required.

Housings of products have to be visually intact. A second coating will be accepted.

Key Points, Exercises, Recommended Further Reading, References/Websites

1. Key Points

- HW will be collected for reuse as required by legislation. The collection rate has to be extended to avoid the loss of valuable materials.
- The energy saving achieved during production runs parallel to the amount of reused components.
- SW can be newly combined with some HW running today without SW to reduce energy consumption.
- SW itself often needs improvement to become energy saving. But metrics for measurement are not always sufficient. Another chance for lower consumption is available by linking the products via networks to achieve an efficiency control.

Exercises

(1) Search the net for models with the target of extending the product lifetime. Find out where and how SW is integrated in these models.
(2) Give some examples how older products or components could become energy saving after a SW program takes control about its use.

Recommended Further Reading

Belli and Quella [3] and Fair [77].

Situation worldwide about WEEE, ecodesign rules, industrial practice: WEEE handbook, Goodship, V., A. Stevels, WEEE Handbook, Woodhead Publishing, Philadephia 2012, ISBN 978-0-85709-089-8.

ERN European Remanufacturing Network: Remanufacturing Market Study https://www.remanufacturing.eu/.

Energy saving with label energy star: https://www.energystar.gov.

References/Websites

Belli and Quella [3, pp. 48–53], European Remanufacturing Network [74], WEEE Handbook Goodship and Stevels [75], Fair [77] New way of manufacturing and using environmentally sound phones: Fairphone: https://www.fairphone.co.

2. Key Points

- Environmental benefits of reuse are focused to HW where the collection rate of waste products has to be strongly increased. In addition durability and useful life should be extended. Further sales, use or marketing models after better eco-design, such as for modularity, will allow industry to move to a circular economy.
- Good take back systems are available for capital goods and in the beginning for consumer products. The latter may have the problem of too high an energy consumption for remanufacturing as-new.
- As SW will now often be reused in the same component or product, the role of SW for energy consumption has to be investigated and improved by some metrics or by the application of tested SW building blocks.

Exercises

(1)   Search some trade-offs withstanding a remanufacturing of household equipment. How could they be overcome?
(2)   Which benefit of reuse will be higher: Lifetime extension, for example like by upgradeability, multi-use, for example by a product rental system, or remanufacturing at the end of life? Make comparable assumptions for the products.

Recommended Further Reading

Belli and Quella [3].

An example for the ICT market is presented in: ICT Remanufacturing in the European B2B market—Questions and answers, www.bitcom.org.

References/Websites

Study for US market for household equipment https://www.web.unit.edu/ebm/www/Publications/Mitei-1-a-2010.pdf.

Study about rare materials: https://www.bgr.bund.de/EN/Themen/Energie/Produkte/energy_study_2015_summary_en.html.

3. Key Points

- Quality testing combined with information about the dependability of components provides information for customers about the state of the product. Usually all tests have to be passed, which an incoming inspection for new components has to fulfill to grant a QAGAN state.
- A company can also give a guarantee different to an as-new state. It can follow different, own standards and conform it by a certificate.

Exercises

(1) Compare the standards in Sect. 11.2. Where do you see problems with comparability, what should additionally be defined that you would make you trust the expertise and a certificate?
(2) Would it be enough to fulfill only ISO 9001 and ISO 14001 to create a reused/remanufactured product of high quality?

Recommended Further Reading

Belli and Quella [3] and IEC 62309. Especially in IEC 62309 several calculation examples are shown how to calculate dependability of components after a certain lifetime.

References/Websites

Belli and Quella [3] and IEC 62309. Especially in IEC 62309 several calculation examples are shown how to calculate dependability of components after a certain lifetime.

Chapter 10
Managerial Aspects

10.1 Reusability Organization

In Figures 1.6, 1.7 and Table 1.2 it is roughly described which tasks have to be organized to manage HW reuse or remanufacturing:

- The take back of a product from the customer, testing, refurbishment of components and products and the organization of the material chain until recovery are both necessary.
- In-detail work procedures from Fig. 1.7 (as an example) have to be integrated into the manufacturing processes.
- Volumes of taken back products and components have to be planned and organized from customers till recycling and sales happen again.
- In addition, processes also have to be optimized, say for an adopted supply chain or a forward/backward logistics process.

Often a department for refurbishment or an affiliation is organizing these processes including all quality testing. The recycling flow from collection up to the sales of refurbished and remanufactured components can be transferred to a specialized third-party company experienced in material recycling.

The reusability organization should also initiate an integration of specific requirements of reuse/remanufacturing into the product development scheme (see also Annex):

- Rules for better reuse like modularity, easy cleaning, recommended materials,
- Materials to be avoided, elimination of those difficult to recycle,
- Experience from quality checks as feedback for new development,
- Cooperation with suppliers and logistics partners to optimize processes,
- Documentation, information for customers about components and products.

The organization of the SW reuse process alone was already described in Sect. 3.5.

F. Belli and F. Quella, *A Holistic View of Software and Hardware Reuse*, Studies in Systems, Decision and Control 315, https://doi.org/10.1007/978-3-030-72261-6_10

During the refurbishment process many processes will deal with SW and SW reuse. It is difficult to organize such processes in general because it will depend on the kind of component or system and what, for example, has to be checked and in which stage of the refurbishment process. Such activities have to be added to the aforementioned steps. Different processes will occur if the models for reuse will be different: If the core of a PC stays the same and only consulting, repair or upgrades are supplied, the processes in a company will look different from those in an area of large industrial goods like plants. Many more different models could be applied. Some companies also sell used products with certificates according to special standards, or with a declaration of quality by themselves, and/or as-new products together with only new products, with or with or without QAGAN components.

A big difference must be seen between integration into the standard manufacturing process, for example with QAGAN components. Their state of technology, age or SW will be identical and all processes of manufacturing will be the same as the manufacturing of new products. It will be an increasingly complicated task if older HW, like military or aerospace products, have to be combined with new or old components, especially those with new SW. Here, individual test procedures have to be defined to guarantee a failure-free function. The testing technique has to be selected from the procedures in Part I.

10.2   Integration into Management Systems

Management systems are required to organize the modern production process. Many aspects have to be considered in order to increase reliability and reproducibility. Environmental, safety, quality, and legal aspects have to be cared for, and especially for reuse many new processes have to be installed.

10.2.1   Required Management Systems

Certified management systems according to ISO 9001 and ISO 14001 will guarantee that quality and environmental aspects are respected and integrated into the management systems. Ecodesign requirements could follow Appendix A of the *Ecodesign Directive* (see Annex 1 of this book). They are a minimum requirement for the content to be integrated into ISO 9001 or 14001. In detail it should become part of the product development scheme.

Important phases and required activities for eco-product design are summarized in Table 5.3 (in accordance with the ISO/TR 14 062) (Table 10.1).

Most of these steps and the tools belonging to them are not yet standardized. There are many solutions available; however, comparability is missing. Nevertheless, this list could be used as a guide for actions. This procedure became known internationally after the publishing of ISO/TR 14 062 [25]. Afterwards, a special solution for

Table 10.1 Phases and tasks of eco-product design

Phase	Task to do for HW	Additional for SW (**new, not in ISO/TR 14062**, suggestion by authors)
Planning	Environmental analysis of the product Benchmarking, determination of environmental aspects and requirements, definition of targets	Environmental and safety analysis of SW components, definition of selection criteria for SW units
Concept phase	Development of design concepts Analysis of alternative products	Nothing mentioned about this phase
Detailed design	Application of design strategies, tools	SW architecture design SW unit design and implementation
Test-/prototyping phase	Verification of conformity with specification	SW unit testing, SW integration and testing, verification of SW safety requirements
Production/putting on the market	Communication and information about materials used, information about environmentally compatible use, take back, disposal, environmental product declaration	Information about state and kind of SW available for customers, enable upgrades
Product review	Analysis of practical experiences (market success, environmental impact of the product)	*Nothing mentioned about this phase here*

ecodesign rules of E&E products was developed (see IEC 62 430 [11]). A SW part was not included; it was added by the authors of this book as a suggestion.

Important in ISO 9001 is the product development scheme, in which all the development rules have to be integrated into the right phase of this scheme. Take back, refurbishment and recycling require many new processes to be defined, organized and integrated. These processes have also to be integrated into the ISO 9001 management system and have to be reviewed. Many companies today have all processes from cradle to cradle described by a SW program often named "process house" as part of ISO 9001. In these processes also marketing, production, and sales processes should be defined. For quality checks reviews have to be planned and metrics for their fulfillment are required. According to ISO 14001 the environmental impacts of products and processes must be determined. These two standards are not in competition, and they could be seen as complementary.

The manufacturer today gets responsibility for the sales of products internationally: responsibility for working conditions, ethical guidelines, safety aspects, technical properties and much more. Rules for Corporate Social Responsibility (CSR) according to ISO 26000 are a new requirement with much influence on products and product development. Therefore, at the moment, the integration of different current management systems such as Quality, Environment, Safety or CSR into a single system is discussed; some companies already applied integrated management systems.

10.2.2 SW Integration/SW Reuse

As shown in Sect. 7.2, integration of reused SW has to be planned in advance. A repository of quality-checked SW modules could be a good basis for that reuse. In the chapter about Green IT (Sect. 7.2) several procedures were suggested to select only such SW which does not contribute to increasing energy consumption, or cause too much waste. In the project SAFE [44] a practical test example for combined HW/SW reuse for the car industry is managed.

One philosophy is that components already in operation without an incident can be reused, if the candidate and the field data fulfill the required criteria. This can be valid for HW and SW (but may not be sufficient for QAGAN). The field data are decisive in determining the dependability of a component. This is normal in quality science.

Prerequisite is the management of functional safety consisting of an overall safety management: Safety management during concept phase and after the components release. In concept phase the safety lifecycle is initiated followed by analysis of hazards, risk assessment and a functional safety concept. HW and SW are developed in parallel. On the SW level the development contains:

- SW architecture design,
- SW unit design and implementation,
- SW unit testing,
- SW integration and testing,
- Verification of SW requirements, and
- The qualification of the SW components as a supporting process.

Accompanying is

- The analysis of dependent failures, and
- Safety analysis.

In documentation of the SW component the following information shall be included:

- Requirements of SW component (unique identification),
- Description of configuration (unique configuration),

- Description of the environment (interfaces),
- Application manual and responsible person,
- Description of integration,
- Definition of reactions under anomalous operating conditions,
- Dependencies with other SW components,
- Description of known anomalies and work-around measures,
- Results of verification measures,
- Maximum target level (SIL[1] level).

It is very important to describe the interfaces and the inter-dependencies in the concept phase. The interfaces of HW/SW are described in the HW/SW specification and in the SW architectural design specification. In a SW component, model integration measures are often included. The known anomalies could be detailed by an error model.

During verification both normal conditions (requirement specification) and behavior in case of failure shall be covered, also metrics at SW unit level for structural coverage. Different types of behavior can be modeled using "Behavior and Fault Behavior" classes. Failure rate reduction targets are classified according to SIL[2] (or ASIL) classes corresponding to failure risks.

The SW component can be considered as qualified if the following criteria are fulfilled:

- Specification of the SW component is available,
- SW component complies with its requirements (evidence available),
- Suitability of SW component for intended use (evidence available),
- SW development process is based on appropriate national or international standard (evidence available).

HW components are checked similarly according to already described criteria (cf. [3], Sects. 1.1.3 and 1.3) or failure curves as required by Fig. 1.2.

As often COTS components from external sources might be required the following should be available:

- Requirement specification of re-used SW component,
- Design specification of re-used SW component,
- Results of previous verification measures,
- Application manual of re-used SW component,
- HW component specification,
- Field data from previous service period.

Supplier declarations should follow ISO 14021 (self-declaration, Type II) or ISO 14025 (certified, including life cycle inventory). In a declaration consistent with ISO/IEC 24700 conformity with all necessary documents has to be confirmed. Also,

---

[1] SIL (= Safety integrity level) acc. to IEC 61500, ASIL (Automotive safety integrity level). Widely used are COTS products (COTS = Commercial off-the-shelf).

[2] See Footnote 1.

conformity with the corresponding European laws is required (cf. Sect. 10.2), if the product is sold in Europe.

10.2.3   Documentation/Information: External/Internal

Documentation is required to control and improve processes. Also, customers like to have a look into the processes to gain confidence in the dependability of the refurbished components and products.

10.2.3.1   External

The external information for customers and the public has to cover the kind of product or component and its liability. The state of the component "as-new" or only "refurbished" is required in order to to explain the state of the complete product. Contracts, service manuals and whatever else is needed should be provided to inform the customer. It should also be explained that several parties gain an advantage: The manufacturer, the customer, the public and very importantly, the environment.

As a customer should not be supplied with defective products, but instead will be provided with new highly valuable ones containing QAGAN components, there is no reason to hide anything. Therefore, unmistakable information about the potential availability of reused components and their state will be necessary. Offers, data sheets, product descriptions, sales literature and contracts are suitable for supplying this information.

It is important to direct the customer to the information about the performance of all important and suitable tests and about product safety, including CE Mark, because the product has been newly put on the market.

> A reference inside the customer documentation to *IEC 62309* makes sense. For example it could read "This new product contains used but" qualified-as-good-as-new "QAGAN components according to *IEC 62309*".

Liability period and liability conditions have to be the same as for products containing only new components. At the end the message will read: *The customer will experience no extended risk but additionally will usually get an economic benefit.*

> **Principle**: Detailed and clear information for the customer about the availability of QAGAN components will be absolutely necessary!

All relevant deviations, properties, characteristic values, liabilities and other important information have to be written into the contracts. Also, an alignment with the contracts of the new products is required.

Environmental aspects are also important: an environmental product declaration according to ISO 14021 or 14025 [47, 48] will inform customers about the potential environmental impact and can also include information about 'Green IT' and 'Green by IT'. Especially for components reused as QAGAN the information about the materials record has to be included. For information about clean production technologies and processes an environmental management system according to ISO 14001 [54] has to be proven. A certified Quality management system according to ISO 9001 [56] is another indispensable requirement.

By a supplier declaration (example see in ISO/IEC 24700 [55]) all information can be summarized and ensured.

10.2.3.2 Internal

If, for example, a customer is interested in seeing the corresponding quality documentation, all processes have to be documented exactly and in a presentable way. The documentation should contain:

- Documentation of processes for verification of lifetime and of dependability. Also processes having to do with the selection of components and products, disassembly and costs if required must be included.
- Records about QAGAN components, their age and state.
- Storage until the end of NDL.
- The documentation will be part of the technical records and of quality documentation according to ISO 9001 (there, Sects. 1.3.2 and 1.5.3).
- Compliance with product liability and product safety legislation and other regulations dealing with the product being put on the market by the responsible persons.
- Safety checks should be newly repeated in case of doubt. It must be tested that all data are removed from a component before reuse of this component or product (data safety).
- If CE-mark is required the documentation has to show that all needed standards are fulfilled. Sometimes a certificate might be required (medical products) otherwise self-declaration is sufficient.

Also, for reasons of product liability it will be required to document the selection process, the FMEA, the disassembly instructions, the test program and the capabilities, if possible.

For products not to be sold as-new after refurbishment or remanufacturing some documentation need not to be done again because they might not be put newly on the market. But anyway it has to be checked whether compatibility with the latest EU directives might be required. If deviations of product properties from requirements of EU directives are found they should be explained, although the products can still be resold because they were not newly put on the market. Also, the SW states have to be documented and potential incompatibilities with the SW of some components. It seems sometimes that too much work has to be done before a reuse is really possible: Indeed, if the documentation of the running products is complete, and if components selected for reuse are not too complicated in structure, the reuse will be rather simple!

> **Principle**: Do not accept potential risks in products to be resold. All procedures and quality characteristics should be recorded exactly and be documented.

Key Points, Exercises, Recommended Further Reading, References/Websites

1. Key Points

- The organization of reuse is a process covering take back, development, remanufacturing and recycling of the residues.
- A special department could organize these subjects, including a recycling specialist. Alternatively, it could be organized by the production site if the components are required there. If a product is only upgraded and will stay in principle the same, then the state of the product could be guaranteed by the sales and marketing department, which could also organize special consulting and potential repair.

Exercises

(1) Draw a hierarchical connection through the tasks of the re-usability organization (see Figs. 1.6 and 1.7). Which department might be responsible for these tasks?

Recommended Further Reading

Criteria for a qualified refurbishment compared with others: https://static.health care.siemens.com/siemens_hwem-hwem_ssxa_websites-context-root/wcm/idc/gro ups/public/@global/@refurb/@imaging/documents/download/mda2/mtcw/~edisp/ rs_brochure_advantages_qualityprocess-03122263.pdf

References/Websites

Belli and Quella [3], IEC 62309

2. Key Points

- Management systems according to ISO 9001 and ISO 14001 are mandatory for every company today. Aspects of sustainability are today also required (ISO 25000).
- Integration of Ecodesign rules into the development scheme is legally required, corresponding to the aspects of the *Ecodesign Directive* (Annex 1).
- A description of all relevant processes is necessary where all processes like take back, recycling, remanufacturing are described. Otherwise these processes cannot be managed.

3. SW integration and SW reuse have to be managed in the same way as part of the development process and part of refurbishment, because these elements are part of the components to be refurbished as-new. For qualification of SW requirements are:

- Specification of the SW component is available,
- SW component complies with its requirements (evidence available),
- Suitability of SW component for intended use (evidence available),
- SW development process is based on appropriate national or international standards (evidence available).

4. Documentation is essential for the sales of remanufactured products.

- It is information for customers, becomes part of the contract and advertisement/marketing.
- Documentation internally requires the technical details of the qualification results (technical file from ISO 9001). Data safety is also needed and all old data have to be removed completely.
- External documentation covers information about liability, availability of QAGAN components in the new product and also information about take back or recycling. Information about the environmental properties should also be given (including *Ecolabel* if possible).
- Internal documentation contains all relevant data necessary to guarantee potential environmental and dependability and which has to be delivered to all potentially interested parties, such as government.

Exercises

(1) Into which management system you would integrate SW reuse and integration. Describe which processes are required over the lifecycle to get a SW program which can later be retro-fitted to a QAGAN product.
(2) Which information about environmental properties should be included in the documentation for customers? Give some examples.

Recommended Further Reading [3]

Qualification of HW and SW together: For example https://www.ivtnetwork.com/art icle/infrastructure-qualification-proposed-standard

Ecolabel schemes for documentation (see https://ec.europa.eu/environment/eco label/).

References/Websites

[EcoD09] EU homepage: https://www.eceee.org/ecodesign/process/

Chapter 11
Legal Aspects, Standardization and Liability

11.1 Laws Concerning Reuse—An Overview

The legal framework concerning reuse of components and systems is sometimes difficult to understand. Fundamentally, every legal standard of a country is binding in that country and also in others if a manufacturer is designing products that will be sold internationally. The following regulations are European-wide and often are very similar in other developed countrsies world-wide. In Europe, such regulations specifically directed to reuse or new entrance into the marketplace have to be considered. These are:

- Waste framework directive [32];
  The Waste Framework Directive sets out legislation addressing the impact of inappropriate Waste, management on greenhouse gas emissions, air pollution and littering.
- End-of-life vehicle directive [33];
  The scope of the directive is limited to passenger cars and light commercial vehicles. The directive covers aspects along the life cycle of a vehicle as well as aspects related to treatment operations. As such it aims at:

  – Preventing the use of certain heavy metals, such as cadmium, lead, mercury and hexavalent chromium,
  – Collection of vehicles at suitable treatment facilities,
  – De-pollution of fluids and specific components,
  – Coding and/or information on parts and components,
  – Ensuring information for consumers and treatment organizations, and
  – Achieving reuse, recycling and recovery performance targets.

- *Ecodesign (ErP) Directive* [1];
  Law about ecodesign of E&E products and related products such as water consuming ones.

F. Belli and F. Quella, *A Holistic View of Software and Hardware Reuse*,
Studies in Systems, Decision and Control 315,
https://doi.org/10.1007/978-3-030-72261-6_11

- RoHS, RoHS2 and RoHS3 [34] and WEEE [4] directives;
- RoHS (Restrictions of Hazardous Substances) in E&E products determines which substances are prohibited in E&E products. Latest versions are counted by numbers 1, 2, 3 …
- WEEE (Waste of Electro and Electronic Products) regulates take back, reuse and recycling of E&E products.
- Dodd-Frank Act about Conflict Minerals
  US law about restrictions for the use of minerals, such as Coltan, sourcing from conflict countries.
- Packaging and packaging waste directive [35];
  The directive deals with the problems of packing waste and the currently permitted heavy
  Metal content in packaging. It

  – sets targets for recovery and recycling,
  – requires the encouragement of the use of recycled packaging materials in the manufacturing of packaging and other products,
  – requires packaging to comply with 'essential requirements' which include the minimisation of packaging volume and weight, and the design of packaging to permit its reuse or recovery,
  – requires the implementation of measures to prevent packaging waste in addition to preventative measures under the 'essential requirements', which may include measures to encourage the reuse of packaging.

- General product safety directive (GPSD) [36];
  Requirements about the safety of products to be put on the market.
- EMC directive (EMC) [37]
  Directive relating to electromagnetic compatibility.
- Battery directive.
  Regulaltion about take back of batteries and their recycling.

– Preventing the use of certain heavy metals, such as cadmium, lead, mercury and hexavalent chromium,
– Collection of vehicles at suitable treatment facilities,
– De-pollution of fluids and specific components,
– Coding and/or information on parts and components,
– Ensuring information for consumers and treatment organizations, and
– Achieving reuse, recycling and recovery performance targets.

Apart from this legislation, general laws also have to be taken into account that may also affect how products containing reused components are put on the market. Liability legislation is important and must be mentioned. Exposure can arise through the criminal code, or in commercial law. It may even be classed as an administrative offence. Civil liability has to be considered, especially related to product liability legislation and the law of obligations.

In this section we consider two European laws, because they strongly influence reuse. In many countries worldwide these laws are valid in a similar manner. Further legislative advice, consequences and background differ, depending on the legislation of the corresponding countries.

11.1.1 RoHS (Restriction of Hazardous Substances, RoHS3) and WEEE (Waste of Electro and Electronic Equipment)

Affected products are at the moment divided into 10 categories (according to Annex 1 of WEEE, and one more (11) for (all) others cf. RoHS3)[1]:

1. Large household appliances
2. Small household appliances
3. IT- and telecommunications equipment
4. Consumer equipment
5. Lighting equipment, discharge lamps
6. Electrical and electronic tools
7. Toys, leisure and sports equipment
8. Medical devices
9. Monitoring and control instruments
10. Automatic dispensers
11. Other EEE not covered by any of the categories above (also cables from 19 July 2019).[2]

The first 10 product groups are allocated to different collection groups for which the same recovery quota are valid (cf. Table 11.4):

- Large household appliances, automatic dispensers,
- IT and telecommunication equipment, consumer equipment,
- Small household appliances, lighting equipment, electrical and electronic tools, toys, leisure and sports equipment, monitoring and control instruments,
- Discharge lamps.

In some countries, such as in Germany, further collection groups were formed because of the special treatment requirements of the E&E waste there.

In Table 11.4 it is distinguished between the total recovery rate including its energy contribution, and a second rate in brackets expressing the sum from materials recovery and the volume of reused components [4]. The split in shares for materials

---

[1] **Category 11** products include all other electronic and electrical equipment not covered under the other categories. Included are 2-wheeled vehicles, electronic nicotine delivery systems (ENDS) such as e-cigarettes, cannabis vaporizers and vape pens. Also included are electrical cables that are less than 250 V working voltage.

[2] See Footnote 1.

to be recovered has been increased in the new edition of WEEE from 2012. In this section the components to be treated selectively are also described.

A new requirement of WEEE in its latest version in §11 urges member states:

"Member states shall take measures, as appropriate, to promote the **reuse of products** and **preparing for reuse activities**, notably ...

– Encouraging the establishment and support of reuse and repair networks,
– The use of economic instruments,
– Procurement criteria,
– Quantitative objectives or other measures."

The promotion of **reuse** is part of WEEE §8 too:

"In order to strengthen the reuse and the prevention, recycling and other recovery of waste, Member states may take legislative or non-legislative measures. Such measures may include...

– Acceptance of returned products,
– Subsequent management of the waste,
– Financial responsibility for such activities,
– Information about re-usability or recyclability, and
– Design of products to reduce their environmental impacts.

And they shall take into account:

"Technical feasibility and economic viability and the overall environmental, human health and social impacts, respecting the need to ensure the proper functioning of the internal market."

Where the recycling is not done by an OEM who would segregate machines and sub-assemblies from within a machine, then the materials collected in recycling containers will be of a vast spread of technologies and ages. Thus, there could be an old wireless radio, built with tubes on an aluminum chassis mixed in with a modern IC and transistor constructed radio.

This mix of technologies precludes any reliable estimate of recyclability or recovery rates of modern products. Therefore, this method of collection does not give either a sustainable source of materials or any confidence that these components have a qualified history. In conclusion, there must be two major changes:

1. All E&E products should be fed back through the OEM as the approved recycling process.
2. Products should be designed with recycling in mind as a key part of the design.
3. Manufacturers should ensure they have adequate records to show the manufacture date and history of the material sources that went into manufacturing of the product.

From the year 2018 the collection groups will change according to the new WEEE [4]. Corrected recovery quota is not published till that date.

RoHS3 directive ISO/IEC 25000 [19] restricts the following hazardous substances: lead (0.1%), cadmium (0.01%), hexavalent chromium (0.1%), mercury (0.1%), brominated biphenyls and diphenyl ethers (0.1%), and there are exemptions. The current details of these exemptions may be downloaded from the RoHS3 guide of the European Commission (https://www.rohsguide.com/rohs3.htm). It will also be available from the corresponding national industry associations. As the exemptions were time sensitive most of them have expired or will expire in the near future. An extension was done in 2015 for 4 Phthalates: Bis(2-Ethylhexyl) phthalate (DEHP): 0.1%; Benzyl butyl phthalate (BBP): 0.1%, Dibutyl phthalate (DBP): 0.1%; Di-isobutyl phthalate (DIBP): 0.1%. DEHP, BBP, DBP and DIBP are primarily used as plasticizers to soften plastics. They will be restricted from **22 July 2019** for all electrical and electronic equipment apart from Category 8 (medical devices) and Category 9 (monitoring and control equipment) that will have an additional two years to comply by **22 July 2021**. In contrast to WEEE (where 10 products groups were affected) **all E&E products are now in the range of RoHS3.**[3]

Be aware: In RoHS3 all E&E products are affected since 2013!

Categories of WEEE (10) and RoHS3 (11) are now independent from each other! Of note is that medical devices have a two-year extension to meet RoHS3 compliance:

The restriction of DEHP, BBP, DBP and DIBP shall apply to medical devices, including in vitro*medical devices, and monitoring and control instruments, including industrial monitoring and control instruments, from* **22 July 2021***.*

With the Dodd-Frank Act [62] from the US new restrictions are coming for so-called conflict minerals Au, Sn, Ta, and W, which require the avoidance of these minerals if they are coming from the Congo and neighboring states. The European Commission is planning a corresponding directive in the next few years. As recycling material is not affected, this legislation could promote recycling material from E&E products.

Besides these restrictions there are many more laws limiting the putting-on-the-market of hazardous chemicals, and the extension of RoHS is planned in the future in REACH legislation Reach [41].

The Restriction of Hazardous Substances RoHS legislation states that from 1 July 2006 no E&E products can be sold with concentrations of hazardous substances above the defined levels of RoHS materials, and this applies to reuse as well. Medical devices and monitoring and control instruments not yet covered by the RoHS were

---

[3]**Category 11** products include all other electronic and electrical equipment not covered under the other categories. Included are 2-wheeled vehicles, electronic nicotine delivery systems (ENDS) such as e-cigarettes, cannabis vaporizers and vape pens. Also included are electrical cables that are less than 250 V working voltage.

integrated into the law in the meantime. The law was put into force step by step until 2017. Additionally, a new group of other products was summarized in a new category 11: This means in the end that all E&E products will be affected by the new RoHS2 Directive [34] but will keep the known exemptions. Categories of RoHS3 and new WEEE will no longer be identical. RoHS conformity has to be confirmed now by the CE mark. New RoHS versions are counted by numbers RoHS, RoHS2 and RoHS3.

> The following rules are valid: Spare parts with substances limited by RoHS can be put on the market for the purpose of repair of equipment that was put on the market before 1 July 2006. For products put on the market afterwards, only such spare parts can be used in which the concentration of the 6 substances affected by RoHS will be below or equal to the corresponding legally defined low limit. In equipment put on the market before 1 July 2016 these parts are allowed for reuse in a closed verifiable system within a company. Consumers must be informed of this fact.

11.1.2 Ecodesign Directive [1]

This framework directive will, in the near future, affect all essential energy-using products produced in a volume of more than 200,000 pieces that are sold in Europe. (This is the sum for all companies put together!) In the first version, only energy-using products were affected. In the recast version, energy-related products are also included. Measures like the setting of energy consumption limits are determined for individual product groups by case studies and set into force one by one. The *Ecodesign Directive* offers only the framework conditions (see https://ec.europa.eu/enterprise/policies/sustainable-business/eco-design/legislation/index_en.htm) for the legislation. Details are or will be defined in product specific measures, so-called implementing measures, after investigation of the state of the art. For the state of the individual studies' planned and finally decided measures see [42]).

In 2019 the EU parliament agreed on extended Implementing Measures (in force from 2021) for several products of the ErP Directive (Ecodesign Directive). Besides energy reduction, the focus is now on repair, spare parts and resources. Manufacturers shall ensure that these spare parts can be replaced with the use of commonly available tools and without permanent damage to the appliance. Information about these parts must be available till the end of required period. For more information see the corresponding Implementing Measure. In Table B.1 for the affected products those components are mentioned which have to be supplied after the end of production (Table 11.1).

Table 11.1 Availability of spare parts after placing the last unit of the model on the market

Product	Availability of spare parts	Components to be made available for repair
Refrigerators [82]	7 years	Thermostats, temperature sensors, printed circuit boards and light sources
	10 years	Door handles, door hinges, trays and baskets for a minimum period of seven years, and door gaskets
Televisions [83]	7 years	Internal power supply, connectors to connect external equipment (cable, antenna, USB, DVD and Blue-Ray), capacitors, batteries and accumulators, DVD/Blue-Ray module if applicable and HD/SSD module
	7 years	External power supply and remote control
	8 years	The latest available security update to the firmware after the placing on the market of the last product of a certain product model, free of charge
Household dishwashers [88]	7 years	Motor; circulation and drain pump; heaters and heating elements, including heat pumps (separately or bundled); piping and related equipment including all hoses, valves, filters and aqua stops; structural and interior parts related to door assemblies (separately or bundled); printed circuit boards; electronic displays; pressure switches; thermostats and sensors
	7 years	SW and firmware including reset SW
	10 years	Door hinge and seals, other seals, spray arms, drain filters, interior racks and plastic peripherals such as baskets and lids

(continued)

Table 11.1 (continued)

Product	Availability of spare parts	Components to be made available for repair
Household washing machine and washer dryers [87]	10 years	Motor and motor brushes; transmission between motor and drum; pumps; shock absorbers and springs; washing drum, drum spider and related ball bearings (separately or bundled); heaters and heating elements, including heat pumps (separately or bundled); piping and related equipment including all hoses, valves, filters and aqua stops (separately or bundled); printed circuit boards; electronic displays; pressure switches; thermostats and sensors
	10 years	SW and firmware including reset SW
Refrigerating appliances with a direct sales function [84]	8 years	Thermostats; starting relays; no-frost heating resistors; temperature sensors; SW and firmware including reset SW; printed circuit boards; and light sources
	8 years	Door handles and door hinges; knobs, dials and buttons; door gaskets; and peripheral trays, baskets and racks for storage
Welding equipment [85]	10 years	Control panel; power source(s); equipment housing; battery(ies); welding torch; gas supply hose(s); gas supply regulator(s); welding wire or filler material drive; fan(s); electricity supply cable
	10 years	SW and firmware including reset SW
Servers [86]	2 years	SW and firmware including reset SW
Electronic displays [90]	7 years	Internal power supply, connectors to connect external equipment (cable, antenna, USB, DVD and Blue-Ray), capacitors, batteries and accumulators, DVD/Blue-Ray module if applicable and HD/SSD module

(continued)

Table 11.1 (continued)

Product	Availability of spare parts	Components to be made available for repair
	7 years	External power supply and remote control

This selection only covers direct requirements for repair. Much more requirements arrive from energy consumption or hazardous substances. They should be part of the repair process (see implementing measures). In parallel ecolabels became law in manufacturer's own implementing measures. The energy requirements might be of interest for products but are not in the focus of repair.

Spare parts must not be new, so this field will be also a subject of as-new components or other suitable ones. Interesting for refurbishment is the requirement for the spare part supplier to supply the latest SW to update the equipment. Such regulations should become standard for other equipment.

In addition to the already known technical measures, it is required that a manufacturer installs an ecodesign concept and a management system including ecodesign (see Appendix A of this book). Conformity of the product and management system with the corresponding standards is confirmed by the CE mark. This also includes conformity with the low voltage directive Low voltage Directive [39] or machinery directive Machinery Directive [40] and with the Ecodesign directive. Technical and environmental conformity is stated by this mark. A self-declaration is sufficient for technical conformity for most of the products affected. This statement is valid for all affected products in regard to environmental conformity.

If reuse of components in products newly put on the market is planned, these regulations have to be considered and the CE mark has to be placed on the product as far as it is legally required.

> **Conclusion**: The CE mark covers not only product safety but also conformity with environmental legislation.

11.2 Industrial Standards—How to Use Them

> A short introduction to this chapter:
> This chapter is not a description of all standards mentioned in this book. These are introduced or explained directly where they are needed! In Annex F a list of all standards is given additional to the literature list.
> Standards can show the character of a law if they become part of a law. Standards can be strong if they are published as a standard. But they can be weak in the role of a technical report (TR) or publicly available specification (PAS). These publications are developed for test applications and can later be converted to a real standard or withdrawn.
> Associations or companies can promote standards if one of their products gains an advantage because some of its properties were standardized. Consequentially, in so far as products of other companies are not covered, they

might become less attractive. So, if many applicants in the market practice or buy standardized goods, competitors might lose business.

In Table 11.3 a comparison is made between different standards for refurbishment and remanufacturing. It shows, although they seem similar, different shades of the same subject. Indeed the quality of the product is different! They all own a certain state of testing. Some companies issue certificates about their quality. The readers of this book should decide which standard of the three they would trust more.

Therefore this chapter is about the understanding of different strategies behind the standards. Readers not so well-acquainted with standardization might get a hint about that. Otherwise questions will arise about the purpose of some standards. The title was therefore selected: How to use them!

Internationally valid standards are developed by international organizations such as ISO or IEC, national organizations or industrial associations. Such standards are developed if, for example, environmental legislation has started in some countries and in others not. Then it is desirable to have the same regulations worldwide and a standard is created. These standards might become part of the new progressing legislation worldwide. Examples are standards for the testing of hazardous substances as it is done by IEC TC 111 (a standardization group initiated by EU Commission) or to give additional assistance to Ecodesign methods with more detailed rules than published in the official *Ecodesign Directive*. By IEC 62430 independent rules are published only with reference to law. As these standards were developed with a mandate from the European Commission, they have a more similar character than legislation produces. They are part of the self-declaration of the CE-mark.

Standards can also be certified, which means that a company asks a certification organization, such as TÜV (Technischer Überwachungsverein: a German technical organization for certification), to check whether all the rules of the standard are sufficiently applied by a company, and grants a certificate if this is true. Many standards are not developed for certification in their industrial areas, but for example, to harmonize definitions, create equal understanding of methods, or for tentative use before the final version is developed.

Other standards such as ISO/TR 14062 recommend general ecodesign rules. But, as the term 'TR' (Technical Report) indicates, it is recommended for application but not for certification, and the authors believe that these will not be the final version of the rules. Other standards are planned for certification such as ISO 9001 and ISO 14001, and every company should possess such a certificate. As environmental and quality aspects might not be correctly realized, customers and the public might not trust that the company is really able to fulfill all of the requirements. In this case, a 3rd party certificate will be presented to provide evidence for an independent control. In other cases, also with laws, self-declaration by industry is allowed for some products, as is the case for most products in the *Ecodesign Directive*.

A standardization process can begin on the highest level, as it did for IEC 62309, and for all E&E products as a general standard, while other electro-technical institutions can develop from that their own standards based on such a standard. From IEC a harmonization with the ISO level can be achieved via a harmonization body. In the case of IEC 62309, a harmonization with ISO was agreed via the JTC 1 (joint implementation committee). So it is valid in both organizations.

All companies can agree and apply the same industrial standard. So together with an order of a product, a certificate for the company to successfully use ISO 9001 and/or ISO 14001 is often required as a precondition. No trade contract between countries is needed, the industrial partners simply apply the same rules. This creates trust by customers and the public. As the requirements are demanded via the whole supply chain, after some years the quality level and the environmental level should be improved for nearly all companies.

For reuse as-new, such a standard was developed for E&E HW with IEC 62309 because the undefined state of as-new was often misused to get too much money for no longer valuable products. Now this state can be treated equally; in addition, a state of the art in the case a legal problem occurs.

With IEC PAS 62814 a similar standard was developed for the reuse of SW together with the reuse of HW. This standard did not get the full approval of a standard but as a publicly available specification (PAS) to enable companies to test the procedures which did not seem finally completed. Later-on, a standard can be created from that state. Both standards, IEC 62309 and IEC PAS 62814 (now withdrawn, but still applicable), were not written for certification because the situation differs in different branches of industry, and products or manufacturing processes can be too different. Nevertheless, a self-declaration is possible: "This product is a new product in accordance with IEC 62309".

Both standards could not be developed together to one single standard because two different subjects, such as HW and SW, were treated, and thus, a unification did not comply with the rules of IEC.

A standard will also be created from associations, such as the Industry standard for medical equipment about "Good refurbishment practice" from COCIR [8], to present and harmonize the policy of their members.

In the following overview in Table 11.2 several standards are compared with the same background. Here one can see that different purposes may require different standards.

National specifications like British standard BS 8887-220 specify requirements for the process of remanufacture. It lists the steps required to change a used product into an as-new product, with at least the equivalent performance and warranty of a comparable new replacement product. This remanufacturing process can include parts or components to be used in subsequent assembly. The validity is limited because it is only a national standard. But it seems it should be transferred to an international level one. Nevertheless, some companies have got a certification, like that of Ricoh confirming to apply the rules of this standard.

In more detail in Table 11.3 a comparison is made in the following overview. The scope is different. Especially some subjects like general quality standards or

Table 11.2 Different standards about remanufacturing

IEC 62307	
IEC 24700	Processes to be defined by company, overall product with used parts can be as-new (avoids exact definition), CE-mark?
ANSI (RIC 2015) 001.12016	Remanufacturing oriented, processes similar to IEC 62309, not so strict dependability requirements, new definitions for the same subject for example reuse comparable with remanufacturing etc., many requirements to be defined by the remanufacturing company, CE-mark?
Industry standard from COCIR "Good refurbishment practice"	Character of recommendation
BS 8887-220	National standard, limited in application

environmental standards like ISO 9001 or 14001 are put in the foreground, although this is today state of the art. More detailed requirements are often not defined in these standards. In IEC 24700 and ANSI RIC 2015 it is pointed out that many definitions should be made by the manufacturer itself. So, it will not be clear whether they are sufficient. This allows the remanufacturer much freedom. On the other hand, strict definitions are often avoided, such as the definition of a new product containing reused parts, or the meaning of an as-new part, because this might be a new legal domain where some lawyers might see a problem with new juridical areas. Nevertheless, the targets of the standards are similar; to sell products with tested reused parts similar to that of the as-new state. The company offering the higher standard might get more trust. With a rather uncertain definition the company might benefit from more freedom of application and might not receive legal complaints. If a manufacturer only gives a guarantee for the as-new product like that for a new product, then dependability could become a problem. On the other hand, a remanufacturer who not only remanufactures its own components or products, but also those of other companies, might have problems to give dependability information; often it can be impossible.

Nevertheless, some companies get certificates to improve the trust of customers, others apply self-declarations. In the following overview the reader can see that differences are located more in the details, but often definitions are not precise or are open for further definitions such as processes in a company. It is recommended to look at the documentation of these companies.

To sum up, IEC 62309 requires the most detailed information about the reused components and the product. Customers should examine the information they get if the remanufacturer sells as-new products.

Table 11.3 Requirements for remanufacturing of products by different standards

Standard	IEC 62309	IEC 24700	ANSI/RIC 001.12016 (RIC 2015), Remanufacturing Industries Council (RIC), UK
Target group	All E&E products	Office equipment	All remanufactured products
Qualification	The functionality of products with Qagan parts shall be verified at least by the same inspection and function testing as the product with only new parts	… the equipment, when comprised of reused parts should have the same cosmetic, functional and performance characteristics as the equipment when comprised of all new parts	Remanufactured product using the remanufacturing process; the as-new/like-new remanufactured finished article results from the recovery and transformation of core
Testing	Products containing reused parts shall meet the same technical standards and, if not otherwise stated, have the same functionality as a product with only new parts. Qagan parts shall have passed the same inspection and functional testing as new parts. The functionality of products with Qagan parts shall be verified at least by the same inspection and function testing as the product with only new parts. Final inspection and testing, including functional testing of products with reused parts, shall be the same as for products with only new parts	The equipment must be tested using the equivalent procedures as defined by the manufacturer regardless of whether or not the product contains all new components or if it contains reused components	Inspection of core The core shall be inspected against documented acceptance criteria to determine whether the core is suitable for remanufacturing Product components shall be inspected and functionally evaluated using documented procedures to determine their eligibility for reuse The product shall undergo performance testing using established, documented test procedures to confirm that its performance meets the technical specifications described in Sect. 6.3

(continued)

Table 11.3 (continued)

Standard	IEC 62309	IEC 24700	ANSI/RIC 001.12016 (RIC 2015), Remanufacturing Industries Council (RIC), UK
Quality	ISO 9001 valid for production Qagan state of a part, which has been put into normal use on one or more occasions but differs from a second-hand part in the respect that it is not just a re-sale, rather reconditioned and subjected to fully defined and documented quality checks prior to re-sale, such that it is in all dependability issues, as good as new	If the equipment is originally manufactured in an ISO 9001 certified factory, then this condition must be maintained. Factory certification to the ISO 9001 may ensure that the product conforms to the product's established design	ISO 9001 valid for production Obtain technical specifications The organization shall obtain or create technical specifications for the remanufactured product to validate the "equivalent, or better, condition and performance compared to the new original product
Warranty	The expected life of a product containing reused parts shall correspond at least to the designed life of a product as stated in specification. The failure intensity/hazard rate of a product containing reused parts shall not exceed given expected values which are also applicable for a product containing only new parts, or shall not exceed given limits. The warranty period and terms of warranty shall not be less than for products with only new parts	The equipment's warranties and guaranties must apply equally to equipment manufactured of all new components and to equipment that contains reused components	The organization may issue a warranty for the product that matches or exceeds that of original or similar articles that have never been placed on the market or operated by an end-user customer

(continued)

Table 11.3 (continued)

Standard	IEC 62309	IEC 24700	ANSI/RIC 001.12016 (RIC 2015), Remanufacturing Industries Council (RIC), UK
Environmental responsibility	ISO 14001 is valid for production For example, special attention should be given to the following environmental aspects: energy consumption, hazardous substances and material recycling	If the equipment is originally manufactured in an ISO 14001 certified factory, then this condition must be maintained. Factory certifi-cation to ISO 14001 may ensure a commitment to continuous improvement in environmental performance	ISO 14001 is valid for production
Supplier Declaration of conformity	Customers shall be made aware in sales literature, through normal documentation (for example quotations, invoices, product brochures), that they are buying a product containing Qagan parts. They should see a statement that this product contains reused parts meeting these standards, (through testing or analysis), and, where required by contract or regulations, listing the reused parts. The kind of declaration may be subject to business conventions	When it is required that a supplier demonstrate conformance with the requirements of this International Standard and the normative references in 2.1, the supplier shall complete Annex A, "Supplier's declaration of conformity in accordance with ISO/IEC 17050-1."	The organization may apply to the standard in order to qualify for self-declaration of conformity to this standard. Such an application shall include detailed, written documentation to establish systematic implementation of the process described in Sect. 2.1 of this standard

(continued)

Table 11.3   (continued)

Standard	IEC 62309	IEC 24700	ANSI/RIC 001.12016 (RIC 2015), Remanufacturing Industries Council (RIC), UK
Documentation	When manufacturing products containing Qagan parts, the processes for verifying their reliability and lifetime shall be documented. This documentation shall list the Qagan parts and describe how their expected reliability and remaining life are verified. (extended see standard)	Documentation prepared in meeting the requirements of the normative references in 2.1 shall be available upon request in accordance with ISO/IEC 17050-2. Other documentation prepared in meeting requirements specifically pertaining to the characteristics of office equipment shall also be available upon request NOTE Characteristics of office equipment are specified by ISO standards such as ISO/IEC 11159, and ISO/IEC 11160	Documentation (for core)shall include the appropriate criteria, such as quality, condition, economic, cosmetic, etc., as well as the techniques to be used to conduct the evaluation, and the disposition of cores that fail to meet the acceptance criteria
Reliability assessment	Through analysis and/or testing it shall be verified that the specified expected failure intensity/hazard rate for the product containing reused parts is not higher than for a product with only new parts, nor does it exceed given limits in the as-new designed life. Reliability stress screening may be omitted on reused parts	No information available	No information available

(continued)

Table 11.3 (continued)

Standard	IEC 62309	IEC 24700	ANSI/RIC 001.12016 (RIC 2015), Remanufacturing Industries Council (RIC), UK
Evaluation of current status	Procedures shall be applied to evaluate the state of products containing Qagan parts. The verification can be made by using the manufacturer's data sheet or by life tests of components or modules. Proper use of quality management systems can support in controlling the process These procedures may be: – visual inspections, measurements and function tests such as those performed on products with only new parts; – evaluations of running time counters, consumption counters, etc. designed to help decide whether products can possibly be reused giving information on the remaining life (for example with characteristic curves): – checking whether qualified-as-good-as-new enclosures/reused parts of products are visually intact; appropriate inspections or testing (that is x-ray or ultra sound testing) shall be included if the mechanical structure is in any way load carrying; – evaluation of the state of SW	No information available	Control of nonconforming product The organization shall ensure that product which does not conform to minimum product technical specifications is identified and addressed to prevent reintroduction into the market, as detailed in the quality management system referenced above. A documented procedure shall be established to define the process for dealing with such a nonconforming product When nonconforming product is corrected such a product shall be subject to reverification to demonstrate conformity to these specifications The organization shall obtain or create technical specifications for the remanufactured product to validate the "equivalent, or better, condition and performance" compared to the new original product The organization shall collect or acquire the core using appropriate documented quality control processes The core shall be inspected against documented acceptance criteria to determine whether the core is suitable for remanufacturing

(continued)

Table 11.3 (continued)

Standard	IEC 62309	IEC 24700	ANSI/RIC 001.12016 (RIC 2015), Remanufacturing Industries Council (RIC), UK
Product safety and control	Whoever puts products containing reused parts into circulation shall be responsible for the product liability and its compliance with product safety laws and other regulations dealing with the circulation and use of products in the same way as for products containing only new parts. Safety tests should be repeated when required and safety analysis has to be updated	No information available	No information available
Summary	More detailed Description, documentation, Test of SW state, Dependability testing, Safety assessment. Result: As-new component, new product with as-new parts, CE-mark required	Processes not clear which should be applied like for new product with extensions Result: As-new equipment, CE-mark?	Processes not clear which should be applied like for new product with extensions Result: As-new equipment, CE-mark?

11.3   Accommodation to Pre-defined Conditions

It doesn't make sense to aim for the recovery of certain high valuable thermoplastics if
the complete product will be thrown into a shredder and there will be no chance to get
this material for high level reuse. Correspondingly, only in such a case, products will
be extracted outside of the legally required separated collection (according to special
product categories from the flow of collection), by which mono-materials could be
gathered or components like printed wiring boards could be extracted. According to
Appendix G of the WEEE, certain components or materials have to be eliminated
from the materials flow, but this could also happen by special processes after the
shredder process, or by a pre-disassembly. As far as the printed wiring boards can be
directly recycled in a copper smelter, they should be mounted in the product, easily
accessible and easy to be dismantled. National treatment regulations must be applied
if required.

But if mono-material fractions will be available, recycling often will be impossible because certain minimum quantities of the same waste have to be collected for
an economic recycling process, that is, if the volume collected is too small for the
volume required by a remanufacturing plant, the refurbishment will not be profitable.
Inside the market there is one group of recyclers dismantling "deeply" forming up to
100 recovery fractions, and another party which prefers to send the waste products
like household appliances to a shredder processes. Which process the development
engineer will prefer depends on their own guideline conditions. Rules for repro-
cessing for recyclers are in some countries, as in Germany, defined by a governmental organization (in Germany: LAGA = Laender-Arbeits-Gemeinschaft Abfall
= waste). Therefore, the possibilities for independent decisions are small. According
to Appendix G of the new WEEE the following have to be treated selectively.

1. As a minimum, the following substances, mixtures and components have to be
    removed from any separately collected WEEE:

    – Polychlorinated biphenyls (PCB) containing capacitors in accordance with
      Council Directive 96/59/EC of 16 September 1996 on the disposal of
      polychlorinated biphenyls and polychlorinated terphenyls (PCB/PCT)
    – Mercury containing components, such as switches or backlighting lamps
    – Batteries
    – Printed circuit boards of mobile phones generally, and of other devices if the
      surface of the printed circuit board is greater than 10 cm^2
    – Toner cartridges, liquid and paste, as well as color toner
    – Plastic containing brominated flame retardants
    – Asbestos waste and components which contain asbestos
    – Cathode ray tubes
    – Chlorofluorocarbons (CFC), hydro chlorofluorocarbons (HCFC) or hydroflu-
      orocarbons (HFC), hydrocarbons (HC)
    – Gas discharge lamps

- Liquid crystal displays (together with their casing where appropriate) of a surface greater than 100 cm^2 and all those which are back-lighted with gas discharge lamps
- External electrical cables
- Components containing refractory ceramic fibers as described in Commission Directive 97/69/EC of 5 December 1997, adapting to technical progress for the 23rd time Council Directive 67/548/EEC on the approximation of the laws, regulations and administrative provisions relating to the classification, packaging and labeling of dangerous substances
- Components containing radioactive substances with the exception of components that are below the exemption thresholds set in Article 3 of Annex I to Council Directive 96/29/Euratom of 13 May 1996, laying down basic safety standards for the protection of the health of workers and the general public against the dangers arising from ionizing radiation
- Electrolyte capacitors containing substances of concern (height > 25 mm, diameter > 25 mm or proportionately similar volume
  These substances, mixtures and components shall be disposed of or recovered in compliance with Directive 2008/98/EC.

2. The following components of WEEE that are separately collected have to be treated as indicated:

- Cathode ray tubes: the fluorescent coating has to be removed,
- Equipment containing gases that are ozone depleting or have a global warming potential (GWP) above 15, such as those contained in foams and refrigeration circuits. The gases must be properly extracted and properly treated. Ozone-depleting gases must be treated in accordance with Regulation (EC) No 1005/2009,
- Gas discharge lamps: the mercury shall be removed.

3. Taking into account environmental considerations and the desirability of preparation for reuse and recycling, points 1 and 2 shall be applied in such a way that environmentally-sound preparation for reuse and recycling of components or whole appliances is not hindered.

The recovery quota, which have to be achieved according to WEEE of 24 July 2012, are also important. At the moment the following quota are valid (Table 11.4).

To understand the values in Table B.2, it first has to be considered that these values have to be achieved as an average for all products of a category. The statement that the required quota were achieved will be taken as fact by the European Commission, the first time there is a summarized reporting by the government of the corresponding country for the different collection groups. An example is the sum of weight of collection containers per category. The designer, however, should achieve the corresponding values for his category with his product already, as a result of his development, and should check the validity by a trial of recovery. This value has only a small influence on the special case in reality, because in the mix from a container the result will be very different. Nevertheless, a manufacturer could acquire a positive

Table 11.4 Recovery quota from 15 August 2018

Category according to Annex III WEEE	Recovery quota[a] to be achieved from 15 August 2018
1. Temperature exchange equipment	85(80)
2. Screens, monitors, and equipment containing screens having a surface greater than 100 cm^2	80(70)
3. Lamps	(80)
4. Large equipment (any external dimension more than 50 cm) including, but not limited to: Household appliances; IT and telecommunication equipment; consumer equipment; luminaires; equipment reproducing sound or images, musical equipment; electrical and electronic tools; toys, leisure and sports equipment; medical devices; monitoring and control instruments; automatic dispensers; equipment for the generation of electric currents. This category does not include equipment included in categories 1–3	85(80)
5. Small equipment (no external dimension more than 50 cm) including, but not limited to: Household appliances; consumer equipment; luminaires; equipment reproducing sound or images, musical equipment; electrical and electronic tools; toys, leisure and sports equipment; medical devices; monitoring and control instruments; automatic dispensers; equipment for the generation of electric currents. This category does not include equipment included in categories 1–3 and 6	75(55)
6. Small IT and telecommunication equipment (no external dimension more than 50 cm)	75(55)

[a]Target value for recovery (preparation for reuse and recycling). In category 3, only recycling, without reuse

image if the company is able to answer questions from the public about whether the recovery quota might be achieved, or exceeded, for its products.

From year 2018 the categories of the products are formed differently. They are summarized in Annex III of the WEEE from 24 July 2012 and in the following Table 11.4.

The increased collection targets, according to WEEE, have already been discussed above. These targets have increased the possibility for gathering bigger volumes for an economically feasible reuse, or for remanufacturing of certain fractions and extracting raw materials.

In the waste framework directive of the European Commission, reuse got a higher ranking within the waste hierarchy. It was put on the second rank. The order therefore is now: Avoidance, **reuse**, material recycling, recovery and deposition. Especially for industrial goods, additional legal regulations about reuse of components are now in political discussion.

In the Waste Framework Directive, legislation mandates the following important points: The manufacturer has to design the products in such a way that reuse will be more practical. Extended product responsibility will affect not only manufacturers but all participants involved in the process chains. They will also have to carry the costs for an inordinate share of waste. Recovery targets were already set, but this was done without naming those conditions by which these have to be achieved. Further interpretations are not possible without these missing conditions.

Many procedures about materials recovery have already been tested and some have been introduced into practice. The integrated material recovery processes, such as the Sicon process [43], which are aligned with the practice of a melt oven, are especially interesting. After a special shredding process the ore rich pellets are directly introduced into the mound of the melt oven, the fiber substances are separated and reapplied for filter purposes and the dust containing elements are put into the melt oven from the top. How far this, and similar procedures, will be really accepted as a material, raw material or energetic process will depend on the legislative discussions still pending about the acceptable recovery processes.

For development engineers, such long decision-making processes for certification are not acceptable, as they have to decide during the design phase which recovery process will be available at the end of product life. Usually designers will contact their recyclers to get their advice. But for some plastics, such procedures are not always on the market in a reliable sustainable form. It is similar for some metals like Indium. For rare earth a process is described in Lorenz et al. [51]. But anyway, there is a big gap between the potential "good will" of the development engineer and a reliable and available process for some material recovery problems.

A further problem will be that these processes after acceptance might be different in countries worldwide! So, what should be done? One potential step might be the international standardization of such accepted processes and what could be recycled by them, hoping the governments will agree too. The basic processes of recycling and related procedures are internationally standardized by IEC TC 111 (drafts in the order by European Commission). In Germany legal requirements are described in their own regulation [49] (only in German).

It seems as if now three different interests have to be brought together:

- The interest of the recyclers who have installed some processes for special optimization,
- the interest of the government to avoid emissions, and
- the interest of the development engineers.

to create a product function with the chance for a recovery of the materials. It might be an idea to define transfer points where every party defines what they will expect so that a discussion could start about how a better recycling process can be achieved together. At the moment, the recycler is the last in line and has to take what it gets. But the thinking should also start from the reverse. Which process will be the best for the recycling of special materials and how could the component be designed to fit better into such a process?

Depending on which guideline conditions will be valid, the optimized design might look different. An information sheet was agreed with the European association of recyclers (EERA) and with the associations of product manufacturers, such as *CECED* and *DigitalEurope*. This information sheet, designed for treatment facilities, relates to certain hazardous substances and components of a product and lets recyclers know how to contact the corresponding product manufacturer, if required. This list contains, in principle, the information required by WEEE and RoHS3 and, additionally, answers questions about the possible contents of asbestos, beryllium and its compounds, as well as information about potentially contained gases, liquids and subjects relevant for safety. However, recyclers of E&E products who were asked by the associations declared that they would know enough about the standard products in the market and that they will no longer require this information.

So far, only certain hazardous substances might be present which are legally restricted by the RoHS3 directive, like the substances lead, cadmium, mercury, chrome(VI), polybrominated biphenyls and diphenyl ethers (PBB, PBDE). These substances can be contained only below definite (low) limits, in products of the aforementioned product categories, that have been newly put on the market since 1 July 2006. All E&E products are affected by RoHS3 now (some exemptions).

Whoever wants to recover **capital goods** will have an advantage. In comparison to consumer products, these components and products will be in a relatively unworn condition so that disassembly has a better chance of being profitable and also the probability of recovery will be higher so that larger amounts of pure mono-material could be formed. The product manufacturer will have the best opportunities with these kinds of products to install a reuse or remanufacturing factory, as are already existing for medical or telecommunication equipment. The European Commission promotes now a project for the reuse of complete old production plants to become new ones [50]. From such projects usually legal requirements will be derived later.

11.4 The Role of Software

SW problems are covered by the general product safety directive [36] if a component or a product is put on the market as QAGAN or as a new product containing QAGAN components. The same will be valid if a product was only repaired. It must be safe.

From the point of the view of the product safety directive the following is valid:

In the absence of specific regulations, and when the European standards established under mandates set by the Commission are not available or recourse is not made to such standards, the safety of products should be assessed taking into account in particular national standards transposing any other relevant European or international standards, Commission recommendations or national standards, international standards, codes of good practice, the state of the art and the safety which consumers may reasonably expect. In this context, the Commission's recommendations may facilitate the consistent and effective application of this Directive pending the introduction of European standards or as regards the risks and/or products for which such standards are deemed not to be possible or appropriate. (from preliminaries of this directive, Chap. 16)

The Commission refers to the application of standards to check the safety of the product. In the case of QAGAN this is the standard for the new product. In other cases one could refer to this standard or to define one's own tests.

Appropriate tests of the SW dependability should be available. Tests must be applied for components and for the whole system. Such system tests were developed for products such as "https://en.wikipedia.org/wiki/Software_of_unknown_p edigree" computers and will be the same as for the new product. If a system will be compiled, completely new such safety tests must be planned for the whole system. A special look should be directed to safety critical SW.

Products as-new newly put on the market have to be checked as to whether copyright, license agreements or SW patents are injured or not. But they should usually be the same as for the new product.

For medical equipment special requirements, for example, from the US Food and Drug Administration (FDA), have to be integrated. An example for a combined testing is given by the FDA for an infusion pump (https://www.fda.gov/med ical-devices/infusion-pumps/infusion-pump-SW-safety-research-fda). The Laboratory of SW Engineering of the FDA recommends to extend SW testing from the development phase to a model-based system testing: "The SW engineering community has been developing tools for modeling SW and its interactions with the system it controls. Safety properties of the model can be systematically examined, and once the model has been verified, the SW derived from it can be proven to conform to the model. The result is SW designs that are far more robust than those developed using traditional methods." Guidance documents are available at the FDA, such as the General Principles of SW Validation Guidance for Industry and FDA Staff January 2002. Cybersafety[4] is another potential test area.

In IEC 62304 (Medical device SW–SW lifecycle processes, 2006) SW lifecycle processes are described and activities are recommended for.

- 3 more or less critical levels A, B, C of safety classification in IEC 62304,
- risk management, and
- SW problem resolution processes.

Such special considerations are interesting for SW of unknown pedigree or provenance.

In our case three points can be extracted for reuse of HW and SW:

---

[4]Cybersafety is the safe and responsible use of Information and Communication Technologies (ICT). NetSafe's approach to cybersafety is founded on:
- Maintaining a positive approach about the many benefits brought by technologies.
- Encouraging the public to identify the risks associated with ICT.
- Putting in place strategies to minimize and manage risks.
- Recognising the importance of effective teaching and learning programs.

Recommendations from the FDA cover legal aspects of a refurbished HW/SW product and a company could take reference of the recommended tools to guarantee safety.
– Model-based tests could enable the testing of complex systems which might be combined from old and new components with different SW.
– Safety classification of SW might be a chance to distinguish between SW and HW risks and the necessity of more intensive testing.
– As a final remark it can be stated that a product as-new will run through the usual standard tests for HW and SW testing of a new product. But if a product is very different from new then more and more testing might be necessary to guarantee a certain state of quality, if there is a lot of unknown history. Cybersafety* might be a new problem which could prevent a company from the reuse of older components of unknown risks.

Key Points, Exercises, Recommended Further Reading, References/Websites

1. Key Points

• Many directives require actions for take-back and recycling including ecodesign. Most important for product development is the *Ecodesign Directive*. In the meantime measures are given for many products which require better environmental compatibility. More will follow. The *Ecodesign Directive* includes not only reuse but also design requirements for the whole lifecycle. Reuse projects are promoted by European Commission and may end in a reuse target.
• Regulations for take back and recycling are part of the WEEE directive. Not only are 10 collection groups formed, but also targets for collection and recovery are set. The volume of reuse will reduce the necessity for recycling rates.
• With the RoHS3 directive 6 substances (Pb, Cd, Cr(VI), Hg, brominated biphenyls and diphenylethers) are prohibited in all E&E products and some phthalates (*DEHP*, BBP, DBP and DiBP) have been added till 2019. Components with such substances cannot be reused in new products. Some more restrictions have to be observed and conformity must be declared by use of the CE mark. Exemptions to the RoHS are granted but with changing validity (see the homepage of the European Commission). Worldwide, similar law is spreading and should be carefully studied.

Exercises

(1) How a company, besides conducting an expensive analysis, can make sure that prohibited substances are not included in a component?
(2) Which legal requirements do you find about reuse of E&E products? Which volumes should be reused?
(3) From where do you get information about legal requirements in other countries?

Recommended Further Reading

Exemptions to RoHS: https://ec.europa.eu/environment/waste/weee/legis_en.htm.

Recommended further reading: Belli and Quella [3],

ReBorn, reuse of factory systems project of EC: www.reborn-eu-project.org.

References/Websites

- Waste framework directive,
- End-of-life vehicle directive,
- *Ecodesign (ErP) Directive* [1],
- RoHS2/RoHS3 [33, 34] and WEEE [4] directives,
- Dodd-Frank Act about Conflict Minerals,
- Packaging and packaging waste directive,
- General product safety directive (GPSD) [36],
- EMC directive: Directive relating to electromagnetic compatibility (EMC) [37],
- Battery directive [38].

2. Key Points

- Standards are the tools enabling industry to do things in the same way. This can happen through acting cooperatively, to save money or to exclude other companies from the market.
- Standards can also be made for the same activity but with less strict requirements, like for all E&E products, or for office equipment. They could also be initiated by governments to persuade industry to unify to do the same things in the same way, such as for environmental protection in ISO 14001, and to increase the pressure on companies to apply the environmental standard. Governments also grant a mandate to a standardization organization to develop a standard, like for an analytical method, as a basis for a chemical law.

- Pressure could be applied on a company to get an independent certification, so that the company fulfills the requirements of the whole standard.
- National standards or publicly available specifications (PAS) are often developed for the test application of a standard before it is presented as a suggestion for an international standard.

Exercises

(1)   What could be the target of the European Commission in giving the IEC TC 111 the task to develop international standards, although the legislation is valid only in the European Union?
(2)   Please describe some differences between a standard, a PAS (publicly available specification) or a TR (technical report). Use the internet for research.

Recommended Further Reading

Standards IEC 62309, ANSI RIC 2015 (Remanufacturing Industries Council, RIC), UK, IEC 24700, IEC/PAS 62814, ISO/TR 14 062.

References/Websites

Unido, Working paper: Role of Standards: https://www.unido.org/sites/default/files/2009-04/Role_of_standards_0.pdf.

ISO: Economic benefits of standards, Volume 1,2, Geneva 2011.

3. Key Points

- Legislation requires different collection and recycling quota for different equipment. In many countries collection procedures are different.
- So often the conditions for the recycling have to be installed or volumes have to be brought together to get the critical volume for an economical recycling system. Engineers from a producing company therefore must find out how the processes work in their country to adapt it to their product. Examples are plastics, which might not be recycled in one country but are in another country.
- They can also install a collection process themselves to which other companies could join.

Exercise

Find out by contacting your local recycler which plastics and which metals they collect separately or which they dismantle for a materials recycling.

4. Key points

The role of software for recycling is still underestimated, especially in legislation

- Safety problems are dealt with in the safety directive.
- Dependability standards are missing, especially for the cooperation of HW and SW. For medical device regulations are available and standardized.
- Best test procedures are available if the product is as-new. In this case all tests for new products are valid.
- SW could extend the life of a product which would otherwise be expired as a result of too high energy consumption.
- Cybersafety is a new task for a reused product.

Exercise

Using the internet identify examples for reuse of HW and SW together. For example, in the automobile industry. What is required?
   Name some products where cyber-safety might play a role during reuse.

Recommended Further Reading

Standard IEC 62304, check standards for reuse in the automobile industry.

References/Websites

Recommendations by the FDA and their standards.
   https://www.fda.gov/medical-devices/digital-health/software-medical-device-samd.

Epilogue—Experiences, Future Research Directions, Conclusions

Fevzi Belli andFerdinand Quella

Epilogue

War and Victory Stories About Reuse

As already mentioned in Sect. 1.2, reuse was one of the primary visions of the young discipline, Software Engineering, that was founded in 1968 by the constitutional NATO conference in Germany [15]. It was stressed that the definition of a macro, or a sub-function, and its reuse does not necessarily form a systematic SW reuse. A reuse-oriented development of the component and composite system, service-based, global offering, and finally a context-based, application-oriented selection of the component-to-be-reused can only lead to a systematic reuse.

Moreover, the decisive importance of assuring the dependability of components built-for-reuse before release and in their application context, integrated into the composite system, built by reuse, cannot be stressed enough. The early history of reuse is full of failed and partly disastrous incidents, such as the careless reuse of existing SW in a new Roentgen system that caused the tragic death of nine people [6], or the crash of the ESA satellite system Ariane 5 that caused damages of over a billion Euro [16].

The sad experiences with amateurish and ill-thought-out reuse soon led to systematic reuse because companies very quickly grasped the importance of the economic impact of dependable reuse in competition and the time-to-market race. Nowadays, any ambitious company has its own program of reuse. Experience reports stress that success requires reuse practices be well integrated into the company's SW processes (and to a certain extent into their company cultures). This also significantly helps

F. Belli and F. Quella, *A Holistic View of Software and Hardware Reuse*, Studies in Systems, Decision and Control 315, https://doi.org/10.1007/978-3-030-72261-6

to improve the product offerings [4]. Literature is full of success stories; see, for example, [1, 4, 7, 9, 14, 17].

The idea of as-new (QAGAN) is one of the success stories for reuse because it helps to create a high value (as it is) and reduces test costs. It is not legally a problem to sell new products with as-new components, if the customer is well informed of the fact. Different definitions are discussed.

At any rate, reuse is meanwhile a success story in nearly all product groups and contributes to the earnings of many companies. These earnings are usually not published, but certainly all readers will be convinced of the success of reuse if they look on the internet and search for reuse. The sheer volume of offers is evidence enough! Reused products are being offered on nearly every major company's home-page. An as-new statement is not required if a company can guarantee certain properties and product life.

This is thanks to capital goods reuse being easier because of their long product lifetime and low damage rate when taken back. Good reuse conditions for consumer products are often not given. They often remain on the market only one year or less. Either users change their minds and use products longer or politics creates suitable conditions. The EU has recently decided to promote the repair of consumer products. Depending on the corresponding forthcoming legislation, it might also intensify reuse in this field. Companies producing consumer products are now forced to offer the latest SW for the repair of certain household equipment to bolster energy saving and to get an as-new like product and not an old-fashioned one.

Sustainability Requires Reuse

17 Sustainability Development Goals (SDGs) have been identified by the UN [22] for the future development of the world. Many companies and associations have adopted these rules and added further detailed aspects they want to convert (for example ZVEI [20] and also many larger companies). Resources will especially be a subject of interest where reuse and recycling are involved. Most important for reuse will be Goal Nr 12: Ensure sustainable consumption and production patterns. Under this goal, some targets are being fast tracked for rapid implementation and will hopefully change the mind of potential customers concerning reused products. The following is a selection of some of the targets:

- Implement the 10-year framework of programs on sustainable consumption and production, all countries taking action, with developed countries taking the lead, taking into account the development and capabilities of developing countries
- By 2030, achieve the sustainable management and efficient use of natural resources

- By 2020, achieve the environmentally sound management of chemicals and all wastes throughout their life cycle, in accordance with agreed international frameworks, and significantly reduce their release to air, water, and soil in order to minimize their adverse impacts on human health and the environment
- By 2030, substantially reduce waste generation through prevention, reduction, recycling, and reuse
- Encourage companies, especially large and transnational companies, to adopt sustainable practices and to integrate sustainability information into their reporting cycle
- Promote public procurement practices that are sustainable, in accordance with national policies and priorities
- By 2030, ensure that people everywhere have the relevant information and awareness for sustainable development and lifestyles in harmony with nature
- Support developing countries to strengthen their scientific and technological capacity to move towards more sustainable patterns of consumption and production
- Develop and implement tools to monitor sustainable development impacts for sustainable tourism that creates jobs and promotes local culture and products.

All other Goals are of relevance for a circular economy and also for the target of reuse. A view on how to apply the goals over the industrial life cycle is presented in the ZVEI Guide about the SDGs.

The path to a circular economy is also a target of EU [18]. While there is a target of 65 % of municipal waste being recycled from 2035, such a target for industrial EEE products is missing. The targets should also be much higher, given that within a few years all the missing resources will disappear!

Recommendations for Future Research on Dependable Reuse

Reuse of materials is still a major problem today, although many plastics and other materials could be reused with no difference to new ones. Japanese E&E products apply only a few plastics so reuse is often directly possible in the recycling factory, also for consumer products. TOMRA sorting systems [21] are a technological breakthrough. They are also capable of sorting out plastics from household and other waste. More such systems will be required. All the targets far below 100 % for collection or recycling are not sufficient and must be improved! Everybody should be aware about what happens to the difference and where it goes: millions of tons of waste straight into the environment!

If SW cannot be upgraded in an HW component, the component becomes waste! It might therefore be important to see directly which SW state is in a component. Some kind of marking would be an idea.

Fault recognition would enable to determine which components to exchange: A widely installed technology. Failure repair is not so common and would enable to bypass an SW containing a failure. Here more research might be necessary.

Consumption of resources, especially energy, is another subject where more research and standardization will be required. Interesting systems are already available, but the ideal path to minimum energy consumption has not yet been found.

Much work has been done to optimize the recovery of rare substances and how to separate the different substances, especially metals. Much of this work looks really successful, but it seems no government cares to promote these technologies into an industrial volume to make them cheaper.

The authors of this book recommend the following research directions for SW reuse, particularly concerning its dependability:

- Model-based development and validation of reusable components and composite systems.
- Application of formal methods.
- Strict, legal structuring of open-source and/or public domain SW and their "harvesting," taking strictly quality and legal aspects into account [3, 5, 10, 19].
- The efforts of the United Nations University International Institute for Software Technology for certification of open-source is a valuable, welcome development that calls for support from academy, industry and politics (https://web.archive.org/web/20071115223300/http://www.iist.unu.edu/).
- Sustainable Edge Computing [8, 12, 13].

And at the end:

The reality is that plastic waste is often still being exported to developing countries without recycling!

Many valuable substances disappear off to here and there and are not recycled because nobody really manages to gather all the collected waste to special recycling factories. Sweden collects 90 % of its waste, volumes in other countries are much lower! After so many years of legislation, the EU is still not very successful in collection.

References

1. Apel, S., Batory, D., Kästner, Ch., Saake, G.: Feature-Oriented Software Product Lines. Springer, Berlin (2013)
2. Avizienis, A., Laprie, J.-C., Randell, B., Landwehr, C.: Basic concepts and taxonomy of dependable and secure computing. IEEE Trans. Depend. Secure Comput. **1**(1) (2004). Available at https://ieeexplore.ieee.org/xpls/abs_all.jsp?arnumber=1335465
3. Antes, G.: Open-source software no longer optional. Comm. ACM 59-8, 15–17 (2016)
4. Ezran, M., Morisio, M., Tully, C.: Practical software reuse. In: Chapter "Two Major Case Histories". Springer Practitioner Series, pp. 155–169 (2002)

5. Meeker, H.: Open (Source) for Business A Practical Guide to Open-Source Software Licensing (2nd edn.). Createspace Independent Publishing Platform (2017)
6. Leveson, N.G., Turner, C.S.: An investigation of the Therac-25 accidents. IEEE Comput. **26**(7), 18–41 (1993)
7. Frakes, W.B., Isoda, S.: Success factors of systematic reuse. IEEE Softw. V **11**(5), 14–19 (1994). Available at https://ieeexplore.ieee.org/stamp/stamp.jsp? tp=&arnumber=311045
8. Gomes, C., et al.: Computational sustainability: computing for a better world and a sustainable future. Commun.Assoc. Comput. Assoc., **62**, 56–65 (2019)
9. Hallsteinsen, S., Paci, M.: Experiences in Software Evolution and Reuse—Twelve Real World Projects. Springer Science and Business Media (1997)
10. Hummel, O., Atkinson, C.: Extreme harvesting: test driven discovery and reuse of software components. In: Proc. IEEE International Conference on Information Reuse and Integration—IRI, pp. 66–72 (2004). https://doi.org/10.1109/IRI.2004.1431438
11. Hummel, O., Janjic, W.: Test-driven reuse: key to improving precision of search engines for software reuse. In: Sim, S.E., Gallardo-Valencia, R.E. (eds.) Finding Source Code on the Web for Remix and Reuse, pp. 65–80. Springer, Berlin (2013)
12. Hamm, A., Willner, A., Schieferdecker, I.: Edge computing: a comprehensive survey of current initiatives and a roadmap for a sustainable edge computing development. In: Proceedings of 15th International Conference on Wirtschaftsinformatik (2020)
13. Kienzle, J., et al.: Toward model-driven sustainability evaluation. In: Communication of the Associative for Computing Machinery, vol. 63, pp. 80–91 (2020)
14. Mili, A., Chmiel, S.F., Gottumukkala, R., Zhang, L.: Managing software reuse economics: an integrated ROI-based model. Annals Softw Eng **11**, 175–218 (2001)
15. Naur, P., Randell, B. (eds.): Software Engineering, Report on a Conference Sponsored by the NATO Science Committee, Garmisch, Germany, 7th to 11th October 1968. Available at https://homepages.cs.ncl.ac.uk/brian.randell/NATO/nato1968.PDF
16. Nuseibeh, B.: Ariane 5: Who Dunnit? IEEE Softw. 15–16 (1997)
17. Poulin, J.S., Caruso, J.M., Hancock, D.R.: Business case for software reuse. IBM Syst. J. **32**(4), 567–594 (1993)
18. Report from the Commission to the European parliament, the Council, the European economic and social committee and the committee of the regions on the implementation of the Circular Economy Action Plan, Brussels, 4.3.2019 COM (2019) 190 final. Sim, S.E., Gallardo-Valencia, R.E. (eds.): Finding Source Code on the Web for Remix and Reuse. Springer, Berlin (2013)

19. Sustainable Development Goals, ZVEI Wegweiser für nachhaltige Entwicklung in der Elektroindustrie (Guide only in German), www.ZVEI.org ZVEI—Zentralverband Elektrotechnik- und Elektronikindustrie e. V. Abteilung Umweltschutzpolitik Lyoner Straße 9 60528 Frankfurt am Main
20. Tomra: Recycling of Household Plastic Waste. https://www.tomra.com/de-de/sorting/recycling
21. United Nations: Department of Economic and Social Affairs Disability. https://www.un.org/development/desa/disabilities/about-us/sustainable-development-goals-sdgs-and-disability.html.

Appendices—Part I

Appendix A—Integrating the System Modeling with Fault Modeling I: Sequential Systems Using Finite State Automata, Event Sequence Graphs, and Regular Expressions

When developing a system, the construction usually starts with creating a model of the system to be built to better understand its "look and feel," including its overall external behavior, which is also important to validate the user requirements. Thus, modeling of a system requires the ability of abstraction, extracting the relevant issues and information from the irrelevant ones, while taking the present stage of the system development into account (see Sects. 1.3.2 and 5.2).

The remaining part of this appendix will explain modeling sequential systems with simple formal means, using examples of graphical user interfaces.

A.1 Graphical User Interfaces as an Example for Sequential Systems

In human-machine systems, graphical user interfaces (GUI) enable the interactions of users with the system. Focusing on a single user with a single system, this interaction usually forms a strictly sequential process, obeying the following order:

(initial) user input – system reaction (output) - user input – system reaction (output) - …
– user input – (concluding) system output

While modeling a GUI, the focus is usually on the correct behavior of the system as *desirable* situations, triggered by correct (*legal*) inputs. Describing the system behavior in undesirable, exceptional situations triggered by *illegal* inputs and other

© The Editor(s) (if applicable) and The Author(s), under exclusive license to Springer Nature Switzerland AG 2021
F. Belli and F. Quella, *A Holistic View of Software and Hardware Reuse*, Studies in Systems, Decision and Control 315, https://doi.org/10.1007/978-3-030-72261-6

undesirable events are likely to be neglected due to the project's time and cost pressures. The precise description of such undesirable situations is, however, of decisive importance for a user-oriented fault handling, because the user has not only a clear understanding how his or her system has to function properly, but also which situations are not in compliance with his or her expectations. In other words, we need a specification to describe the system behavior *both* in legal and illegal situations in accordance with the expectations of the user. Once we have such a complete description, we can then also precisely specify our hypotheses to detect undesirable situations and determine the due steps to localize and correct the faults that cause these situations.

Summarizing the discussion about a good modeling, we need a formal specification tool with the following capabilities (see also Sect. 1.3.2.1):

- *Generic,* that is, describing both legal and illegal situations;
- *Recognizing,* that is, distinguishing between legal and illegal situations;
- *Operable,* that is, enabling calculations based on efficient, verifiable algorithms for a quantitative view, for example, through the enumeration of the generated test cases, assigning them weights in the order of their importance for test coverage, etc. The operability is best given by an algebra, consisting of well-defined operations according to a calculus, an order relation, and neutral element(s).

These requirements are fulfilled by Finite-State Automata (FSA) and Regular Expressions (RegEx), having equivalent recognition and generation capabilities, in the sense of regular grammars and languages (type-3 level of Chomsky hierarchy), and building an event algebra [6, 15].

A.2 Finite-State Modeling of GUI

Deterministic finite-state automata (FSA), also called finite-state, sequential machines have been successfully used already for many decades to model sequential systems; for example, logic design of both combinatorial and sequential circuits [7, 12], protocol conformance of open systems [17], compiler construction [2], but also for UI specification and testing [13]. FSA are broadly accepted for the design and specification of sequential systems because of their recognition capabilities to effectively distinguish between correct and faulty events/situations.

A FSM can be represented by:

- a set of inputs
- a set of outputs
- a set of states
- an output function that maps pairs of inputs and states to outputs
- a next-state function that maps pairs of inputs and states to next states.

This is an informal but nevertheless sufficiently precise definition that will be used in this appendix; for a formal definition, see, for example, [201]. For representing GUI, we will interpret the elements of FSA as follows:

- Input set: Identifiable objects that can be perceived and controlled by input/output devices; they are elements of WIMPs (Windows, Icons, Menus, and Pointers).
- Output set has two distinct subsets:

  - desirable events: Outcomes that the user expects and wants to have, that is, correct, legal responses, and
  - undesirable events: Outcomes that the user does not want, that is, a faulty result or an unexpected result that surprises the user.

Note our following assumptions that do not constrain the generality:

- We use FSA and its state transition diagram (STD) synonymously.
- STDs are directed graphs having an *entry* node and an *exit* node, and there is at least one path from entry to exit (we will use the notions "node" and "vertex" synonymously).
- Outputs are neglected, in the sense of Moore Automata.
- We will merge the inputs and states, assigning them to the vertices of the STD of the FSA.
- Next-state function will be interpreted accordingly, that is, inducing the next input that will be merged with the due state.

Thus, we use the notions "state" and "input" on the one hand and "state," "system response," and "output" on the other synonymously, because the user is interested in the external behavior of the system that will be manifested by *events*, which can be perceived visually or acoustically and not in its internal states and mechanisms that are hidden from the user [10, 11, 16]. This view leads to a simplified graphical representation that is called *Event Sequence Graph (ESG)* [3]. Any chain of events (represented by the nodes of ESG) from one vertex to another one, materialized by sequences of user inputs-states-triggered out-puts, defines an *event sequence (ES)* traversing the ESG model from one vertex to another.

Basically, an *event* is an externally observable phenomenon, such as an environmental or a user stimulus, or a system response, punctuating different stages of the system activity. It is clear that such a representation disregards the detailed internal behavior of the system, which is given by means of its different states and, hence, an ESG is a more abstract representation compared to, for example, a state transition diagram (STD) or finite-state automaton (FSA) [14]. Figure A.1 compares representation by FSA and ESG.

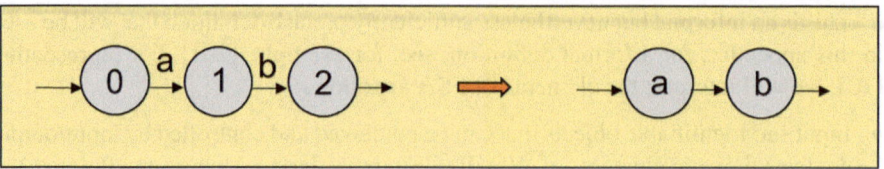

Fig. A.1 On the equivalence of FAS and ESG, both representing the sequence *ab* (based on [11])

A.3 Regular Expressions

Efficient algorithms exist for converting an FSA, and, therefore, the corresponding ESG, into an equivalent regular expression (RegEx), and v.v. [8, 14, 15]. RegEx, on the other hand, are traditional means to systematically generate and recognize legal (desirable, correct) and illegal (undesirable, incorrect) situations and events, building an event algebra [204, 8].

To introduce informally, we assume that a *regular expression* RegEx consists of symbols *a, b, c, ...* of an alphabet that can be connected by operations:

- *Sequence* (usually no explicit operation symbol, for example, "ab" means "b follows a");
- *Selection* ("+", for example, "*a+b*" means "*a* (exclusive) or *b*");
- *Iteration* ("*", Kleene's Star Operation, for example, "*a**" means "*a* will be repeated arbitrarily often"; "+": at least one occurrence of "*a*").

Example 1 $T = [(ab(a + c)*)*]$.

Figure A.2 represents the regular expression given in Example 1 as an ESG. Note that the brackets "[" and "]" are pseudo-events and symbolize begin and end of the ESG.

The symbols of a RegEx (and ESG) can be atomic/terminal symbols or also regular expressions. Accordingly, they can be interpreted as single actions or an aggregation of actions. An action can represent a command, a system response, etc. in the sense of an event.

A.4 An Example for a GUI

Figure A.3 represents a small part of an MS WordPad-like word processing system (see also [1]). This GUI will usually be active when text is to be loaded from a file or to be manipulated by cutting and pasting or copying. The GUI will be used also for saving the text in the file (or, in another one). At the top level, the GUI has a pull-down menu with the options `File` and `Edit` that invoke other components, for example, `File` event opens a sub-menu with `Save As` and `Open` as sub-options. These sub-options have further sub-options. `Select` can invoke sub-directories or select

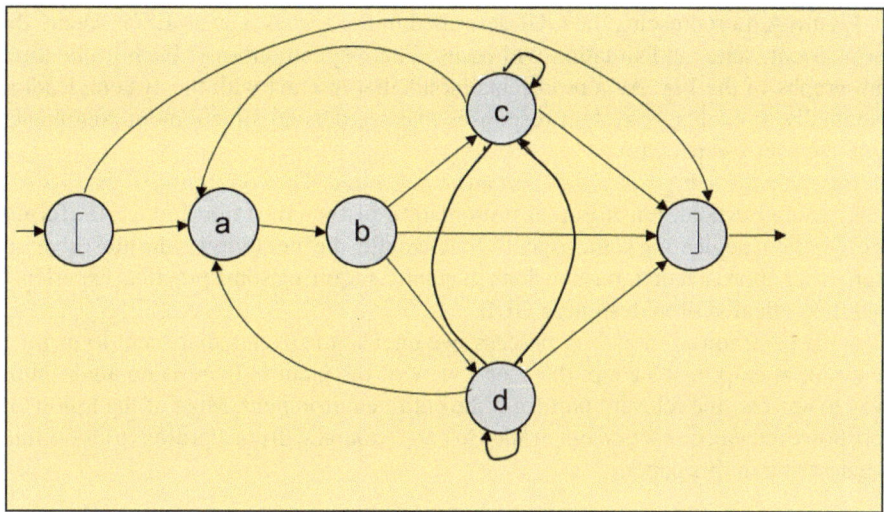

Fig. A.2 ESG of the Example 1

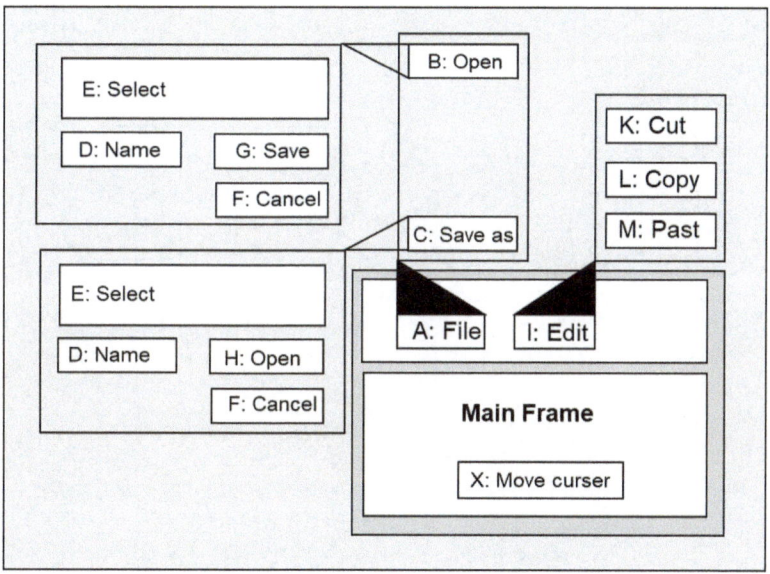

Fig. A.3 Example of a GUI

files. There are still more window components, which will not be described further. The window can be closed by selecting either Open or Cancel. The described components are used to traverse through the sequences of the menus and sub-menus, creating many different combinations and accordingly, many applications.

Figure A.4a represents the GUI described in the Fig. A.3 as an ESG. Again, the terms event, state, and situation will be used here synonymously. Each of the three sub-graphs of the Fig. A.4a represents inputs that interact with the system, leading eventually to events as system responses that are desired situations in compliance with the user's expectation.

Based on the sub-graphs, the ESs can be generated. The conversion of the Fig. A.3 (easy to understand, but informal presentation of the GUI) into Fig. A.4a (formal presentation, neglecting some aspects, for example, the hierarchy) is the most abstract step in our approach that must be done manually, requiring some practical experience and theoretical skill in designing GUIs.

As is common in modeling process, we choose the events that seem to us most relevant, attempting to adopt the user's view of the picture; there is no algorithmic way to abstract the relevant part from the entire environment. Most of the following job, however, can be carried out at least in part automatically, according to algorithms we describe in this paper.

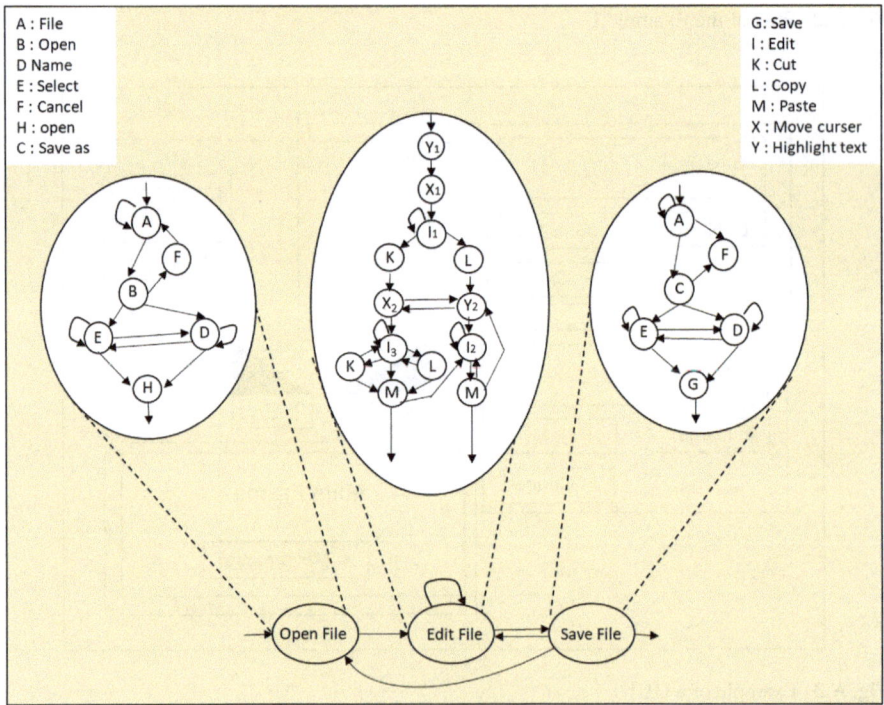

Fig. A.4 a ESG of the GUI. Represented in Fig. A.3. **b** Sub-graph open file of the (**a**)

Fig. A.4 (continued)

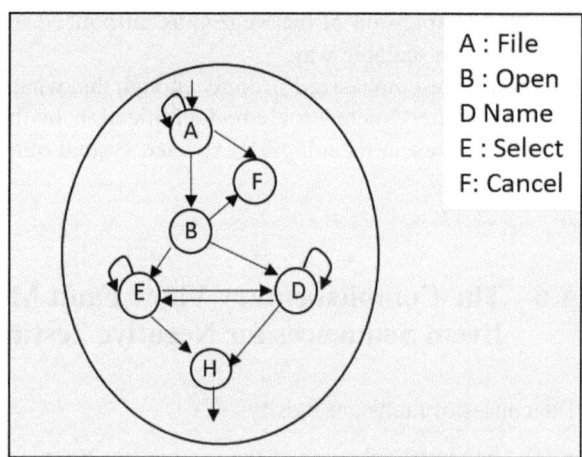

A : File
B : Open
D Name
E : Select
F: Cancel

Table A.1 (**a**) EPs of the sub-graph open file of the Fig. A.4b. (**b**) RegEx of the sub-graph open file of the Fig. A.4b

(a)	
Sub-graph	EPs
Open File	AA, AB, BD, BE, BF, EH, FA, ED, EE, DD, DE, DH
(b)	
Sub-graph	RegEx
Open File	$A^+B(FA^+B)^* (E^+D^*+D^+E^*)^+H$

A.5   Event Sequences (ES) and Complete ES (CES) for Positive Testing

Once the ESG has been constructed, more information can be gained by analyzing its structure. First, we can now identify all legal sequences of user-system interactions as event sequences (ES). An ES that traverse the ESG from beginning to end is a *complete event sequence (CES)*. Thus, a CES is a well-defined system response that the user expects the system to carry out. An *incomplete* event sequence is a sub-sequence of a complete even sequence. Second, we can identify the entire set of the *compatible*, that is, legal, *event pairs (EP)* as the nodes of the ESG (Table A.1a, EPs based on the sub-graph in Fig. A.4b). This is the key issue of the test approach, since it will enable to define the *event coverage* notion as a test termination criterion. Validating that the system under consideration (SUC) works the way the user expects will be called *positive testing*.

The generation of the CESs and EPs can be based either on the ESG, or more elegantly, on the corresponding RegEx, whichever is more convenient for the test engineer (Table A.1b, RegEx for the sub-graph in Fig. A.4b). For example, the systematic expansion of the RegEx, as introduced in [26], can be used to generate test cases in a scalable way.

It cannot be emphasized strongly enough that what we are doing here is an elegant solution of the Oracle Problem: Identification of the Complete Event Sequences (CES) to represent meaningful, expected system outputs which will be constructed systematically.

A.6   The Complementary View: Fault Modeling through Event Sequences for Negative Testing

The causes of faults are mostly:

- The expected behavior of the system has been wrongly specified (*Specification Errors*), or
- the implementation is not in compliance with the specification (*Implementation Errors*).

In our approach, we will exclude the *User Errors*, suggesting that the user is *always* right, that is, we suggest that there are no user errors. We require that the system must detect all inputs that cannot lead to a desired event, inform the user, and navigate him or her properly in order to reach a desired situation. This requirement is called *input completeness.*

One consequence of this requirement is that we need a view that is complementary to the modeling of the system, in the sense of a *holistic* approach. This can be done by systematical and stepwise manipulation of the ESG that models the system. For this purpose, we introduce the notion *Faulty/Incompatible Event Pairs (FEP)*, which consist of inputs that are not legal in the sense of the specification.

The faulty events pairs for the sub-graph open file of Fig. A.5.

Figure A.5 systematically generates all potential faulty event pairs for the sub-graph open file (Fig. A.4b) by threefold manipulations:

- Add edges in opposite direction wherever only one-way edges exists (Fig. A.5a).
- Add loops to vertices wherever none exists in the specification (Fig. A.5b).
- Add edges between vertices wherever none exists (Fig. A.5c).

Adding all manipulations to the ESG defines the *Completed ESG* (*CESG*, Fig. A.5d).

Now we can construct all potential interaction faults systematically, building all illegal combinations of symbols that are not in compliance with the specification (FEPs in Table A.2a). Once we have generated a FEP, we can extend it through an ES that starts with entry and ends with the first symbol of this FEP; we then have a *faulty/illegal complete event sequence* (*FCES*), bringing the system into a faulty situation (Table A.2b).

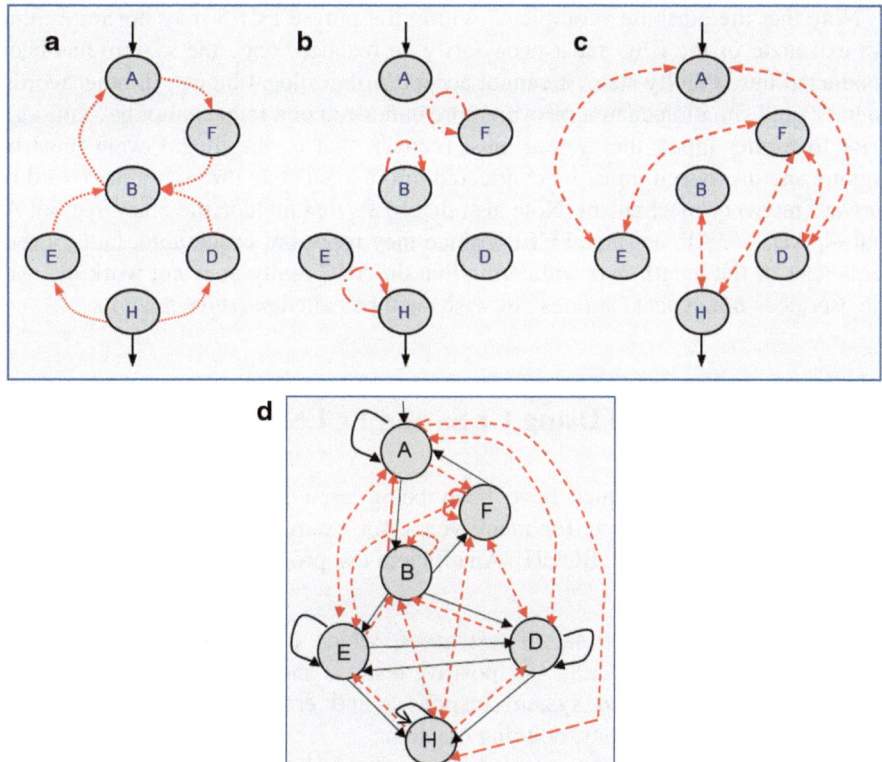

Fig. A.5 **a** Reversing connections. **b** Loops. **c** Networking connections. **d** CESG (*Completed* ESG)

Table A.2 (**a**) The set of FEPs (faulty event Pairs), (**b**) the set of FCESs (faulty complete event sequences), which transfer the system into a faulty state

(a)

Sub-graph	FIPs
File Open	AD, AE, AF, AH, BA, BB, BH, DA, DB, EA, EB, FB, FF, HA, HB, HD, HE, HH

(b)

Sub-graph	FCISs
File Open	AD, AE, AF, AH, ABA, ABB, ABH, ABDA, ABDB, ABEA, ABEB, ABFB, ABFF, AB(E+D)HA, AB(E+D)HB, AB(E+D)HD, AB(E+D)HE

Note The total number of the arcs of an ESG with n nodes is n^2, including the loops (circles) on the nodes. Thus, there are n^2-m FEPs, given that the number of EPs is m

Note that the attribute "complete" within the phrase FCES may not imply that the exit node of the ESG must necessarily be reached; once the system has been conducted into a faulty state, it cannot accept further illegal inputs; in other words, an undesired situation cannot be even more undesired or a fault cannot be "faultier." Prior to further input, the system must recover, that is, the illegal event must be undone and the system must be conducted into a legal state through a backward or forward recovery mechanism. Note also that FEPs that include the entry symbol A, that is, AD, ..., AF, are also FCEPs, since they represent executable, faulty event sequences of the length two. Validating that the SUC really does not work the way the user does not expect (or does not wish) will be called *negative testing*.

A.7 Test Procedure Using CESs and FCESs

Finite-state-based techniques have been being used successfully by numerous research works and projects for many years, for example, for testing the conformance of protocols [SARI, BOCH]. An efficient test process using the ESG concept is summarized below:

- Construct the complete set of test cases, which includes all types of event sequences, that is, all CESs (for positive testing) and FCESs (negative testing) to produce the desired system responses and error messages, respectively (*Predictability* of the tests, defining oracles).
- Input CESs and FCESs to transfer the system into a legal or illegal state, respectively (*Controllability* of the tests through test cases).
- Observe the system output that enables a unique decision whether the output leads to a desired system response or an undesired, faulty event occurs that invokes an error message/warning, provided that a good exception handling mechanism has been implemented (*Observability* of the tests for determining pass/fail) (Sects. 1.3.3.4, 1.3.3.16.2, [9]).

If the steps 1–3 can be carried out effectively, we have a monitoring capability of testing process that leads to a high grade of testability. Monitoring requires a special structure of software, which must be designed carefully, considering the methods and principles of the modern Software Engineering (*Design for Testability*).

Note that in the test procedure represented in Sects. A.6 and A.7, event sequences of the length 2, that is, event pairs, have been considered to be covered. For a more comprehensive test, however, longer event sequences, for example, event triples, event quadruples, etc. can be considered. A comparison of the impacts of the sequence length to be covered on the test performance and aspects of costs and cost optimization of the approach can be found in the literature, for example, [3, 5].

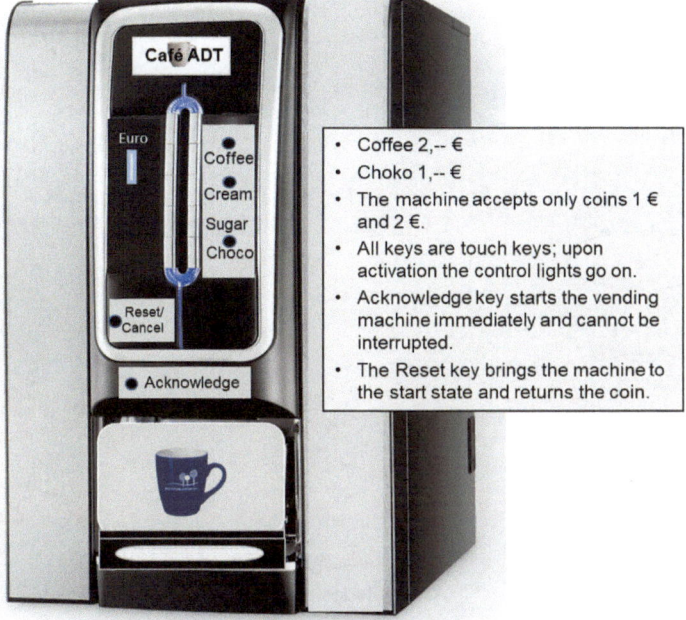

Fig. A.6 A simplified vending machine

A.8 An Experiment

The approach we described here has been used in different environments. The following simple example is meant to demonstrate its applicability to reactive systems.

The vending machine in Fig. A.6 operates with coins. The operation/transaction is initiated by the insertion of a 1- or 2-euro coin. The client is then able to make a selection (coffee or chocolate). In addition, the vending machine enables the client to add extra sugar and/or cream before the hot drink is poured into a cup.

The simplified functionality of the vending machine in Fig. A.6 is modeled in Fig. A.7 by an ESG. Test sequences can be generated by applying the test procedures to this ESG as outlined in Sect. A.7 of this Appendix.

Detected faults are listed in Table A.3. They represent potential faults because the tests were run hypothetically on the model and not on the real system that is under construction, and thus not yet available. This is a good thing, given that the costs of fixing these design flaws in later stages—let alone after implementation—would be several orders of magnitude higher than the costs of immediately correcting them while modeling the system and processes.

The reader might wonder about the multitude of dotted arcs in Fig. A.7 that represent illegal interactions. Those are possible faults. The ESG in Fig. A.7 has 12 nodes, thus 144 arcs (see the note at the end of Sect. A.6 of this Appendix). 29 of

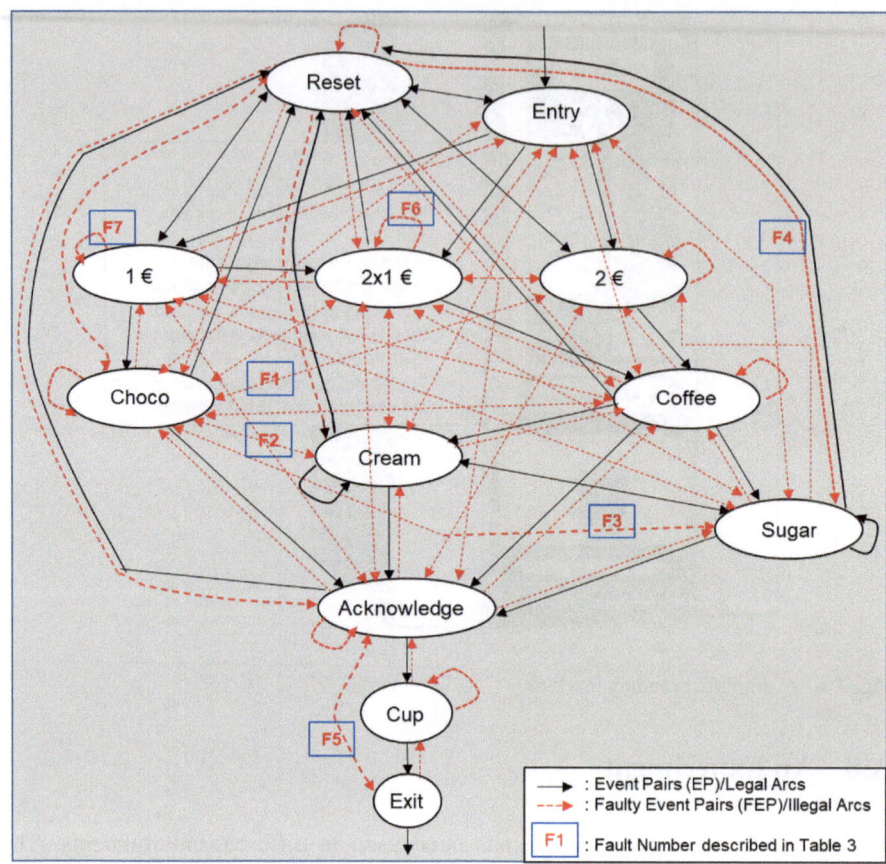

Fig. A.7 Completed ESG of the vending machine represented in Fig. A.6

Table A.3 Faults detected in vending machine

Fault 1	The construction omitted the case that chocolate should also be delivered when 2 Euros have been inserted. The user should then receive either one cup of coffee, or two consecutive cups of chocolate, or 1 Euro in return, dependent upon an additional selection that is missing in this design
Faults 2, 3	The machine should warn the user in case he/she wants cream (FEP on arc 2) and/or sugar (FEP on arc 3) with his/her chocolate, without knowing that sugar and cream are already included in the chocolate. The result often ruins its taste and flavour
Fault 4	The selection keys should be kept locked before a coin has been inserted. A display should inform the user appropriately
Fault 5	An additional sensor should ensure that a cup has been placed before the filling process starts. The sensor should also stop the filling process if the cup is removed before the process concludes
Faults 6, 7	A lock should exclude the multiple insertion of 1- and 2-Euro coins, except 1 Euro twice, which can be alternatively inserted instead of a single 2-Euro coin

them represent legal interactions, thus are EPs (The reader will hopefully accept the author's apologies for omitting some of the FEPs that make little sense). As can be seen, there are almost five times more chances of misusing this vending machine than operating it properly. This clearly explains why, in real life situations, clients of similar machines struggle and even box with them to get the commodity they desire and have already paid for!

This also clearly explains why it is so expensive to design and implement smart systems and processes that tolerate the faults made by their patrons and patiently help them get the goods they desire and reach an end that satisfies all. Note that handling each and every dotted line (FEP) causes additional programming effort. Nevertheless, this fault-tolerating behavior can be of crucial importance for software to be reused in critical applications.

References

1. Aho, A., Ullman, J.: The theory of languages. Math. Systems Theory **2**, 97–125 (1968)
2. Aho, A.V., Hopcroft, J.E., Ullman, J.D.: *Principles of Compiler Design*. Addison-Wesley (1977)
3. Belli, F., Budnik, Ch.B., White, L.: Event based modelling, analysis and testing of user interactions: approach and case study. In: Software Testing, Verification and Reliability 16/1, pp. 3–32 (2006)
4. Belli, F., Dreyer, J.: Program segmentation for controlling test coverage. In: Proceedings of the Eighth International Symposium on Software Reliability Engineering ISSRE '97, pp. 72–83 (1997)
5. Belli, F.: Finite state testing and analysis of graphical user interfaces. In: Proceedings 12th International Symposium on Software Reliability Engineering, pp. 34–43 (2001)
6. Chomsky, N.: On certain formal properties of grammars. Inform. Control **2**, 137–167 (1959)
7. David, R., Thevenod-Fosse, P.: Detecting transition sequences: application to random testing of sequential circuits. In: Proceedings of International Symposium Fault-Tolerant Computing FTCS-9, pp. 121–124 (1979)
8. Gluschkow, W.M.: *Theory of Abstract Automata* (in German), VEB Verlag der Wissensch., Berlin (1963)
9. Goodenough, J.B.: Exception handling—issues and a proposed notation. Comm. ACM 18/12, 683–696 (1975)
10. Kleene, S.C.: Representation of events in nerve nets and finite automata. In: Shannon, C.E., McCarthy, J. (eds.) Automata Studies, pp. 3–42. Princeton University Press, Princeton, New Jersey (1956)
11. Myhill, J.: Finite automata and representation of events. In: Fundamental Concepts in the Theory of Systems, Tech. Report No. 57-624, ASTIA Document No. AD 1557 41, Wright Air Development Center, Cincinnati-Ohio (1957)

12. Naito, S., Tsunoyama, M.: Fault detection for sequential machines by transition tours. In: Proceedings of FTCS, pp. 238–243 (1981)
13. Parnas, D.L.: On the use of transition diagrams in the design of user interface for an interactive computer system. In: 24th ACM National Conference, pp. 379–385 (1969)
14. Salomaa, A.: Theory of Automata. Pergamon Press, New York (1969)
15. Salomaa, A.: Formal Languages. Academic Press (1973)
16. Salomaa, A.: Events and languages. In: Calude, C.S. (ed.) People and Ideas in Theoretical Computer Science, pp. 253–274 (1999)
17. Shen, Y.N., Lombardi, F., Dahbura, A.T.: Protocol conformance testing using multiple UIO sequences. IEEE Trans. Commun. **40**, 1282–1287 (1992)
18. Shehady, R.K., Siewiorek, D.P.: A method to automate user interface testing using finite state machines. In: International Symposium on Fault-Tolerant Computing FTCS-27, pp. 80–88 (1997)

Appendix B—Integrating the System Modeling with Fault Modeling II: Concurrent Systems Using General Net Theory (Petri Nets)

In many situations, several processes can run at the same time without causing a critical situation. However, if these processes share some resources, critical situations can emerge, for example, bottlenecks in resource sharing, or more critical, deadlocks through mutual exclusions. Finite-state and sequential techniques, as reviewed in Appendix A, cannot model such processes and situations.

The General Net Theory (GNT), also called Petri Net Theory, was founded by Carl Adam Petri (1961, Bonn) to model and analyze concurrent systems and processes considering their critical features [10, 14]. GNT is not an interesting research area, but has become more and more attractive for industrial and commercial applications [5]. It presents a combination of algebraic and graphical methods, and a generalization of automata theory and formal languages. This appendix summarizes GNT and illustrates some kinds of Peri nets that have become popular in academia and practice.

This brief introduction is meant to help the reader in understanding the basics, especially in correctly and easily interpreting the figures and examples where Petri net notation is used in this book. It is not meant to replace a "crash course" in Petri nets.

B.1   Introductory Example

Imagine you have to write a report and need a special book. The consequence is (Fig. B.1): (i) You need to buy it. While taking a walk, (ii) you see this book in a

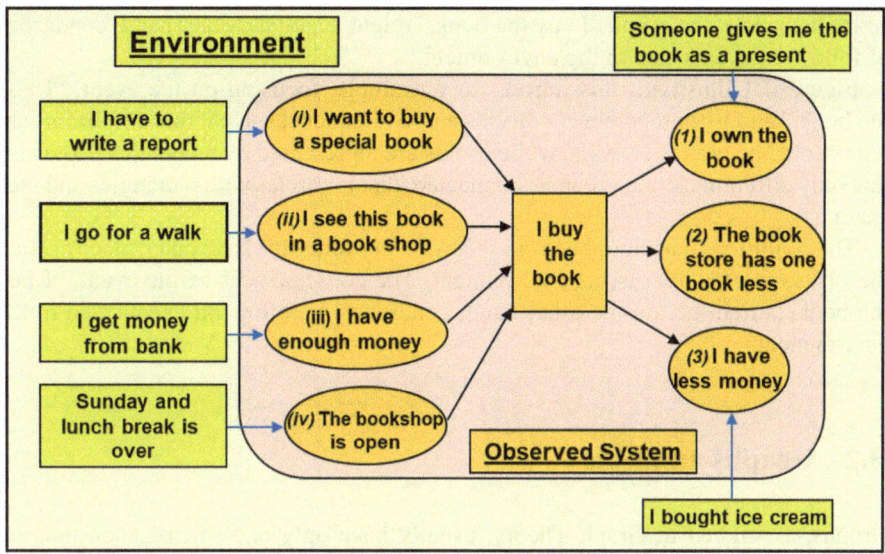

Fig. B.1 Introductory example for a condition-event Petri net

book shop you pass by. Since you have just been to the bank, (iii) you have enough cash with you. Furthermore, it is not lunch time or the weekend, so (iv) the bookshop is open. Given if and when all of these four conditions, (i)–(iv), are fulfilled, you can go into the shop and buy the book. As a consequence of this transaction, (1) you are now the new owner of this book; (2) the book shop has one book less; and, last but not least, (3) your stock of money has been reduced. Note that these three aftermaths of your action, (1)–(3), "I buy the book," emerge instantaneously at the same time. The conditions (i) to (iv), on the other hand, need not be fulfilled at the *same* time, that is, simultaneously. Each of them can occur at different points in time; we say, they occur "concurrently". To trigger your action, "I buy the book", they only need to be completed at *any* point of time.

You can, of course, own this book because someone, your aunt, for example, has given it to you as a present. If this really happened, you would then possess two copies of the book. Another reason why you have less money could be that you have bought some ice cream, which might have happened before, after, or at the same time you buy the book. If this really happened, then you will have an even less amount of money.

Figure B.1 illustrates this introductory example focusing on the event, "I buy the book," and its immediate conditions and consequences. Now, critical situations become visible. For example, if you do not have enough money, you cannot buy the book.

Note that conditions are represented by circles or ovals, while events are represented by rectangles. The conditions could have been caused by other events that have taken place outside the observed area, that is, the action's environment. The consequences of the event, "I buy the book," might be, on the other hand, conditions of follow-on events also in the environment.

Figure B.1 illustrates this introductory example, focusing on the event, "I buy the book," and its immediate conditions and consequences. Note that conditions are represented by circles or ovals, while events are represented by rectangles. Note also that only different elements can be connected, that is, circles with rectangles and vice versa.

The conditions may have been caused by other events that have taken place outside the observed area, that is, its environment. The consequences of the event, "I buy the book," might be, on the other hand, conditions of follow-on events also in the environment.

B.2   Graphs and Nets

Graphs, as studied in Graph Theory, usually have only one sort of node that are connected with each other by arcs (edges). As pointed out in the introductory example (Fig. B.1), General Net Theory introduces two different dual nodes (two-sorted nets): circles and rectangles [9, 13, 212]. These basic elements will be termed:

- s-elements of the set S (*states*), represented by circles, represented by a circle ○)
- t-elements of the set T (*transitions*), represented by rectangles,).

To form a directed net, we need a relation (called *flow*) between s-elements and t-elements, that is,

$f \in F: (S \times T) \cup (T \times S)$, represented by an arrow between s- and t-elements.

So, a *directed net* is a triple $N = (S, T, F)$ with

$S \cap T = \emptyset \wedge S \times T \neq \emptyset \wedge F \subseteq (S \times T) \cup (T \times S)$ and

$Domain(F) = S \cup T$ (no isolated elements are allowed).

B.3   Variety of Interpretations of the s- and t-Elements

The syntax of a specific Petri net was introduced in the last section. This is, briefly speaking, a calculation scheme without content. In the course of the research activities in General Net theory, many interpretations have been introduced that helped to formalize the semantic of some scientific areas ([1, 8, 9]; Table B.1).

It can be concluded that General Net Theory offers a graphical calculus that makes all these interpretations of Table B.1 coherent.

The next sections will review the 0th and 9th interpretation of GNT.

Table B.1 Interpretations of the s- and t-elements

#	s-element	t-element
0	State (place)	Transition
1	Condition	Event
2	Condition	Fact
3	Statement	Dependency
4	Model domain	Specification
5	Chemical compound	Chemical reaction
6	Channel	Office
7	Language	Translator
8	Product	Production activity
9	Predicate	Transition
...	...	...

Fig. B.2 A state-transition Petri net

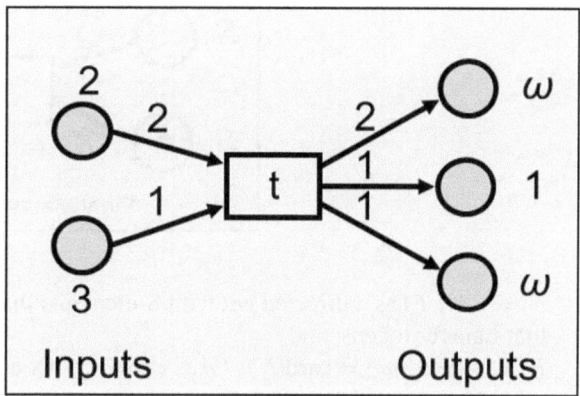

B.4   Interpretation of GNT: Place-Transition Nets

Place-transition (PT) or state-transition nets are one of the most popular interpretations of the GNT. They are also called (basic) Petri nets. The introductory example (Fig. B.1) represents, actually, a PT net, whereby the states are to be interpreted as conditions and transitions as events, bringing us to the 1st interpretation of GNT (Table B.1). Nevertheless, the introductory example lacks a very essential element that will be introduced as next.

A PT net includes "tokens" (or marks) placed in s-elements that enable the net to "live" by transferring those tokens from one place to another via corresponding transitions.

Formally, PT extends the notion of the directed net as follows (Fig. B.2):

$PT = (N, K, W, M_0)$ whereby.

Fig. B.3 A state-transition
Petri net, ready to be fired

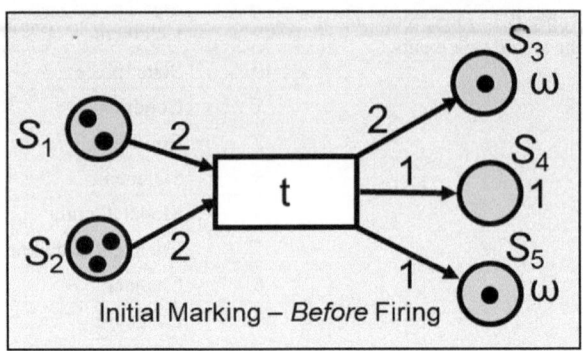

Fig. B.4 A state-transition
Petri Net of Fig. B.3, after
yielding a firing

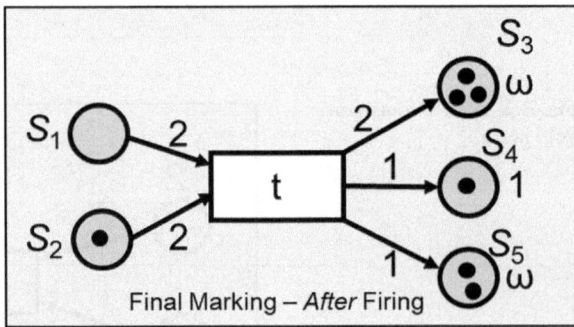

$N = (S, T; F)$ as a directed net with S-elements that carry tokens and T-elements
that transfer tokens, and
$K: S \rightarrow N \cup \{ \omega \} = \text{card}(N)$: *Token capacity* of s-elements with ω as "don't care,"
as many as you like, to resemble with the infinite symbol "∞", and
$W: F \rightarrow N$: *Transfer width* (channel width)
$M: S \rightarrow N \cup \{ \omega \}$: *marking* (distribution of the tokens in s-elements) with M_0: S
$\rightarrow N \cup \{ \omega \}$: *initial marking.*

By a "switch" ("trigger," "firing"), a token will be moved from an input place via
a transition to an output place, if and only:

- if all other (if existing) inputs have enough tokens (at least as many as required
  by the transfer or channel or width), and
- if all outputs still have enough space; that is, their token capacity will not be
  injured by the transition under consideration.

By "firing," the number of tokens in the inputs are reduced by the respective
transmission width and the number of tokens in the outputs are increased by the
respective transmission width. Thus, a switch transfers a marking to a succeeding
follow-on marking. See Figs. B.3 and B.4.

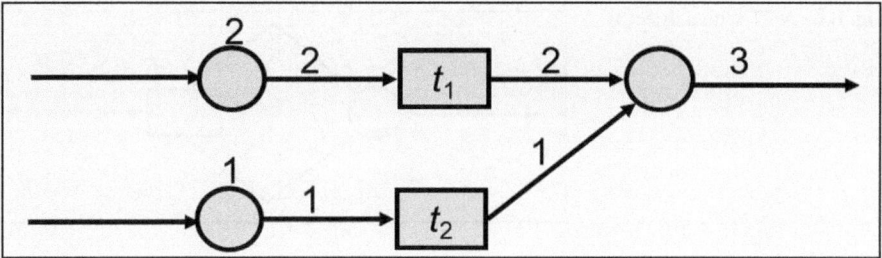

Fig. B.5 A PT with the concurrent processes t1 and t2

Note that a PT is a *condition event* net when the capacity of the *s*-elements and transfer (channel) width is fixed to one (*Single Stone Nets*).

Higher Petri nets have tokens that are distinguishable and enable algebraic notations and operations [6, 11].

Configuration of the tokens in the example represented in Fig. B.3 as *initial marking*:

$$c1 = \{(S_1, 2), (S_2, 3), (S_3, 1), (S_4, 0), (S_5, 1)\}$$

Configuration of the tokens in the example represented in Fig. B.4 as *end marking*:

$$c2 = \{(S_1, 0), (S_2, 1), (S_3, 3), (S_4, 1), (S_5, 2)\}$$

$c2 = \{ (S_1, 0), (S_2, 1), (S_3, 3), (S_4, 1), (S_5, 2) \}$

The marking $c2$ is *reachable* from $c1$ in one *step*, keeping the net *alive*.

B.5   Concurrency, Loops

PT enable to comfortably represent concurrent processes. Figure B.5 contains two processes, t1 and t2, that run concurrently. Note that t1 and t2 need not take place at the same time. All temporal combinations are possible; for example, t1 can run and then come to an end before t2 starts, or vice versa. It is also possible that t1 and/or will never take place because their places will not be "fed", causing the net to "die".

A *direct loop* (or *cycle*) can occur if a place is the initial and end place of a transition (Fig. B.6). Loops over a sequence of transitions help a PT to "live," that is, to switch a transition.

If the net contains no loops, it is "pure" (*loop-free*).

Fig. B.6 A PT with a direct
loop

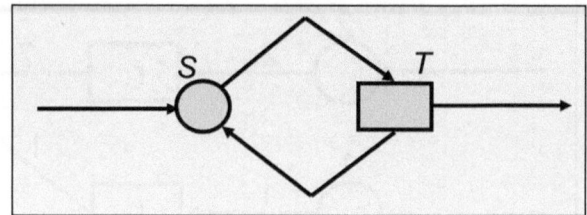

Fig. B.7 A PT with
potential for unsafety

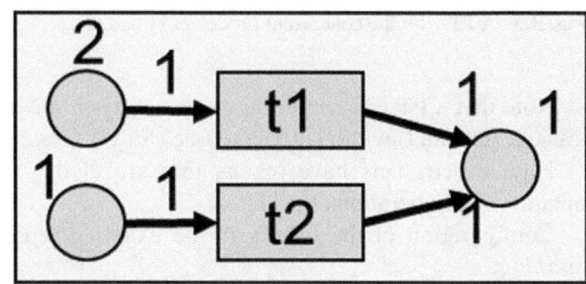

B.6 Critical Features of PT

The following properties are of interest for analyzing nets:

- A net is *safe* if the capacity of the *s*-elements is not exceeded by a firing. That means, the firing does not cause an overriding of the maximum number of tokens in any place (Fig. B.7).
- A *conflict* exists when firing a transition excludes another one.
- A *confusion* exists when firing a transition causes a second and a third transition to be in conflict.
- A net is *live* when the succeeding marking is also fireable ("deadlock free") after each switch.

Figure B.7 illustrates a situation where t1 and t2 are concurrent transitions. If t1 switches, t2 cannot switch since the output place is occupied, that is, already carries a token, and a second token would injure its max. capability of one.

Figure B.8 illustrates a situation where t1 and t2 are again concurrent transitions. They mutually exclude each other, because if t1 switches, t2 cannot switch, since the only input place has only one token that can fire t1 (exclusive) or t2.

Figure B.9 illustrates a situation where t1 and t2 are again concurrent transitions. While t1 has only one input, t2 has two inputs. T1 can fire if its input has a token. T2, on the other hand, needs one token from the input it shares with t1 and another two tokens from its input below the first one. So, t1 and t2 are not in conflict as long as t2's second input has no token. However, if t3 switches, this input will be fed and now carries the token t2 needs to be fired. Thus, a conflict of t1 and t2 emerges if and only if t3 switches.

Fig. B.8 A PT with potential for T-conflict

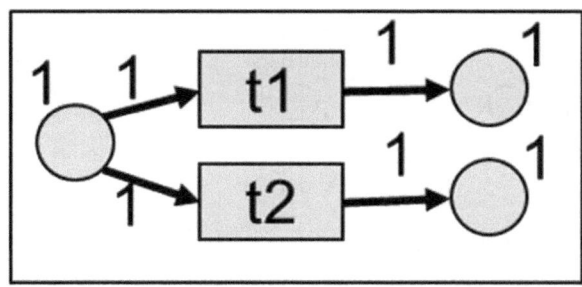

Fig. B.9 A PT with potential for P-conflict

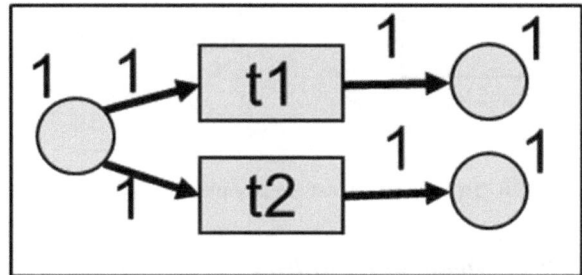

B.7 Some Theoretical Results

GNT has been the object of intensive research for many decades. Accordingly, there is a rich theory focusing on critical features. Some of them are summarized below:

- Liveliness and reachability are (recursively) equivalent properties. However, it is not known whether they are decidable.
- If they are decidable, the decision time increases exponentially with the input (components + marking). They are decidable for conflict-free networks.
- The languages represented by Petri nets are only closed under intersection (not under union, complement, K-cover, homomorphism).

B.8 Examples for PT

In this section, some well-known examples are selected to demonstrate the expressional power of PT.

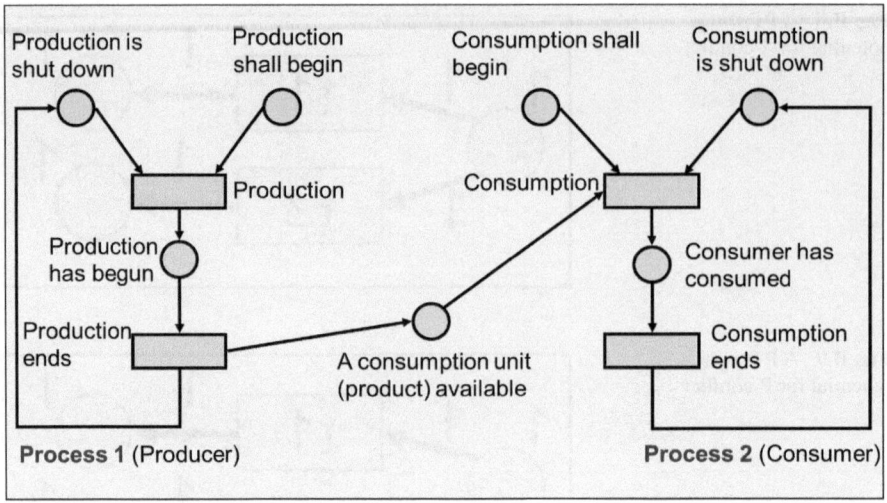

Fig. B.10 Two processes that are "coupled"

B.8.1 Process Coupling

Figure B.10 illustrates two processes that are coupled by a unit that enables an interaction.

The processes in Figure B.10 can be interpreted as *producer* and *consumer* that will be coupled by a product container for the consumption unit, which is available at the end of the production process.

B.8.2 Processes with "Buffer" and "Distributer"

Figure B.11 models two processes that are, as in Fig. B.10, coupled by a product container that can be large enough to store many products that will be distributed. Each of the entry transitions produce *m* products that can be buffered at a maximum of 2*m*. These goods will then be distributed equally by the "Distribute" transition to the following consumer transitions.

B.8.3 Two Concurrent Processes—Possible Interactions

The following example (Fig. B.12) represents again two processes; this time with an additional s-unit to visualize their interaction. Depending on which t-elements of

Fig. B.11 Processes with a "buffer" and "distributer"

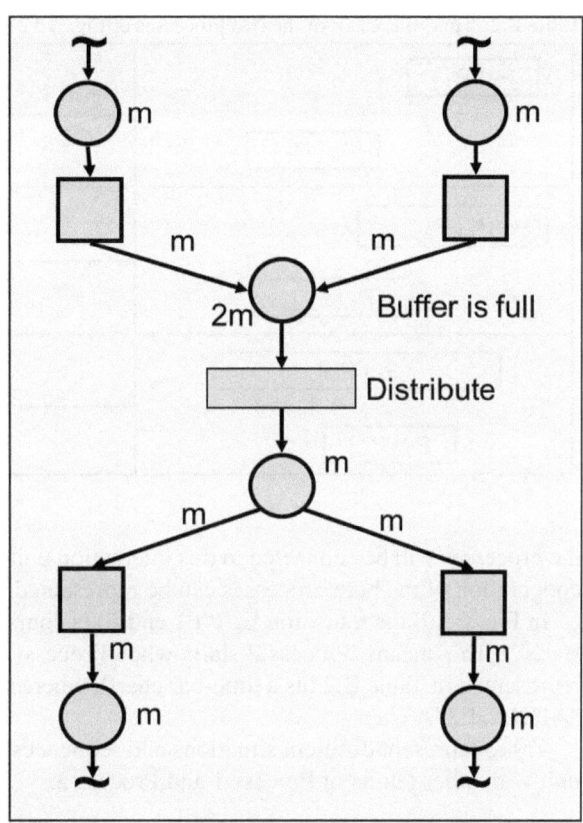

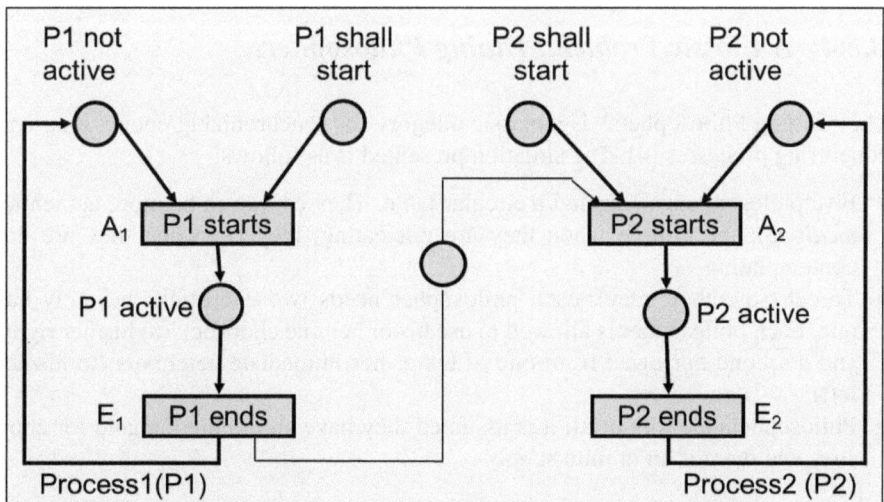

Fig. B.12 Two concurrent process

Table B.2   Time-Bar Chart of the two Processes of Fig. B.12

$P_1[P_2]$	$A_1\,E_1\,A_2\,E_2$	**X**
$P_2[P_1]$	$[A_2\,E_2\,A_1\,E_1]$	
$P_1[P_2]$	$A_1\,A_2\,E_1\,E_2$	
$P_2[P_1]$	$[A_2\,A_1\,E_2\,E_1]$	
$P_1[P_2]$	$A_1\,A_2\,E_2\,E_1$	
$P_2[P_1]$	$[A_2\,A_1\,E_1\,E_2]$	

the processes will be connected to this interaction unit, different constellations of the cooperation of the both processes can be represented.

In Fig. B.12, the transition E1 ("P1 ends") is connected to the transition A2 ("P2 starts"). This means, Process 2 starts when Process 1 ends. This event sequence is represented in Table B.2 (as a time-bar chart), where the first row is checked, that is, "A1E1A2E2".

Try to represent different situations and sequences by connecting the additional s unit with other t units of Process 1 and Process 2.

B.8.4   A Classic Problem: Dining Philosophers

The "Dining Philosophers" is a classic allegory for synchronizing cooperating and concurring processes [4]. The situation presented is as follows:

- Five philosophers sit around a circular table. They eat (or, to be more authentic, *meditate*); the infinite when they are not eating; they eat when they are not contemplating.
- To eat (say, chop suey), each philosopher needs two chopsticks but only has one. Each philosopher is allowed to use his or her one chopstick (to his/her right) and a second borrowed from one of his or her immediate neighbors (to his/her left).
- Philosophers are immortal; it is assumed they have an infinite demand for chop suey and there is an infinite supply.

Now, the task is to coordinate the eating/thinking process, so that

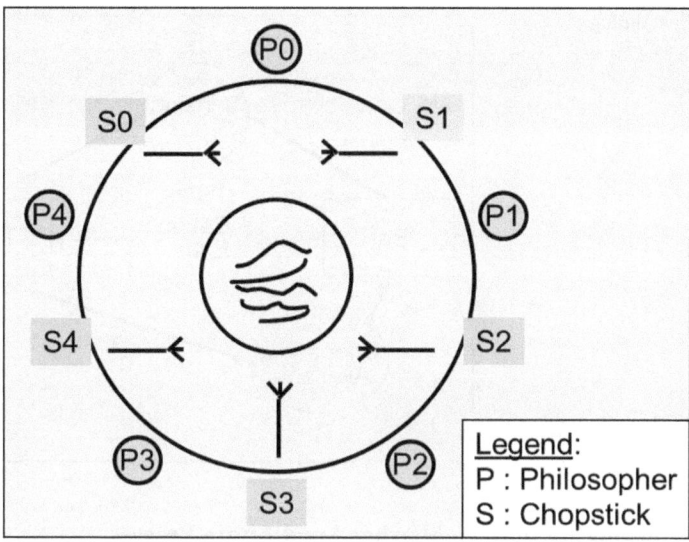

Fig. B.13 P1–P5 and the dish they share [7]

- No philosopher will starve, because, for example, the other two never put down their chopsticks to make them available to another philosopher.
- Two philosophers never obstruct each other. For example, one holds a chopstick and the other does not make his or her chopstick available (with the consequence they both starve).
- A philosopher is only allowed to take one of his or her immediately adjacent neighbors' chopsticks.

Figure B.13 illustrates the constellation of philosophers and the food they share.

B.8.4.1   Eating Process of an Individual Philosopher

The PT in Fig. B.14 represents the dining process in two cycles: one for eating, one for thinking. If a philosopher is thinking, he or she needs no chopstick. Two chopsticks are necessary to start eating. When done eating, the chopsticks are made available to other philosophers.

However, this representation has some flaws:

- The philosophers cannot be identified individually.
- Therefore, the assignment of philosophers to chopsticks is not unique. As a consequence, a philosopher can pick up any of the other philosophers' chopstick, not solely the chopstick of the neighbor to his or her left. (This will cause a synchronization problem.)

Fig. B.14 Five dining
philosophers as a PT [7]

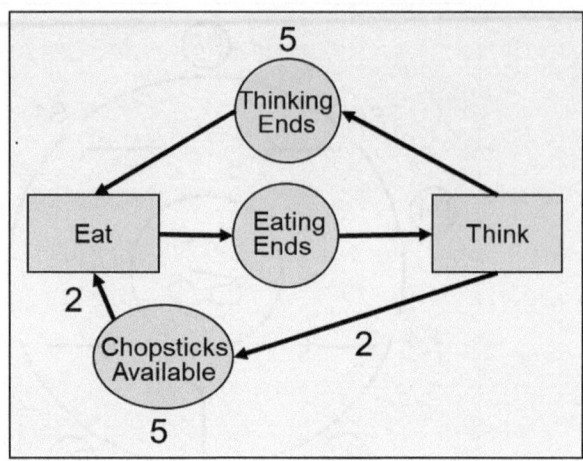

B.8.4.2 Considering the Neighborhood and Entire Process

The PT in Fig. B.15 can aid in representing the situation from the aspect of a single philosopher. As in Fig. B.14, if a philosopher under consideration (PUC) is thinking, no chopstick is necessary and the PUC puts his or her chopstick aside. Two chopsticks are necessary to re-start eating at the end of thinking: one of his or her own chopsticks and one set aside by the neighbor to the left. When done eating, both chopsticks are made available. In contrast to Fig.B.14, the interaction with other philosophers is realized by the chopstick of the neighbor to the right.

Finally, Fig. B.16 represents all five philosophers in action, considering also their interactions.

Note that the PTs depicted in Figs. B.14, B.15 and B.16 only represent, or describe, the configuration of the situation and the constellation of actors (philosophers) and the resources (chop suey) they share. These PTs are not meant to solve the following synchronization problems:

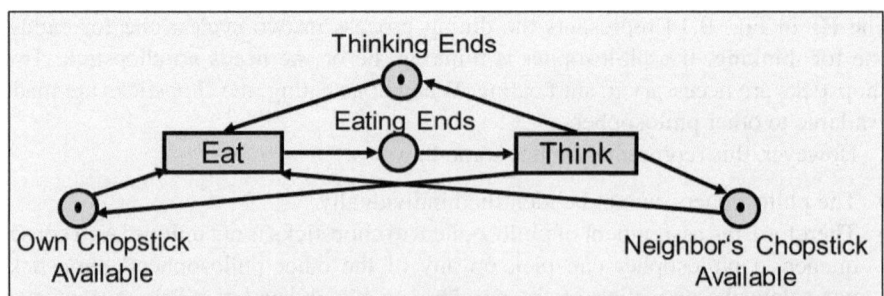

Fig. B.15 Eating process of an individual philosopher—considering chopsticks of neighbors at either sides

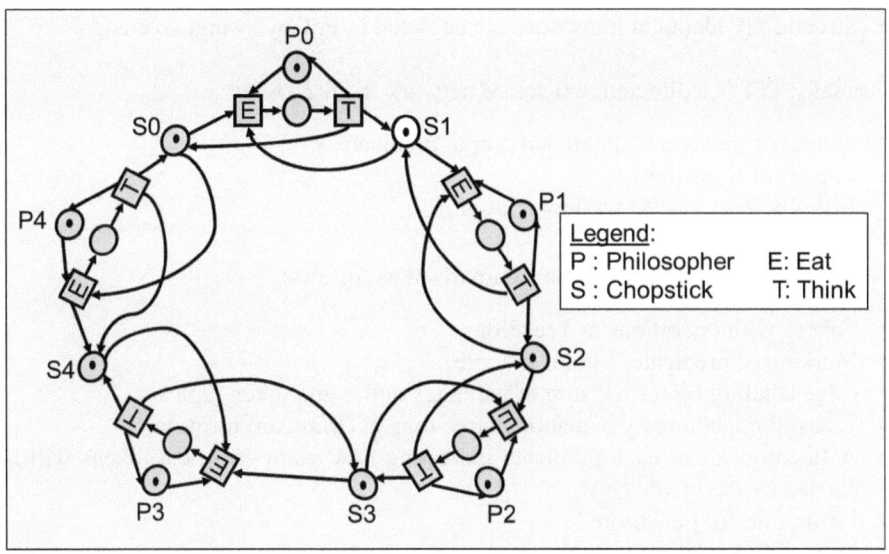

Fig. B.16 Five dining philosophers in action and interaction [7]

- Single-sided synchronization: Each philosopher must wait until a chopstick on a predefined side (right or left) has been put on the table. He/she is not allowed to pick up any available stick.
- Multi-Sided Synchronization: Several philosophers are not allowed to simultaneously access the same stick.

In addition, a good solution has to fulfill the following features:

- Avoid the mutual exclusion that arises when a philosopher holds a stick and never puts it down (releases).
- Fairness in the sense that all philosophers should be equally well-fed.

  Solutions using different strategies can be found in the literature [3, 4].

B.9 Predicate/Transition Nets (PrT)

The tokens used in PT are all the same, only their number in s-elements and channels is variable. Furthermore, transitions are only supposed to transfer the tokens from one place (location) to another. Genrich and Lautenbach added features that considerably extended the interpretation of PT [6]:

- In PT, tokens are indistinguishable. In PrT, tokens can be distinguished.
- Algebraic relationships can be formed between the input/output places to fire a transition in PrT.

- Structurally identical transitions are depicted in PrT by a single rectangle.

Formally, PrT is a directed two-sorted network (S, T, F) with:

S: a set of predicates (input and output predicates)
T: a set of transitions
$F \subseteq S \times T \cup T \times S$ is a flow relation.

Further properties of PrT can be summarized as follows:

- Tokens with operations and relations
- Marking of predicates via token tuple
- Edge labeling by formal sum of variables tuples and token tuples
- Transition labeling by a quantifier-free logical (Boolean) formula
- A function K for each predicate indicating how many identical tokens will be carried by the predicate
- Firing rule for transition T
- Logical formula (if available) in T
- Input predicates contain enough tokens
- No "overflow" in output predicates.

B.9.1 Example for PrT-Net

Figures B.17 and B.18 illustrate a simple PrT before and after firing [2].

The transition inputs consist of tokens a, b, and c in different constellations, represented by vectors (or tuples). Additionally, an anonymous token is also included in the upper left input. Transition requires the input tokens be grouped in lexical order, such as "a before b." If this requirement can be fulfilled, the transition can be "fired" by transferring tokens from the inputs in a specific order:

- First, two tokens of the same kind, x and an anonymous token $\cent$ ("+" does not mean "addition" but "putting together").
- Second, two tokens of different types, y and z, that also differ from the x token above.
- The outputs will be formed grouping the tokens as required by the labels on the output arcs.

B.9.2 Dining Philosophers as PrT

The example given in Sect. B.8.4 is represented in Fig. B.19 as a PrT.

Fig. B.17 Example for
PrT—initial marking [2]

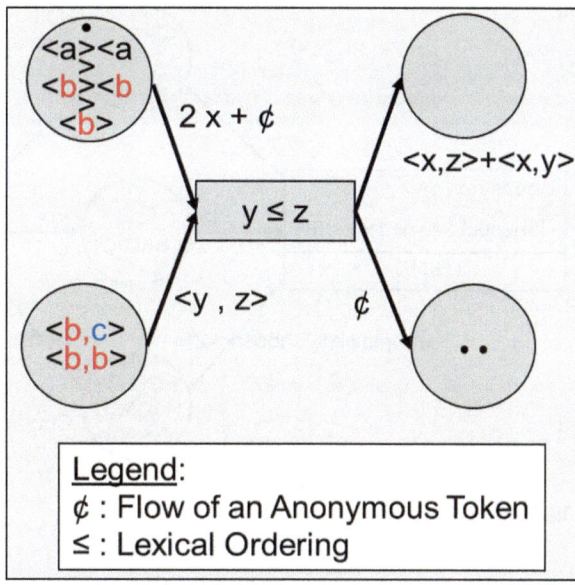

Fig. B.18 Example for
PrT—end marking [2]

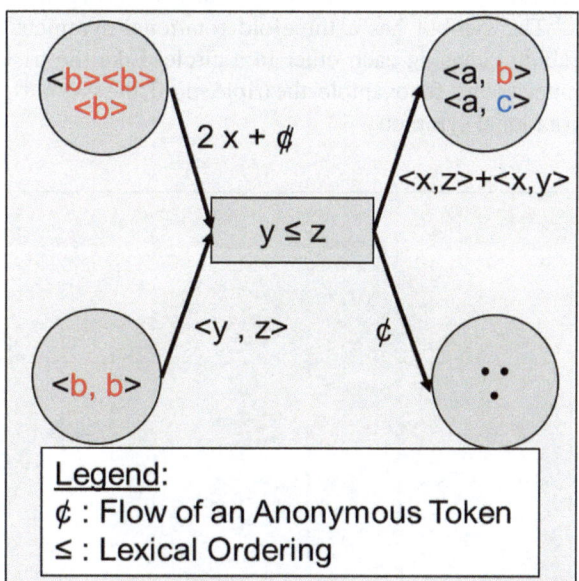

B.9.3 A Folkloristic Remark—The Three Hares

The three hares (or three rabbits) is a circular motif that originated in China, appearing
also in Mongolia and Iran. It has an ancient, mystical background and can be seen in

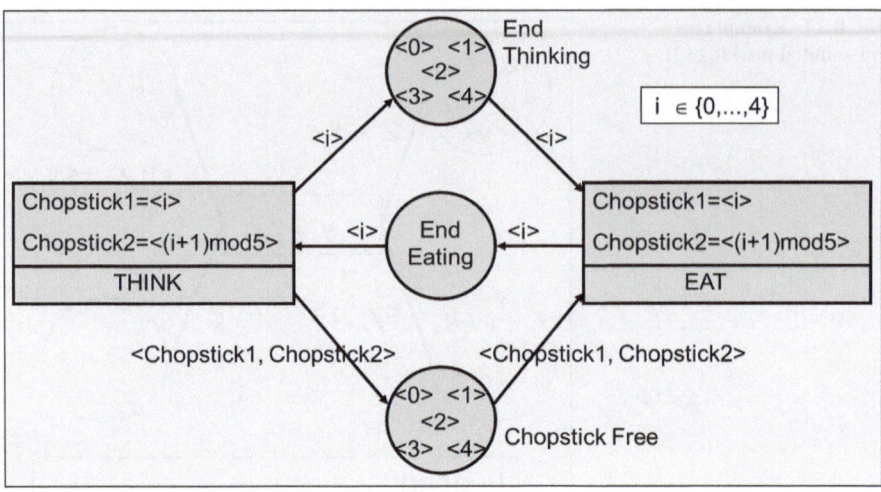

Fig. B.19 Dining philosophers as PrT [7]

sacred venues in some European churches scattered across Europe and Great Britain, for example, in the Cathedral of Paderborn (Fig. B.20).

The symbol has a threefold rotational symmetry and features three hares (or rabbits) chasing each other in a circle. Like the triskelion, the triquetra, and their antecedents, for example, the triple spiral, the symbol of the three hares has a threefold rotational symmetry.

Fig. B.20 Three hares (simplified version, based on three hares window in the Cathedral of Paderborn)

Each of the ears is shared by two hares, so that only three ears are actually shown. There are several interpretations of this motif, such as the Holy Trinity or the virtue of sharing precious possessions with others. The reader might excuse the author for his (mis)use of this motif as another example of coordination and cooperation that needs a precise synchronization, given that the hares depicted are in motion.

It is left to the reader to model the situation utilizing two Petri nets (i), one from the point of the view of one hare, and (ii) a second from the point of the view of an organizer, who seeks to synchronize the hares so that each always has two ears (see Figs. B.15 and B.16).

References

1. Bernardinello, L., De Cindio, F.: A survey of Basic Net Models and Modular Net Classes. LNCS vol. 609, pp. 304–351. Springer. (1992)
2. Belli, F., Großpietsch, K.-E.: Specification of fault-tolerant system issues by predicate/transition nets and regular expressions—approach and case study. IEEE Trans. Softw. Eng. **17**, 513–526 (1991)
3. Chandy, K.M., Misra, M.: The drinking philosophers problem. ACM Trans. Prog. Lang. Syst. **6**(4), 632–646 (1984)
4. Dijkstra, E.W.: Hierarchical ordering of sequential processes. Acta Informatica **1**(2), 115–138 (1971)
5. Fettke, P., Reisig, W.: Modelling service-oriented systems and cloud services with HERAKLIT. In: Proceedings of 16th International Workshop on Engineering Service-Oriented Applications and Cloud Services, pp. 28–30 (2020)
6. Genrich, H.J., Lautenbach, K.: System modelling with high-level Petri nets. Theor Comput Sci **13**(1), 109–135 (1981). https://doi.org/10.1016/0304-397 5(81)90113-4
7. Herrtwich, R.G., Hommel, G.: Nebenläufige Programme (Concurrent Programs, in German), Springer, Berlin (1994)
8. Murata, T.: Petri Nets: properties, analysis and applications. Proc IEEE **77**(4), 541–558 (1989). https://doi.org/10.1109/5.24143
9. Petri, C.A., Reisig, W.: Petri net. Scholarpedia 3(4), 6477 (2008). https://doi.org/10.4249/scholarpedia.6477
10. Petri, C.A.: Communication with Automata (in German), 1962, Ph.D. Dissertation, Univ. Darmstadt, 1962 for English version, *see Technical report RADC-TR-65-377*, Volume I, Final Report, Supplement I (1966), or *DTIC Research Report AD0630125*, or visit https://edoc.sub.uni-hamburg.de/informatik/vollte xte/2010/155/pdf/diss_petri_engl.pdf
11. Reisig, W.: Petri nets and algebraic specifications. Theor Comput Sci **80**(1), 1–34 (1991). https://doi.org/10.1016/0304-3975(91)90203-e
12. Reisig, W.: *Understanding Petri Nets*. Springer, Berlin (2013)
13. Rozenburg, G., Engelfriet, J.: Elementary net systems. In: Reisig, W., Rozenberg, G. (eds.) Lectures on Petri Nets I: Basic Models—Advances in Petri Nets, Lecture Notes in Computer Science, 1491, pp. 12–121. Springer, Berlin (1998)

14.   Silva, M.: 50 years after the Ph.D. thesis of Carl Adam Petri: a perspective. In: IFAC Proceedings Volumes, vol. 45, issue 29, pp. 13–20 (2012). https://doi. org/10.3182/20121003-3-MX-4033.00006

Appendices—Part II

Appendix A—Method for Setting Generic Ecodesign Requirements (Acc. to *Ecodesign Directive* [1])

Generic ecodesign requirements aim at improving the environmental performance of products, focusing on significant environmental aspects thereof without setting limit values. The method referred to in this Annex must be applied when it is not appropriate to set limit values for the product group under examination. The Commission must, when preparing a draft implementing measure to be submitted to the Committee referred to in Article 19(1), identify significant environmental aspects which must be specified in the implementing measure. In preparing implementing measures laying down generic ecodesign requirements pursuant to Article 15, the Commission must identify, as appropriate to the product covered by the implementing measure, the relevant ecodesign parameters from among those listed in Part 1, the information supply requirements from among those listed in Part 2 and the requirements for the manufacturer listed in Part 3.

Part 1. Ecodesign parameters for products

1.1.   In so far as they relate to product design, significant environmental aspects must be identified with reference to the following phases of the life cycle of the product: (a) raw material selection and use; (b) manufacturing; (c) packaging, transport, and distribution; (d) installation and maintenance; (e) use; and (f) end-of-life, meaning the state of a product having reached the end of its first use until its final disposal.

1.2.   For each phase, the following environmental aspects must be assessed where relevant: (a) predicted consumption of materials, of energy and of other resources such as fresh water; (b) anticipated emissions to air, water or soil; (c) anticipated pollution through physical effects such as noise, vibration, radiation, electromagnetic fields; (d) expected generation of waste material; and (e)

© The Editor(s) (if applicable) and The Author(s), under exclusive license
to Springer Nature Switzerland AG 2021
F. Belli and F. Quella, *A Holistic View of Software and Hardware Reuse*,
Studies in Systems, Decision and Control 315,
https://doi.org/10.1007/978-3-030-72261-6

possibilities for reuse, recycling and recovery of materials and/or of energy, taking into account Directive 2002/96/EC.

1.3. In particular, the following parameters must be used, as appropriate, and supplemented by others, where necessary, for evaluating the potential for improving the environmental aspects referred to in point 1.2:

(a) weight and volume of the product;

(b) use of materials issued from recycling activities;

(c) consumption of energy, water and other resources throughout the life cycle; EN 31.10.2009 Official Journal of the European Union L 285/23

(d) use of substances classified as hazardous to health and/or the environment according to Council Directive 67/548/EEC of 27 June 1967 on the approximation of laws, regulations and administrative provisions relating to the classification, packaging and labelling of dangerous substances (1) and taking into account legislation on the marketing and use of specific substances, such as Council Directive 76/769/EEC of 27 July 1976 on the approximation of the laws, regulations and administrative provisions of the Member States relating to restrictions on the marketing and use of certain dangerous substances and preparations (2) or Directive 2002/95/EC;

(e) quantity and nature of consumables needed for proper use and maintenance;

(f) ease for reuse and recycling as expressed through: number of materials and components used, use of standard components, time necessary for disassembly, complexity of tools necessary for disassembly, use of component and material coding standards for the identification of components and materials suitable for reuse and recycling (including marking of plastic parts in accordance with ISO standards), use of easily recyclable materials, easy access to valuable and other recyclable components and materials; easy access to components and materials containing hazardous substances;

(g) incorporation of used components;

(h) avoidance of technical solutions detrimental to reuse and recycling of components and whole appliances;

(i) extension of lifetime as expressed through: minimum guaranteed lifetime, minimum time for availability of spare parts, modularity, upgradeability, reparability;

(j) amounts of waste generated and amounts of hazardous waste generated;

(k) emissions to air (greenhouse gases, acidifying agents, volatile organic compounds, ozone depleting substances, persistent organic pollutants, heavy metals, fine particulate and suspended particulate matter) without prejudice to Directive 97/68/EC of the European Parliament and of the Council of 16 December 1997 on the approximation of the laws of the Member States relating to measures against the emission of gaseous and particulate pollutants from internal combustion engines to be installed in non-road mobile machinery (3);

(l)   emissions to water (heavy metals, substances with an adverse effect on the oxygen balance, persistent organic pollutants); and (m) emissions to soil (especially leakage and spills of dangerous substances during the use phase of the product, and the potential for leaching upon its disposal as waste).

Part 2. Requirements Relating to the Supply of Information

Implementing measures may require information to be supplied by the manufacturer that may influence the way the product is handled, used or recycled by parties other than the manufacturer. This information may include, as applicable:

(a)   information from the designer relating to the manufacturing process;
(b)   information for consumers on the significant environmental characteristics and performance of a product, accompanying the product when it is placed on the market to allow consumers to compare these aspects of the products;
(c)   information for consumers on how to install, use and maintain the product in order to minimise its impact on the environment and to ensure optimal life expectancy, as well as on how to return the product at end-of-life, and, where appropriate, information on the period of availability of spare parts and the possibilities of upgrading products; and
(d)   information for treatment facilities concerning disassembly, recycling, or disposal at end-of-life. Information should be given on the product itself wherever possible. This information must take into account obligations under other Community legislation, such as Directive 2002/96/EC. EN L 285/24.

Official Journal of the European Union 31.10.2009; (1) OJ 196, 16.8.1967, p. 1.; (2) OJ L 262, 27.9.1976, p. 201.; (3) OJ L 59, 27.2.1998, p. 1.

Part 3. Requirements for the Manufacturer

1.   Addressing the environmental aspects identified in the implementing measure as capable of being influenced in a substantial manner through product design, manufacturers of products must perform an assessment of the product model throughout its lifecycle, based upon realistic assumptions about normal conditions and purposes of use. Other environmental aspects may be examined on a voluntary basis.

On the basis of this assessment, manufacturers must establish the product's ecological profile. It must be based on environmentally relevant product characteristics and inputs/outputs throughout the product life cycle expressed in physical quantities that can be measured.

2.   Manufacturers must make use of this assessment to evaluate alternative design solutions and the achieved environmental performance of the product against

benchmarks. The benchmarks must be identified by the Commission in the implementing measure on the basis of information gathered during the preparation of the measure. The choice of a specific design solution must achieve a reasonable balance between the various environmental aspects and between environmental aspects and other relevant considerations, such as safety and health, technical requirements for functionality, quality, and performance, and economic aspects, including manufacturing costs and marketability, while complying with all relevant legislation. EN 31.10.2009 Official Journal of the European Union L 285/25.

Appendix B—Practical Rules for Recycling-Oriented Product Design (Extended from VDI 2243)

Generally the following rules for example from VDI 2243 will be found in nearly all guidelines or standards of companies although the validity of some of them could be discussed. Many companies already have developed more general rules from these including their own requirements for environmentally compatible product design which might be better suited for the manufacturing company and especially will focus on the experience of the manufacturers' engineers over time.

Building structure

- Provide functional modular structure.
- Prefer horizontal structures.
- Place components suitable for circulation and/or components and materials to be disassembled, accessible and easy to be disassembled.
- Guarantee easy removal and accessibility of auxiliaries like oils and their recovery. Construct and arrange cable harnesses and electrical nets that are easily removable.
- E&E components/boards easy accessible/removable and, if possible, in the highest level of disassembly.

Add-ons from the authors:

- Avoid batteries, if possible substitute by capacitors.
- Design worn elements easy exchangeable, if possible exchange only the worn element.
- Potentially mark those components which will be reusable.

Materials and surfaces

- Marking of materials according to relevant standards.
- Avoidance of substances critical for recycling, and of hazardous and harmful substances.[1]

---

[1]Remark of the authors: This will depend on the specific application, see checklist of the E&E associations together with EERA.

- Application of materials economically recyclable as materials.[2]
- Application of materials compatible with others for recycling following instructions of the manufacturer in modules/boards especially with composite materials.[3]
- Reduction of material diversity and material harmonization in modules/boards.
- Design surface coatings if possible compatible for recycling together with carrier material/substrate.[4]
- Application of coatings and paints compatible for recycling.[5]
- With plastic parts plan for use of recycled plastics and avoid metallic inserts.[6]
- Apply halogen-free printed circuit boards.[7]

Add-ons by the authors:

- By laser inscription recyclability will not be influenced.
- Design surfaces so that they don´t become unattractive after aging or so they could be cleaned easily for reuse.
- Reduction of the mass of materials (optimization of thickness, combination of parts, substitution of mechanical elements by electronic functions).
- Avoidance of unnecessary packaging.
- Application of standard components could increase the chances for recovery for certain materials.
- Avoid labels or select them to be compatible for recycling together with the plastic they are attached to.

Disassembly and connection technology

- Minimize number and diversity of connecting elements.[8]
- Harmonize connecting elements.
- Provide consistent direction of disassembly, preferably axial in direction of disassembly.
- Design nondestructively planned connections simply solvable, also after the designed working life still identifiable and accessible.
- Prefer snapping connections if possible ahead of screws.

---

[2]In E&E industry this is often not profitable. In case of low volumes there will be no market. Many components in E&E industry cannot be bought compatible if only one manufacturer exists worldwide and doesn't like to change the product.

[3]See Footnote 2.

[4]Avoid completely if possible! Stainless steel has advantages and can be inscribable by laser.

[5]See Footnote 4.

[6]For mass products sufficient volumes are not available at the market.

[7]There are many problems (available volumes might be too small, material might not be applicable for high frequency purposes, manufacturing processes not identical in Europe and in Japan) to be solved, although substitution in most cases should be possible!

[8]Remark of the authors: There were already products like PCs which could be produced without having tight connection technology. The components were laid into a form and pressed simultaneously on the contacts situated underneath.

- Minimize unsolvable connection technologies for example welding, riveting, bonding—and apply if possible only material compatible for recycling.
- Provide application of standard disassembly tools and guarantee accessibility.
- For assembled printed circuit boards apply snapping connections situated outside and avoid screw connections.
- Design fixing elements for electromechanical components, accessible without existing power supply.

Removal of the auxiliaries

- Operating liquids have to be removable independently from each other: Simple, quick and complete.
- Care for drain openings and their good visibility, guarantee accessibility.
- If no possibility for draining exists, a mark should be set showing how and where the component containing the liquid could be opened. Then it can be destroyed.

Appendix C—Checklist for the Reuse and Update of SW in a HW Component or Product for Resale, esp. for "Qualified-as-Good-as-New" Components of IEC 62309 (According to IEC/PAS 62814:2012-12 [26])

Background: The SW to be reused might be necessary to run on HW equipment which should be remanufactured. Without knowing the necessary old and new functions an upgrade of the equipment to the modern state of the art will be difficult.

The following checklist should be used to qualify the reused and updated SW and should be seen in addition to IEC 62309. Of course the preconditions in the main part of this SW PAS should also be valid.

3.1 State of the SW in the old equipment
3.2 Which SW upgrades are available and which would fit to the old and to the upgraded equipment?
3.3 Which updates have been done with the old SW?
3.4 Is compatibility of upgrades or reused SW with the remanufactured product already tested?
3.5 Which HW components (new/reused) should be integrated and which new functions should be available in the new HW product?
3.6 Is there a need to develop new SW program steps for some HW components which are new in the upgraded/remanufactured HW product?
3.7 Is the number of new HW components to be used with the remanufactured product limited, should it be limited, because problems with SW could occur?
3.8 Which HW components cannot be controlled by the old, upgraded or reused SW in the new HW product?
3.9 Are there different standards, for example, transfer rates etc. for data in components used?

3.10   In a network for example in a production line the product might have worked in a network. Is the reused/upgraded product and its SW compatible to the network of the, for example, production line?

3.11   Is testability of the whole product/system possible?

3.12   If the HW product/component will be used in a network in a plant or with for example internet what are frame conditions for the SW reuse?

3.13   Can installation media for old data also be used with the updated HW product (CD, etc.)?

Appendix D—Checklists for SW Reuse and Environmentally Compatible SW (According to IEC/PAS 62814:2012-12 [26])

In IEC 62309 all those components were excluded which contained, for example, hazardous substances, or had a too high energy consumption in comparison to the new product in the market. Additionally, legal trends worldwide will require to reduce especially energy consumption. Also for customers the benefit from reused, cheaper equipment could become a real disadvantage if energy consumption was too high.

Therefore, a second checklist was developed to evaluate the state of energy consumption by the application reused SW. Also this checklist should be used in addition to a qualification of HW components for new applications according to IEC 62309. Anyway the operational conditions of the HW need to be known. So some of the questions below might not be valid or others had to be put.

Potential influence on energy consumption of reused SW in new HW components and products and in "qualified-as-good-as-new" components

4.1   Is it necessary or possible to integrate energy saving elements in the SW, for example, automatic pull down to stand-by?

4.2   Are loading commands to batteries, capacitors etc. checked for energy consumption?

4.3   Which standards are available to estimate comparable modern energy consumption for equipment such as, for example Energy Star?

4.4   Is the energy consumption of the upgraded equipment much higher with the upgraded reused SW and how can it be reduced?

4.5   Is the run time for some tasks too long?

4.6   Energy consumption in comparison with new HW equipment—Can a too high consumption be reduced by a new SW approach?

4.7   Is a combination of new HW and SW necessary to reduce energy consumption, for example, a combination of switchable power supplies?

4.8   Are the impacts by the components, like printers, checked for energy consumption?

4.9   Can HW in the upgraded product be substituted by SW, for example external fax, now integrated?

4.10   Can HW be simplified, for example, batteries substituted with capacitors?

4.11   Can the system (product + SW) be simply tested?

4.12   Is it easy to use the product and its SW?

4.13   Are there recommendations available for an energy saving mode of the product?

4.14   Can other consumables besides energy like paper consumption, water consumption or ink consumption be reduced by changes in SW?

4.15   Which energy consumption is caused by the running of the SW itself? How can it be reduced?

Appendix E—Aspects Regarding the Harmonization of the State of the HW and Potential States of the SW During the Process of Refurbishment of a Product or System (According to IEC/PAS 62814:2012-12 [26])

5.1   Is it tested which state of the HW is available in the system to be reused? Usually the "Device Master Record" of the manufacturer defines the requirements.

5.2   Which versions of the SW fit to the actual state of the HW? Usually the "Device History Records" of the manufacturer explain which version of the SW fits to which state of the HW.

5.3   Is the update process for the SW defined? Usually there will be a recommendation of the manufacturer for this purpose. Remark: Sometimes the SW updates have to be implemented in a special order.

5.4.   Was the system upgrade tested?

5.5   Were all actions documented that are connected with the update of the SW?

5.6   Is a deposit available for tested (quality, energy consumption or optimized runtime) SW components (also reused)?

Appendix F—List of Standards

Safety

IEC 61508: 2010-4 Functional Safety of Electrical/Electronic/Programmable Electronic Safety-related Systems (E/E/PE, or E/E/PES.

IEC/PAS 62814:2012-12 Dependability of software products containing reusable components—Guidance for functionality and tests. Geneva/Switzerland: Bureau Central de la Commission Electrotechnique Internationale—IEC 62628:2012 Guidance on software aspects of dependability.

ISO 26262: 2011-1 Road vehicles—Functional safety.

EN 50128: 2012-03 Railway applications—Communication, signalling and processing systems—Software for railway control and protection systems.

IEEE 1633-2008 Recommended Practice on Software Reliability.

IEEE 610.12-1990—IEEE Standard Glossary of Software Engineering Terminology.

ANSI AIAA-013-1992 Software Reliability.

Software Development

IEC 62304: 2006-05 Medical device software—Software life cycle processes.

ISO/IEC 7498-1: 1994 Information technology—Open Systems Interconnection—Basic Reference Model: The Basic Model.

ISO/IEC 12207: 2017 Systems and software engineering—Software life cycle processes.

Ecodesign rules.

IEC 62430:2009-02 Environmentally conscious design for electrical and electronic products and systems. Geneva/Switzerland: Bureau Central de la Commission Electrotechnique Internationale. ISSN 2-8318-1032-7.

VDI 2243:2002-07 Recycling-oriented product development. VDI guideline. Düsseldorf: VDI—Association of German Engineers.

ISO/TR 14062:2002-11 Environmental management—Integrating environmental aspects into product design and development. Geneva/Switzerland: ISO—International Organization for Standardization.

Labels

ISO 14021:2016 Environmental labels and declarations—Self declared environmental claims (Type II environmental labeling). Geneva /Switzerland: ISO—International Organization of Standardization.

ISO 14025:2006 Environmental labels and declarations—Type III environmental declarations—Principles and procedures. Geneva /Switzerland: ISO—International Organization of Standardization.

Management

ISO 9001:2015-09 Quality management systems—Requirements. Geneva/Switzerland: ISO—International Standardization Organization.

ISO 14001:2015-11 Environmental management systems—requirements with guidance for use. Geneva/Switzerland: ISO—International Standardization Organization.

ISO 26000: 2010 Guidance on social responsibility, Geneva/Switzerland: ISO—International Standardization Organization.

Reuse

ANSI/RIC 001.12016: Specifications for the process of remanufacturing https://www.remancouncil.org/files/5fdeCD/RIC2015-Specifications-for-the-Process-of-Remanufacturing-Draft-11-25-151.pdf, Remanufacturing Industries Council.

1335 Jefferson Road #20157, Rochester, NY 14602-0157.

BS 8887-220:2010-03 Design for manufacture, disassembly and end-of-life processing (MADE), Design for manufacture, assembly, disassembly and end-of-life processing (MADE). The process of remanufacture. Specification. BSI London/UK.

COCIR (www.cocir.org, Association of manufacturers of medical equipment). Cocir Industrial Standard: Good Refurbishment Practice (GRP), Brussels/Belgium, 6/2009.

IEC 62309:2004-07 Dependability of products containing reused parts—Requirements for functionality and tests. Geneva/Switzerland: Bureau Central de la Commission Electrotechnique Internationale. ISBN 2-8167-7668-0.

ISO 8887-1:2017 Technical product documentation—Design for manufacturing, assembling, disassembling and end-of-life processing—Part 1: General concepts and requirements, International Organization for Standardization, Geneva, Switzerland.

ISO/TR 14062:2002-11 Environmental management—Integrating environmental aspects into product design and development. Geneva/Switzerland: ISO—International Organization for Standardization.

ISO/IEC 24700: 2005-07 Quality and Performance of Office Equipment that contains reused components. Geneva/Switzerland: ISO—International Standardization Organization.

VDI 2343 Sheet 4:2012-01 Recycling of electrical and electronic products—marketing. VDI guideline. Düsseldorf/Germany: VDI—Association of German Engineers.

Quality, dependability

ISO/IEC 25000:2014 International Standardization Organization, Software engineering systems and Software quality requirements and evaluation (SQuaRE)—Guide to Systems and to SQuaRE.

IEC/PAS 62814:2012-12 Dependability of software products containing reusable components. Guidance for functionality and tests. Geneva/Switzerland: Bureau Central de la Commission Electrotechnique Internationale. ISBN 978-2-83220.

IEC 24700:2005-07-01 Quality and performance of office equipment that contains reused components, Geneva/Switzerland: Bureau Centrale de la Commission Electrotechnique Internationale.

IEC 60300-1: 2014, Dependability management—Part 1: Guidance for management and application Geneva/Switzerland: Bureau Central de la Commission Electrotechnique Internationale. ISBN 978-2-8322-1777-1.

Energy consumption

ISO/IEC 14 756:1999 Information technology—Measurement and rating of performance of computer-based software systems, ISO copyright office, CH 1211 Geneve 20, Casa Portale 56. 1999.

Literature Part I

1. de Almeida, A., Lucredio, D., Garcia, V.C., de Lemos Meira, S.R.: A survey on software reuse processes. In: Proceedings of IEEE International Conference on Information Reuse and Integration (IRI), pp. 66–71 (2005). https://doi.org/10.1109/IRI-05.2005.1506451
2. Acar, H., Benfenatki, H., Gelas, J.-P., da Silva, C.F., Alptekin, G., Benharkat, A.-N., Parisa Ghodous, P.: Software greenability: a case study of cloud-based business applications provisioning. In: Proceedings IEEE 11th International Conference on Cloud Computing (CLOUD), pp. 875–878 (2018). https://doi.org/10.1109/CLOUD.2018.00125.hal-01887065
3. Apel, S., Batory, D., Kästner, Ch., Saake, G.: Feature-Oriented Software Product Lines. Springer, Berlin, Heidelberg (2013)
4. Addy, E.A.: A framework for performing verification and validation in reuse-based software engineering. Annals Softw. Eng. **5**, 279–292 (1998). https://doi.org/10.1023/A:1018968222862
5. Aho, A.V., Dahbura, A.T., Lee, D., Uyar, M.Ü.: An optimization technique for protocol conformance test generation based on UIO sequences and rural chinese postman tours. IEEE Trans. Commun. **39**, 1604–1615 (1991)
6. Aho, A., Ullman, J.: The theory of languages. Math. Systems Theory **2**, 97–125 (1968)
7. Aho, A.V., Hopcroft, J.E., Ullman, J.D.: Principles of Compiler Design. Addison-Wesley (1977)
8. AIAA R-013-1992 Recommended. Practice: Software. Reliability (1992)
9. Aiello, R.: Configuration Management Best Practices: Practical Methods that Work in the Real World. Addison-Wesley (2010)
10. Avizienis, A., Laprie, J.-C., Randell, B., Landwehr, C.: Basic concepts and taxonomy of dependable and secure computing. IEEE Trans. Depend. Secure Comput. **1**(1) (2004). Available at https://ieeexplore.ieee.org/xpls/abs_all.jsp?arnumber=1335465
11. Ammann, P., Offutt, J.: Introduction to Software Testing. Cambridge University Press, UK (2008)
12. Anderson, T., Randell, B.: Computing Systems Reliability—An Advanced Course. University of Newcastle upon Tyne, Cambridge University Press (1979)
13. Antes, G.: Open-source software no longer optional. Comm. ACM **59–8**, 15–17 (2016)
14. Ayav, T., Belli, F.: Boolean differentiation for formalizing myers' cause-effect graph testing technique. In: 2015 IEEE International Conference on Software Quality, Reliability and Security—Companion, pp. 138–143 (2015)
15. Bashari, M., Bagheri, E., Du, W.: Dynamic software product line engineering: a reference framework. Int. J. Softw. Eng. Knowl. Eng. **27**(2), 191–234 (2017). https://doi.org/10.1142/S0218194017500085

F. Belli and F. Quella, *A Holistic View of Software and Hardware Reuse*, Studies in Systems, Decision and Control 315, https://doi.org/10.1007/978-3-030-72261-6

16. Bertolino, A., Fantechi, A., Gnesi, S., Lami, G.: Product line use cases: scenario based specification and testing of requirements. In: Käkölä, T., Duenas, J.C. (eds.) Proceedings of Software Product Lines Research Issues in Engineering and Management, Chap. 11, pp. 425–445. Springer, Berlin (2006)
17. Bach, J., Schroeder, P.J.: Pairwise Testing: A Best Practice That Isn't. Available at http://www.testingeducation.org/wtst5/PairwisePNSQC2004.pdf
18. Badareen, A.B., Selamat, M.H., Jabar, M.A., Din, J., Turaev, S.: Reusable software component life cycle. Int. J. Comput. 5(2), 191–199 (2011). Available at www.naun.org/journals/computers/19-863.pdf
19. Belli, F., Budnik, ChJ, Hollmann, A., Tuglular, T., Wong, W.E.: Model-based mutation testing—approach and case studies. Sci. Comput. Program. 120, 25–48 (2016)
20. Belli, F., Budnik, Ch.J., White, L.: Event-based modelling, analysis and testing of user interactions: approach and case study. In: Software Testing, Verification and Reliability, pp. 3–32 (2006)
21. Boehm, B., Brown, A., Madachy, R., Yang, Y.: A software product line life cycle cost estimation model. In: Proceedings of International Symposium on Empirical Software Engineering (ISESE), pp. 156–164 (2004). https://doi.org/10.1109/ISESE.2004.1334903
22. Bass, L., Campbell, G., Clemens, O., Northrop, L., Smith, D.: Third Product Line Practice Workshop, Techn. Report CMU/SEI-99-TR-003, ESC-TR-99-03 (1999)
23. Beck, K., Andres, C.: Extreme Programming Explained, 2nd edn. Addison-Wesley Professional (2004)
24. Bernardinello, L., De Cindio, F.: A Survey of Basic Net Models and Modular Net Classes, LNCS vol. 609, pp. 304–351. Springer, Berlin (1992)
25. Beck, K.: Test Driven Development: By Example. Addison-Wesley Professional (2002)
26. Belli, F., Dreyer, J.: Program segmentation for controlling test coverage. In: Proceedings of the Eighth International Symposium on Software Reliability Engineering ISSRE '97, pp. 72–83 (1997)
27. Belli, F., Echtle, K., Görke, W.: Methoden und Modelle der Fehlertoleranz, Informatik Spektrum, vol. 9, No. 2, pp. 68–81. Springer, Berlin (1986)
28. Bencomo, N., Hallsteinsen, S., Almeida, E.: A view of the landscape of dynamic software product lines. Computer 45(10), 36–41 (2012)
29. Belli, F., Hollmann, A.: Test generation and minimization with "basic" statecharts. In: Proceedings of ACM Symposium on Applied Computing SAC '08, pp. 718–723 (2008). https://doi.org/10.1145/1363686.1363856
30. Bayer, J., Flege, O., Knauber, P., Laqua, R., Muthig, D., Schmid, K., Widen, T., DeBaud', J.-M.: PuLSE: a methodology to develop software product lines. In: Proceedings of ACM Symposium on Software Reusability, pp. 122–131 (1999)
31. Beydeda, S., Gruhn, S., Mayer, J., Reussner, R., Schweiggert, F. (eds.): Testing of component-based systems and software quality. In: GI Edition—Lecture Notes in Informatics, Series of the Gesellschaft für Informatik (GI), vol. P-58 (2004)
32. Belli, F., Großpietsch, K.-E.: Specification of fault-tolerant system issues by predicate/transition nets and regular expressions -approach and case study. IEEE Trans. Softw. Eng. 17, 513–526 (1991)
33. Beizer, B.: Software Testing Techniques, 2 edn. International Thomson Computer Press and Dreamtech (1990)
34. Belli, F., Jedrzejowicz, P.: Fault-tolerant programs and their reliability. IEEE Trans. Reliab. 39(2), 184–192 (1990)
35. Belli, F.: Finite state testing and analysis of graphical user interfaces. In: Proceedings 12th International Symposium on Software Reliability Engineering, pp. 34–43 (2001)
36. Benington, H.D.: Production of large computer programs. IEEE Annals Hist. Comput. IEEE Educ. Activ. Depart. 5(4), 350–361 (1983). https://doi.org/10.1109/MAHC.10102. Retrieved 2011-03-21
37. Binder, R.V.: Testing Object-Oriented Systems: Models, Patterns, and Tools. Addison-Wesley (2006)

38. James Blair, J., Batory, D.: A Comparison of Generative Approaches: XVCL and GenVoca (2004). Available at https://www.cs.utexas.edu/ftp/predator/xvcl-compare.pdf
39. Bosch, J., Krueger, C. (eds.) Proceedings of 8th International Conference—Software Reuse: Methods, Techniques, and Tools (ICSR), LNCS 3107 (2004)
40. Boehm, B., Lane, J.A., Koolmanojwong, S.: The Incremental Commitment—Spiral Model: Principles and Practices for Successful Systems and Software. Addison Wesley (2014)
41. Cohen, D.M., Dalal, S.R., Fredman, M.L., Patton, G.C.: The AETG system: an approach to testing based on combinatorial design. IEEE Trans. Softw. Eng. **23**(7), 437–444 (1997)
42. Chandy, K.M., Misra, M.: The drinking philosophers problem. ACM Trans. Program. Lang. Syst. **6**(4), 632–646 (1984)
43. Carbonnel, J., Huchard, M., Miralles, A., Nebut, C.: Feature Model Composition Assisted by Formal Concept Analysis. In: Proceedings ENASE: Evaluation of Novel Approaches to Software Engineering, pp. 27–37 (2017)
44. Chomsky, N.: On certain formal properties of grammars. Inf. Control **2**, 137–167 (1959)
45. Cleaveland, C.T.: Building application generators. IEEE Softw. 25–33 (1988)
46. Clarke, E.M., Orna Grumberg, O., Peled, D.A.: Model Checking. MIT Press (1999)
47. Clements, P., Northrop, L.: Software Product Lines—Practices and Patterns, SEI Series in Software Engineering. Addison-Wesley (2002)
48. O'Connor, J., Mansour, C., Jerri Turner-Harris, J., Campbell, Jr., G.H.: Reuse in Command-and-Control Systems. IEEE Softw. **11**, 70–79 (1994). https://doi.org/10.1109/52.311065
49. Cichos, H., Oster, S., Lochau, M., Schürr, A.: Model-based coverage-driven test suite generation for software product lines. In: Proceedings of MODELS 2011, LNCS 6981, pp. 425–439 (2011). Springer, Berlin, and Extended Version of Model-based Coverage-Driven Test Suite Generation for Software Product Lines, Informatik-Bericht Nr. 2011-07, Technische Universität Braunschweig, 32 pp (2011)
50. Cichos, H., Lochau, M., Oster, S., Schürr, A.: Reduktion von Testsuiten für Software-Produktlinien. In: Proceedings of Software Engineering, Gesellschaft für Informatik e.V., pp. 143–154 (2012)
51. Cox, B.J.: Planning the software revolution. IEEE Softw. **7**(6), 25–35 (1990)
52. Cox, C.: Surviving software dependencies. Commun. Assoc. Comput. Assoc. **62**, 36–43 (2019)
53. Cohen, S.G., Stanley, Jr., J.L., Peterson, A.S., Krut, Jr., R.W.: Application of Feature-Oriented Domain Analysis to the Army Movement Control Domain. Technical Report CMU/SEI-91-TR-028, ESD-91-TR-028 (1992)
54. Czarnecki, K., Eisenecker, U.: Generative Programming: Methods, Tools, and Applications. Addison-Wesley (2000)
55. Czarnecki, K., Wasowski, A.: Feature Diagrams and Logics: There and Back Again. In: 11th International Software Product Line Conference (SPLC 2007), pp. 23–34 (2007)
56. David, R., Thevenod-Fosse, P.: Detecting Transition Sequences: Application to Random Testing of Sequential Circuits. In: Proceedings of International Symposium on Fault-Tolerant Computing FTCS-9, pp. 121–124 (1979)
57. DeRemer, F., Kron, H.: Programming-in-the large versus programming-in-the-small. In: Proceedings of ACM International Conference on Reliable Software, pp. 114–121 (1975). https://doi.org/10.1145/800027.808431
58. DeMillo, R.A., Lipton, R.J., Sayward, F.G.: Hints on test data selection: help for the practicing programmer. IEEE Comput. **11**(4), 34–41 (1978)
59. German Association For Quality, Deutsche Gesellschaft für Qualitätssicherung: Software Qualitätssicherung, DGQ-NTG-Schrift Nr. 12-51. Frankfurt (1986).
60. Dijkstra, E.W.: Notes on Structured Programming, Section 3 in On the Reliability of Mechanisms. T.H.-Report 70-WSK-03, Technology University of Eindhoven, Department of Mathematics (1970)
61. Dijkstra, E.W.: Hierarchical ordering of sequential processes. Acta Informatica **1**(2), 115–138 (1971)

62. Beydeda, S., Gruhn, V. (eds.): Testing Commercial-off-the-Shelf Components and Systems. Springer, Berlin (2005)
63. Devroey, X.: Behavioural Model-Based Testing of Software Product Lines, Ph.D. Thesis. University of Namur, PReCISE Research Center (2017)
64. Many Relevant Papers and Books. see http://ivknet.de/index.php/en/publications
65. Dimov, A., Punnekkat, A.: On the estimation of software reliability of component-based dependable distributed systems. In: Proceedings of QoSA-SOQUA 2005, LNCS 3712, pp. 171–187. Springer, Berlin (2005)
66. Dintzner, N., Kulesza, U., Deursen, A.V, Pinzger, M.: Evaluating Feature Change Impact on Multi-product Line Configurations Using Partial Information. In: Schaefer, I., Stamelos, I. (eds.) Proceedings of Software Reuse for Dynamic Systems in the Cloud and Beyond, 14th International Conference on Software Reuse (ICSR), Lecture Notes in Computer Science, vol. 8919, pp. 1–16. Springer, Berlin (2015)
67. DO-178C ED-12C:2011: Software Considerations in Airborne Systems and Equipment Certification (available 2012)
68. Edwards, S.H. Kulczycki, G.: Formal foundation of reuse and domain engineering. In: Proceedings of 11th International Conference on Software Reuse (ICSR). Springer Science and Business Media (2009)
69. El-Fakih, K., Hierons, R.M., Turker, U.C.: K-branching UIO sequences for partially specified observable non-deterministic FSMs. IEEE Trans. Softw. Eng. (2019). https://doi.org/10.1109/TSE.2019.2911076
70. Eichelberger, H., Kröher, Ch., Schmid, K.: An Analysis of Variability Modeling Concepts: Expressiveness versus Analyzability. In: Favaro, J., Morisio, M. (eds.) Proceedings of 13th International Conference on Software Reuse—ICSR, LNCS 7925, pp. 32–48. Springer, Berlin (2013)
71. Ezran, M., Morisio, M., Tully, C.: Practical Software Reuse. Springer Practitioner Series (2002)
72. Ezran, M., Morisio, M., Tully, C.: Practical Software Reuse, Chapter "Two Major Case Histories", pp. 155–169. Springer Practitioner Series (2002)
73. EN 50128:2012: Railway Applications—Communication, Signalling and Processing Systems—Software for Railway Control and Protection Systems
74. Engström, E., Runeson, P.: Software product line testing—a systematic mapping study. Inf. Softw. Technol. **53**, 2–13 (2011)
75. Fafchamps, D.: Organizational factors and reuse. IEEE Softw. **11**(5), 31–41 (1994). https://doi.org/10.1109/52.311049
76. Fagan, M.E.: Design and code inspections to reduce errors in program development. IBM Syst. J. **15**(3), 182–211 (1976). https://doi.org/10.1147/sj.153.0182
77. Favaro, J., Morisio, M. (eds.): Proceedings of 13th International Conference on Software Reuse—ICSR, LNCS 7925. Springer, Berlin (2013)
78. Fettke, P., Reisig, W.: Modelling Service-Oriented Systems and Cloudservices with Heraklit. In: Proceedings of 16th International Workshop on Engineering Service-Oriented Applications and Cloud Services (2020). http://heraklit.dfki.de/assets/documents/fettke_2020_service_modeling.pdf
79. Ferreira, J.M., Vergilio, S.R., Quinaia, M.: Software product line testing based on feature model mutation. Inte. J. Softw. Eng. Knowl. Eng. **27**(05), 817–839 (2017). https://doi.org/10.1142/S0218194017500309
80. Filho, R.A.M., Vergilio, S.R.: A mutation and multi-objective test data generation approach for feature testing of software product lines. In: Proceedings of 29th Brazilian Symposium on Software Engineering, pp. 21–30 (2015). https://doi.org/10.1109/SBES.2015.17
81. Frakes, W.B., Isoda, S.: Success factors of systematic reuse. IEEE Softw. V **11**(5), 14–19 (1994). Available at http://ieeexplore.ieee.org/stamp/stamp.jsp?tp=&arnumber=311045
82. Frakes, W.B., Kang, K.: Software reuse research: status and future. IEEE Trans. Softw. Eng. **31**(7), 529–536 (2005)

83. Fragal, V.H., Simao, A., Mousavi, M.R.: Validated test models for software product lines: featured finite state machines. In: Proceedings of 13th International Conference on Formal Aspects of Component Software (FACS), Revised Selected Papers, pp. 210–227 (2016). https://doi.org/10.1007/978-3-319-57666-4_13

84. Frakes, W.B., Terry, C.: Software reuse: metrics and models. ACM Comput. Surv. **28**(2), 415–435 (1996). https://doi.org/10.1145/234528.234531

85. Frankl, P.G., Weyuker, E.J.: An applicable family of data flow testing criteria. IEEE Trans. Softw. Eng. **14**(10), 1483–1498 (1988)

86. Fragal, V.H., Simao, A., Mousavi, M.R., Turker, U.C.: Extending HSI test generation method for software product lines. Comput. J. **62**(1), 109–129 (2019). https://doi.org/10.1093/com jnl/bxy046

87. Ganter, B., Stumme, G., Wille, R.: Formal Concept Analysis: Foundations and Applications. Springer, Berlin (2005)

88. Genrich, H. J., Lautenbach, K.: System modelling with high-level Petri nets. Theor. Comput. Sci. **13**(1), 109–135 (1981). https://doi.org/10.1016/0304-3975(81)90113-4

89. Geppert, B., Li, J., Rößler, F., Weiss, D.M.: Towards Generating Acceptance Tests for Product Lines. In: Bosch, J., Krueger, C. (eds.) Proceedings of 8th International Conference—Software Reuse: Methods, Techniques, and Tools (ICSR), LNCS 3107, pp. 35–48 (2004)

90. Gazzola, L., Goldstein, M., Mariani, L., Segall, I., Ussi, L.: Automatic ex-vivo regression testing of microservices. In: Proceedings of 1st IEEE/ACM International Conference on Automation of Software Test (2020)

91. Gamma, E., Helm, R., Johnson, R.E., Vlissides, J.: Design Patterns. Elements of Reusable Object-Oriented Software. Addison-Wesley Professional Computing Series, USA (1994)

92. Grönniger, H., Krahn, H., Pinkernell, C., Rumpe, B.: Modeling variants of automotive systems using views. In: Proceedings of Modellierungs-Workshop MBEFF, Berlin, Informatik-Bericht 2008-01 (2008). Available at www.se-rwth.de/publications

93. Gluschkow, W.M.: Theory of Abstract Automata (in German). VEB Verlag der Wissensch, Berlin (1963)

94. Gomes, C., et al.: Computational sustainability: computing for a better world and a sustainable future. Commun Assoc. Comput. Assoc. **62**, 56–65 (2019)

95. Goodenough, J., Gerhart, S.L.: Toward a theory of test data selection. ACM SIGPLAN Notices **1**(6), 156–173 (1975). https://doi.org/10.1145/390016.808473

96. Goodenough, J.B.: Exception handling—issues and a proposed notation. Comm. ACM **18**(12), 683–696 (1975)

97. Gorgonia, K.C., Perkusich, A.: Adaptation of coloured petri nets models of software artifacts for reuse. In: Bosch, J., Krueger, Ch. (eds.) Proceedings of 8th International Conference on Software Reuse: Methods, Techniques, and Tools (ICSR), pp. 240–254. Springer, Berlin (2004)

98. Groher, I., Voelter, M.: Expressing feature-based variability in structural models. In: Proceedings of Workshop on Managing Variability for Software Product Lines at SPLC (2007). Available at http://citeseerx.ist.psu.edu/viewdoc/download?doi=10.1.1.571.593&rep=rep1& type=pdf

99. Hallsteinsen, S., Paci, M.: Experiences in Software Evolution and Reuse—Twelve Real World Projects. Springer Science and Business Media, 18.09.1997

100. Hennell, M.A., Woodward, M.R., Hedley, D.: On program analysis. Inf. Process. Lett. **5**(5), 136–140 (1976)

101. Herrtwich, R.G., Hommel, G.: Nebenläufige Programme (Concurrent Programs, in German). Springer, Berlin (1994)

102. Hooper, J.W., Chester, R.O.: Software Reuse Guidelines and Methods. Springer, Berlin (1991)

103. Hopcroft, J.E., Motwani, R., Ullman, J.D.: Introduction to Automata Theory, Languages, and Computation, 2 edn. Pearson Education (2000)

104. Haugen, Ø., Møller-Pedersen, B., Oldevik, J, Olsen, G.K., Svendsen, A.: Adding standardized variability to domain specific languages. In: Proceedings of International Software Product Line Conference (SPLC '08), pp 139–148 (2008). https://doi.org/10.1109/SPLC.2008.25

105. Hossain, Sh.: Rework and reuse effects in software economy. Glob. J. Comput. Sci. Technol. C Softw. Data Eng. **1**(4) (2018)
106. Howden, W.E.: Theoretical and empirical studies of program testing. In: Proceedings of 3rd International Conference on Software Engineering—ICSE '78, pp. 305–311 (1978)
107. Henard, C., Papadakis, M., Perrouiny, G., Klein, J., Le Traon, Y.: Assessing software product line testing via model-based mutation: an application to similarity testing. In: IEEE Sixth International Conference on Software Testing, Verification and Validation Workshops, Luxembourg, pp. 188–197 (2013). https://doi.org/10.1109/ICSTW.2013.30
108. Hristov, D., Hummel, O., Huq, M., Janjic, W.: Structuring software reusability metrics for component-based software development. In: Proceedings of The Seventh International Conference on Software Engineering Advances, pp. 421–429 (2012)
109. Heuer, A., Stricker, V., Budnik, C.J., Konrad, S., Lauenroth, K., Pohl, K.: Defining variability in activity diagrams and petri nets. Sci. Comput. Program. **78**, 2414–2432 (2013)
110. Hummel, O., Atkinson, C.: The managed adapter pattern: facilitating glue code generation for component reuse. In: Proceedings of 11th International Conference on Software Reuse (ICSR), pp. 211–224. Springer, LNCS 5791 (2009)
111. Hamm, A., Willner, A., Schieferdecker, I.: Edge computing: a comprehensive survey of current initiatives and a roadmap for a sustainable edge computing development. In: Proceedings of 15th International Conference on Wirtschaftsinformatik (2020)
112. Schaefer, I., Stamelos, I. (eds.) Proceedings of software reuse for dynamic systems in the cloud and beyond. In: 14th International Conference on Software Reuse (ICSR), Lecture Notes in Computer Science, vol. 8919. Springer, Berlin (2015)
113. IEC 62309:2004: Dependability of Products Containing Reused Parts—Requirements for Functionality and Tests
114. IEC 62304:2006 (amended 2015), Medical Device Software—Software Life Cycle Processes
115. IEC 61508:2010: Functional Safety of Electrical/Electronic/Programmable Electronic Safety-Related Systems: Part 1: General Requirements Part 2: Requirements for Electrical/Electronic/Programmable Electronic Safety-Related Systems Part 3: Software Requirements Part 3-1: Software Requirements—Reuse of Pre-existing Software Elements to Implement all or Part of a Safety Function (2016) Part 4: Definitions and Abbreviations
116. IEC/PAS 62814-2012: Dependability of Software Products Containing Reusable Components—Guidance for Functionality and Tests (withdrawn, however available)
117. IEC 62628:2012: Guidance on Software Aspects of Dependability
118. IEC/PAS 62814: Dependability of Software Products Containing Reusable Components—Guidance for Functionality and Tests (2012)
119. ISO/IEC 12207: 2017: Systems and Software Engineering—Software Life Cycle Processes
120. IEEE 1517-2010: IEEE Standard for Information Technology, System and Software Life Cycle Processes, Reuse Processes (2010)
121. IEEE 1633: 2016: Recommended Practice for Software Reliability
122. IEEE Std 610.12-1990: IEEE Standard Glossary of Software Engineering Terminology (Reaffirmed 2002)
123. ISO/IEC/IEEE 42010:2011: Systems and Software Engineering—Architecture Description
124. ISO/IEC/IEEE 12207: 2017: Systems and Software Engineering—Software Life Cycle Processes
125. ISO 26262 (1 to 12): 2018: Road Vehicles—Functional Safety
126. ISO/IEC 7498: 1994: Information Technology—Open Systems Interconnection—Basic Reference Model: The Basic Model
127. Jatain, A., Goel, S.: Comparison of domain analysis methods in software reuse. Int. J. Inf. Technol. Knowl. Manag. **2**(2), 347–352 (2009)
128. Jacobson, I, Griss, M., Jonsson, P.: Software Reuse—Architecture, Process and Organization for Business Success. ACM Press, Addison Wesley Longman (1997)
129. Jensen, R.W.: Improving Software Quality. Prentice-Hall (2014)
130. Jensen, R.W., Tonies, Ch.: Software Engineering. Prentice-Hall (1979)

131. Jamshidi, P., Pahl, C., Mendonça, N.C., Lewis, J., Tilkov, S.: Microservices: the journey so far and challenges ahead. IEEE Softw. **35**(3), 24–35 (2018). https://doi.org/10.1109/MS.2018.2141039. ISSN 0740-7459

132. Karlsson, E.A., Brantestam, J.: Generic Reuse Development Processes. In: Karlsson, E.A. (ed.) Software Reuse: A Holistic Approach, pp. 253–270. Wiley (1995)

133. Kang, K., Cohen, S., Hess, J., Nowak, W., Peterson, S.: Feature-Oriented Domain Analysis (FODA) Feasibility Study (Report), Software Engineering Institute, Carnegie Mellon University (1990). Available at http://www.floppybunny.org/robin/web/virtualclassroom/chap12/s4/articles/foda_1990.pdf

134. Kamischke, J., Lochau, M., Baller, H.: Conditioned model slicing of feature-annotated state machines. In; Proceedings of 4th International Workshop on Feature-Oriented Software Development (FOSD), pp. 9–16 (2012). https://doi.org/10.1145/2377816.2377818

135. Kaner, C., Falk, J., Nguyen, H.Q.: Testing Computer Software 2nd edn. Wiley (1999)

136. Kaindl, H., Mannion, M.: A Feature-Similarity Model for Product Line Engineering. In: 14th International Conference on Software Reuse (ICSR), Lecture Notes in Computer Science, vol. 8919, pp 34–41. Springer, Berlin (2015). https://doi.org/10.1007/978-3-319-14130-5_3

137. King, J.C.: Symbolic Execution and Program Testing. Comm. ACM, 385–394 (1976)

138. Kim, Y., Stohr.: Software reuse: survey and research directions. J. Manag. Inf. Syst. **14**, 113–147 (1998)

139. Kleene, S.C.: Representation of events in nerve nets and finite automata. In: Shannon, C.E., McCarthy, J. (eds.) Automata Studies, pp. 3–42. Princeton University Press, Princeton, New Jersey (1956)

140. Kienzle, J., et al.: Toward model-driven sustainability evaluation. Commun. Assoc. Comput. Mach. **63**, 80–91 (2020)

141. Kreowski, H.-J., Montanari, U., Orejas, F., Rozenberg, G., Taentzer, G.: Formal Methods in Software and Systems Modeling. Springer, Berlin (2005)

142. Krueger, C.W.: Software reuse. ACM Comp. Surv. **24**(2), 131–183 (1992)

143. Lackner, H.: Model-based product line testing: sampling configurations for optimal fault detection. In: Proceedings of 17th International SDL Forum on SDL 2015: Model-Driven Engineering for Smart Cities, vol. 9369, pp. 238–251 (2015). https://doi.org/10.1007/978-3-319-24912-4_17

144. Lackner, H.: Domain-Centered Product Line Testing, Ph.D. Thesis, Humboldt-University, Berlin (2016)

145. Laprie, J.-C. (ed.): Dependability: Basic Concepts and Terminology in English, French, German, Italian and Japanese. Springer, Berlin (1992)

146. Lackner, H., Schmidt, M.: Towards the assessment of software product line tests: a mutation system for variable systems. In: Software Product Analysis Tools (ACM SPLC), vol. 1, pp. 62–69 (2014)

147. Lackner, H., Schmidt, M.: Potential errors and test assessment in software product line engineering. In: Proceedings of Tenth Workshop on Model-Based Testing, EPTCS 180, pp. 57–72. Springer, Berlin (2015). https://doi.org/10.4204/EPTCS.180.4

148. Lackner, H., Schlingloff, B.-H.: Advances in testing software product lines. In: Memon, A.M. (ed.) Advances in Computers, vol. 107, pp. 157–217. Elsevier (2017). https://doi.org/10.1016/bs.adcom.2017.07.001

149. Luo, Y., Brand, M., Engelen, L., Favaro, J., Klabbers, M., Sartori, G.: Extracting Models from ISO 26262 for Reusable Safety Assurance. In: Favaro, J., Morisio, M. (eds.) Proceedings of 13th International Conference on Software Reuse—ICSR, LNCS 7925, pp. 192–207. Springer, Berlin (2013)

150. Leach, R.J.: Software Reuse, Second Edition: Methods, Models, Costs, 2nd edn. (2012)

151. van der Linden, F.J, Schmid, K., Rommes, E.: Software Product Lines in Action. Springer, Berlin (2007)

152. Lochau, M., Schaefer, I., Kamischke, J., Lity, S.: Incremental model-based testing of delta-oriented software product lines. In: Brucker, A.D., Julliand, J. (eds.) Tests and Proofs. TAP 2012, Lecture Notes in Computer Science, vol. 7305, pp. 67–82. Springer, Berlin. https://doi.org/10.1007/978-3-642-30473-6_7

153. Leveson, N.G., Turner, C.S.: An investigation of the therac-25 accidents. IEEE Comput. **26**(7), 18–41 (1993)
154. Levy, L.S.: A metaprogramming method and its economic justification. IEEE Trans. Softw. Eng., **SE-12**(2), 272–277 (1986)
155. Lim, W.C.: Managing Software Reuse. Prentice Hall PTR (1998)
156. Lisboa, L.B., Li, J.J., Morreale, P., Heer, D., Weiss, D.M.: An Evaluation to Compare Software Product Line Decision Model and Feature Model. In: Proceedings of 9th International Conference on Evaluation of Novel Approaches to Software Engineering (ENASE), pp. 1–8 (2014)
157. Lity, S., Nahrendorf, S., Thüm, T., Seidl, C., Schaefer, I.: 175% modeling for product-line evolution of domain artifacts. In: Proceedings of 12th International Workshop on Variability Modelling of Software-Intensive Systems (VAMOS), pp. 27–34 (2018). https://doi.org/10.1145/3168365.3168369
158. Lochau, M., Oster, S., Goltz, U., Schürr, A.: Model-based pairwise testing for feature interaction coverage in software product line engineering. Softw. Q. J. **20**, 567–604. https://doi.org/10.1007/s11219-011-9165-4
159. Lackner, H., Thomas, M., Wartenberg, F., Weißleder, S.: Model-based test design of product lines: raising test design to the product line level. In: Proceedings of 7th International Conference on Software Testing, Verification, and Validation (ICST), pp. 51–60 (2014)
160. Luckham, D.: The Power of Events. Addison Wesley (2005)
161. Martínez-Fernández, S., Ayala, C.P, Franch, X., Marques, H.M.: REARM: A reuse-based economic model for software reference architectures. In: Favaro, J., Morisio, M. (eds.) Proceedings of 13th International Conference on Software Reuse—ICSR, LNCS 7925, pp. 97–112. Springer, Berlin (2013)
162. Mathur, A.P.: Foundations of Software Testing, 2 edn. Pearson (2013)
163. McClure, C.: Software Reuse: A Standards-Based Guide. Wiley (2001)
164. Mili, A., Chmiel, S.F., Gottumukkala, R., Zhang, L.: Managing software reuse economics: an integrated ROI-based model. Annals Softw. Eng. **11**, 175–218 (2001)
165. Meeker, H.: Open (Source) for Business A Practical Guide to Open-Source Software Licensing, 2nd edn. Createspace Independent Publishing Platform (2017)
166. Microsoft, Accenture and WSP Environment and Energy Study. https://www.news.microsoft.com/2010/11/04/microsoft-accenture-and-wsp-environment-energy-study-shows-significant-energy-and-carbon-emissions-reduction-potential-from-cloud-computing/
167. Miller, E.F., Howden, W.E.: Tutorial: Software Testing and Validation Technique. IEEE Computer Society Press, IEEE Catalog No. EHO 180-0 (1981)
168. Miller, E.F.: Structurally based automatic program testing. In: Proceedings of EASCON-74 (1974)
169. Mili, H., Mili, A., Yacoub, S., Addy, E.: Reuse-Based Software Engineering, Techniques, Organization, and Controls. Wiley (2002)
170. Maglio, P.P, Weske, M., Yang, J., Fantinato, M.: Service-oriented computing. In: Proceedigs of 8th International Conference (ICSOC) (2010)
171. Meinicke, J., Thüm, T., Schröter, R., Benduhn, F., Leich, T., Saake, G.: Mastering Software Variability with FeatureIDE. Springer, Berlin (2017)
172. Muthig, D., Atkinson, C.: Model-driven product line architectures. In: Proceedings of Software Product Lines (SPLC 2002), Lecture Notes in Computer Science, vol. 2379, pp. 110–129. Springer, Berlin (2002)
173. Murata, T.: Petri nets: properties, analysis and applications. Proc. IEEE **77**(4), 541–558 (1989). https://doi.org/10.1109/5.24143
174. Myers, G.: The Art of Software Testing, 3 edn. Wiley (2012)
175. Myhill, J.: Finite Automata and Representation of Events in Fundamental Concepts in the Theory of Systems, Tech. Report No. 57-624, ASTIA Document No. AD 1557 41, Wright Air Development Center, Cincinnati-Ohio (1957)
176. Naito, S., Tsunoyama, M.: Fault detection for sequential machines by transition tours. In: Proceedings of FTCS, pp. 238–243 (1981)

177. Naur, P., Randell, B. (eds.): Software Engineering, Report on a conference sponsored by the NATO Science Committee, Garmisch, Germany, 7th to 11th October 1968. Available at https://homepages.cs.ncl.ac.uk/brian.randell/NATO/nato1968.PDF
178. Nelson, V.P.: Fault-tolerant computing: fundamental concepts. IEEE Comput. **23**(7) (1990)
179. Newman, S.: Building Microservices: Designing Fine-Grained Systems. O'Reilly Media (2015)
180. Noback, M.: Principles of Package Design: Creating Reusable Software Components Technology. Springer, Berlin (2018)
181. Linda, M., Northrop, L.M., Clement, P.C.: A Framework for Software Product Line Practice, Version 5.0, Software Engineering Institute (SEI), Carnegie Mellon University (REV-03.18.2016) (2012). Also available at https://resources.sei.cmu.edu/asset_files/WhitePaper/2012_019_001_495381.pdf
182. Northrop, L.M.: SEI's software product lines tenets. IEEE Softw. **19**, 32–40 (2002)
183. Nurolahzade, M., Walker, R.J., Maurer, F.: An assessment of test-driven reuse: promises and pitfalls. In: Favaro, J., Morisio, M. (eds.) Proceedings of 13th International Conference on Software Reuse—ICSR, LNCS 7925, pp. 227–252. Springer, Berlin (2013)
184. Object Management Group (http://www.uml.org), Unified Modeling Language (UML®), Version 2.5.1 (2017)
185. Oracle Practitioner Guide, Determining ROI of SOA through Reuse (2012)
186. Orrego, A., Mundy, G.: SRAE: An Integrated Framework for Aiding in the Verification and Validation of Legacy Artifacts in NASA Flight Control Systems. In: Proceedings 31st Annual International Computer Software and Applications Conference (COMPSAC). IEEE Computer Press (2007). https://doi.org/10.1109/COMPSAC.2007.199
187. Parnas, D.L.: On the use of transition diagrams in the design of user interface for an interactive computer system. In: Proceedings of 24th ACM National Conference, pp. 379–385 (1969)
188. Paulk, M.C.: How ISO 9001 compares with the CMM. IEEE Softw. **12**(1), 74–83 (1995). https://doi.org/10.1109/52.363163, see also https://resources.sei.cmu.edu/asset_files/TechnicalReport/1994_005_001_435267.pdf
189. Pohl, K., Böckle, G., Linden, F.: Software Product Line Engineering: Foundations, Principles, and Techniques. Springer, Berlin (2005)
190. Petersen, K., Badampudi, D., Syed, M.A.S., Wnuk, K., Gorschek, T., Papatheocharous, E., Axelsson, J., Sentilles, S., Crnkovic, I., Cicchetti, A.: Choosing component origins for software intensive systems: in-house, COTS, OSS or outsourcing? A case survey. IEEE Trans. Softw. Eng. **44** (2018). https://doi.org/10.1109/TSE.2017.2677909
191. Petri, C.A., Reisig, W.: Petri net. Scholarpedia **3**(4), 6477 (2008). https://doi.org/10.4249/scholarpedia.6477
192. Petri, C.A.: Communication with Automata (in German), 1962, Ph.D. Dissertation, University Darmstadt, 1962 For English version, see Technical report RADC-TR-65-377, Volume I, Final Report, Supplement I (1966), or visit DTIC Research Report AD0630125, or visit https://edoc.sub.uni-hamburg.de/informatik/volltexte/2010/155/pdf/diss_petri_engl.pdf
193. Pfleeger, S.L.: Measuring reuse: a cautionary tale. IEEE Softw. **13**(4), 118–127 (1996). https://doi.org/10.1109/52.526839
194. Poulin, J.S., Caruso, J.M., Hancock, D.R.: Business case for software reuse. IBM Syst. J. **32**(4), 567–594 (1993)
195. Pressman, R.S.: Software Engineering: A Practitioner's Approach, 9 edn. Palgrave Macmillan (2020)
196. Rapps, S., Weyuker, E.J.: Selecting software test data using data flow information. IEEE Trans. Softw. Eng. **11**(4), 367–375 (1985)
197. Reifer, D.J.: Practical Software Reuse—Strategies for Introducing Reuse Concepts in Your Organization. Wiley (1997)
198. Reuys, A., Kamsties, E., Pohl, K., Reis, S.: Model-based system testing of software product families. In: Pastor, O., Cunha, J.F. (eds.) Proceedings of International Conference on Advanced Information Systems Engineering (CAiSE), LNCS 3520, pp. 519–534 (2005)

199. Rozenburg, G., Engelfriet, J.: Elementary net systems. In: Reisig, W., Rozenberg, G. (eds.) Lectures on Petri Nets I: Basic Models—Advances in Petri Nets, Lecture Notes in Computer Science, vol. 1491, pp. 12–121. Springer, Berlin (1998)

200. Rocha, C.R., Martins, E.: A strategy to improve component testability without source code. In: Beydeda, S., Gruhn, V. (eds.) Testing Commercial-off-the-Shelf Components and Systems, pp. 47–62. Springer, Berlin (2005)

201. Salomaa, A.: Theory of Automata. Pergamon Press, New York (1969)

202. Salomaa, A.: Formal Languages. Academic Press (1973)

203. Salomaa, A.: Events and languages. In: Calude, C.S. (ed.) People and Ideas in Theoretical Computer Science, pp. 253–274 (1999)

204. Salomaa, A.: Two Complete Axiom Systems for the Algebra of Regular Events, 1966, Journal of the ACM **13**(1), pp. 158–169, (1966). https://doi.org/10.1145/321312.321326.

205. Sametinger, J.: Software Engineering with Reusable Components. Springer, Berlin (1997)

206. Schmietendorf, A., Dimitrov, E., Dumke, R., Foltin, E., Wipprecht, M.: Conception and experience of metrics-based software reuse. In: Proceedimgs of International Workshop on Software Measurement (IWSM'99), pp. 178–189 (1999)

207. CMMI® for Services, Version 1.2, CMMI-SVC, V1.2 CMMI Product Team, Improving processes for better services, Technical Report CMU/SEI-2009-TR-001 ESC-TR-2009-00 (2009). Available at https://resources.sei.cmu.edu/asset_files/TechnicalReport/2009_005_001_15092.pdf see also https://www.sei.cmu.edu/search.cfm#stq=cmm&stp=1

208. Seidl, C., Wille, D., Schaefer, I.: Software reuse: from cloned variants to managed software product lines. In: Automotive Systems and Software Engineering: State of the Art and Future Trends, pp. 77–108. Springer International Publishing (2019)

209. Shen, Y.N., Lombardi, F., Dahbura, A.T.: Protocol conformance testing using multiple UIO sequences. IEEE Trans. Commun. **40**, 1282–1287 (1992)

210. Shehady, R.K., Siewiorek, D.P.: A method to automate user interface testing using finite state machines. In: Proceedings of International Symposium Fault-Tolerant Computing FTCS-27, pp. 80–88 (1997)

211. Sim, S.E., Gallardo-Valencia, R.E. (eds.): Finding Source Code on the Web for Remix and Reuse. Springer, Berlin (2013)

212. Silva, M.: 50 years after the PhD thesis of Carl Adam Petri: A perspective, IFAC Proceedings Volumes, vol. 45, issue 29, pp. 13–20 (2012). https://doi.org/10.3182/20121003-3-MX-4033.00006

213. Schmid, K., John, I., Kolb, R., Meier, G.: Introducing the PuLSE approach to an embedded system population at Testo AG. In: Proceedings of 27th International Conference on Software Engineering, (ICSE), pp. 544–552 (2005). https://doi.org/10.1109/ICSE.2005.1553600

214. Sommerville, I.: Software Engineering, 10th edn. Pearson (2015)

215. Software Productivity Consortium (SPC), Reuse-Driven Software Process Guidebook, Herndon, VA (1993)

216. Specht, J.: Terminology Proposal: Redundancy for Fault Tolerance, University of Duisburg—Essen available at http://www.ieee802.org/1/files/public/docs2013/new-tsn-specht-redundancy-terminology-20130115-v01.pdf

217. Shatnawi, A., Seriai, A., Sahraoui, H.: Recovering architectural variability of a family of product variants. In: Proceedings of ICSR, LNCS 8919, pp. 17–33. Springer, Berlin (2015)

218. Salman, H.E., Seriai, A.-D., Dony, Ch.: Feature-level change impact analysis using formal concept analysis. Int. J. Softw. Eng. Knowl. Eng. **25**(1), 69–92 (2015)

219. Schweigert, T., Vohwinkel, D., Korsaa, M., Nevalainen, R.: Agile Maturity model: analysing agile maturity characteristics from the SPICE perspective. J. Softw. Evol. Proc. **26**, 513–520 (2014)

220. Tracz, W.: Confessions of a used-program salesman: lessons learned. In: Proceedings of the Symposium on Software Reusability, pp. 11–13 (1995)

221. Vouk, M.A.: Back-to-back testing. Inf. Softw. Technol. **32**(1), 34–45 (1990)

222. Weißleder, S.: Influencing factors in model-based testing with UML state machines: report on an industrial cooperation. In: Proceedings on Model Driven Engineering Languages and

Systems (MODELS), Lecture Notes in Computer Science, vol. 5795, pp. 211–225. Springer, Berlin (2009). https://doi.org/10.1007/978-3-642-04425-0_16

223. Weißleder, S., Lackner, H.: Top-down and bottom-up approach for model-based testing of product lines. In: Proceedings MBT 2013, EPTCS 111, pp. 82–94. Springer, Berlin (2013). https://doi.org/10.4204/EPTCS.111.7

224. Whittaker, J.A.: Statistical testing for cleanroom software engineering. In: Proceedngs of Twenty-Fifth Hawaii International Conference on System Sciences, vol. 2, pp. 428–436 (1992)

225. Wirth, N.: A plea for lean software. IEEE Comput. **28**(2), 64–68 (1995). https://doi.org/10.1109/2.348001

226. Weißleder, S., Wartenberg, F., Lackner, H.: Automated test design for boundaries of product line variants. In: Proceedings of ICTSS 2015, LNCS 9447, pp. 86–101 (2015). https://doi.org/10.1007/978-3-319-25945-16

227. Zander, J., Schieferdecker, I., Pieter, J. (eds.): Model-Based Testing for Embedded Systems. CRC Press (2012). Available at https://www.researchgate.net/publication/288984372_Model-Based_Testing_for_Embedded_Systems

228. Zschaler, S., Sánchez, P., Santos, J., Alférez, M., Rashid, A., Fuentes, L., Moreira, A., Araújo, J., Kulesza, U.: VML*—a family of languages for variability management in software product lines. In: Proceedings of Second International Conference Software Language Engineering (SLE), pp. 82–102 (2009)

229. Zuse, H.: Software Complexity—Measures and Methods, Programming Complex Systems, vol. 4. De Gruyter (1991). https://doi.org/10.1515/9783110866087

Literature Part II

1. Belli, F., Beyazıt, M., Budnik, CH. J., Tuglular, T.: Advances in Model-Based Testing of Graphical User Interfaces, in: Atif M. Memon (ed.), Advances in Computers, Vol. 107, Burlington: Academic Press, pp. 219–280 (2017)
2. Ecodesign Directive: Directive 2009/125/EC of the European Parliament and of the Council of October 2009 establishing a framework for the setting of ecodesign requirements for energy-related products. Official Journal of the European Union 53 (2009) L285 of October 31, 2009; p.34-43. ISSN 0378-6978. For Implenting Measures see homepage of DG Energy: https://ec.europa.eu/info/energy-climate-change-environment/standards-tools-and-labels/ products-labelling-rules-and-requirements/energy-label-and-ecodesign/energy-efficient-pro ducts_en. For plastics see: Ecos report: For better not worse: Applying ecodesign principles to plastics in the circular economy, Brussels, June 2019. For labeling see: Energy efficiency directive: Directive 2012/27/EU of the European Parliament and of the Council of 25 October 2012 on energy efficiency. Off. J. L **315**, 1–56 (2012)
3. Metri,G., Sabharwal, M., Iyer,S., Agrawal, A.: HW/SW codesign to optimize SoC device battery life. Computer 89–92 (2013)
4. Belli, F., Quella, F.: Utilization of Used Components. VDE Verlag, Berlin-Offenbach (2015). ISBN 978-3-8007-3898-4
5. WEEE Directive: Directive 2012/19/EC of the European parliament and the council of July 4, 2012 on waste electrical and electronic equipment. Off. J. Eur. Union **55**, 38–67 (2012). ISSN 1977-0677
6. Innovative Technologies for Resource Efficiency—Research for the Provision of Raw Materials of Strategic Importance. http://www.bgr.bund.de/EN/Themen/Min_rohstoffe/Projekte/ Rohstoffverfuegbarkeit_laufend_en/INTRAr4_en.html?nn=1560240
7. IEC 62309:2004-07: Dependability of Products Containing Reused Parts—Requirements for Functionality and Tests. Bureau Central de la Commission Electrotechnique Internationale, Geneva/Switzerland. ISBN 2-8167-7668-0
8. IEC 60300-1: 2014: Dependability Management—Part 1: Guidance for Management and Application. Bureau Central de la Commission Electrotechnique Internationale, Geneva/Switzerland. ISBN 978-2-8322-1777-1
9. COCIR: (www.cocir.org, Association of manufacturers of medical equipment). Cocir Industrial Standard: Good Refurbishment Practice (GRP), Brussels/Belgium, 6/2009
10. Braun, M., Arglebe, C., Plumeyer, M.: Gebraucht und doch wie neu. Medizinprodukte J. **14**(4), 199–2004 (2007). ISSN 978-3-8167-7668-0
11. DERA (Deutsche Rohstoffagentur): Rohstoffinformationen 30 (2016): Wachstumsraten-Monitor, https://www.deutsche-rohstoffagentur.de/DERA/DE/Publikationen (growth rate

© The Editor(s) (if applicable) and The Author(s), under exclusive license to Springer Nature Switzerland AG 2021
F. Belli and F. Quella, *A Holistic View of Software and Hardware Reuse*, Studies in Systems, Decision and Control 315, https://doi.org/10.1007/978-3-030-72261-6

monitor); availability: https://www.bgr.bund.de/EN/Themen/Min_rohstoffe/Rohstoffverfueg barkeit/rohstoffverfuegbarkeit_node_en.html

12. IEC 62430:2009-02: Environmentally Conscious Design for Electrical and Electronic Products and Systems. Bureau Central de la Commission Electrotechnique Internationale, Geneva/Switzerland. ISSN 2-8318-1032-7

13. Sabharwal, M., Agrawal, A., Metri, G.: Enabling Green IT through Energy Aware Software, IT Pro, pp. 19–27 (2013)

14. Chabukswar, R.: Creating Energy-Efficient Software (part 1). https://software.intel.com/en-us/user/336455, Intel, pp. 1–6 (2011)

15. Steigerwald, B., Chabukswar, R., Krishnan, K., De Vega, J.: Creating Energy-Efficient Software, Intel: White Paper on Energy Efficient Platforms—Consideration for Application of Software and Services 2/2011, Rev.2.0. https://www.content/dam/develop/external/us/en/doc uments/green-hill-sw-20-185393.pdf

16. Chabukswar, R.: Creating Energy-Efficient Software (part 4). https://software.intel.com/en-us/user/336455, pp. 1–5 (2012)

17. Kern, E., Dick, M., Naumann, S., Guldner, A., Johann, T.: Green Software and Green Software Engineering—Definitions, Measurements, and Quality Aspects. In: Hilty, L., Aebischer, B., Andersson,G., Lohmann, W. (eds.) ICT4S 2013: Proceedings of the First International Conference on Information and Communication Technologies for Sustainabilty, ETH Zurich, pp. 87–94 (2013). https://doi.org/10.3929/ethz-a-007337628

18. ISO/IEC 14756:1999 Information Technology—Measurement and Rating of Performance of Computer-Based Software Systems. ISO copyright office, CH 1211 Geneve 20, Casa Portale 56 (1999)

19. Dirlewanger, W.: Proceedings of SOSP 2014, pp. 91–104. Stuttgart, Germany (2014)

20. ISO/IEC 25000:2014: International Standardization Organization. In: Software Engineering Systems and Software Quality Requirements and Evaluation (SQuaRE)—Guide to Systems and to SQuaRE

21. Chen, F.F., Schneider, J.-G., Yang, Y., Grundy, J., He, Q.: An energy consumption model and analysis tool for cloud computing environments. In: Proceeding of the First International Workshop on Green and Sustainable Software Engineering (Greens), Zurich, Switzerland, pp. 45–50. IEEE, Piscataway, NJ (2012)

22. Quella, F.: Ecodesign strategies: A missing link in ecodesign. In: Lee, K.-M., Kauffmann, J. (eds.) Handbook of Sustainable Engineering, pp. 269–284. Springer, Dordrecht/Netherlands (2013). ISBN 978-1-402-08940-4

23. Züst, R.: Ressourceneffizienz ganzer Prozessketten am Beispiel „hochfeste Stähle für neue Anwendungen", Bericht aus einer Arbeitsgruppe, Schlussbericht, Mai 2015, Auftraggeber: Bundesamt für Umwelt (BAFU), Abteilung Abfall und Rohstoffe, CH-3003 Bern

25. VDI 2243:2002-07: Recycling-Oriented Product Development. VDI guideline. Düsseldorf: VDI—Association of German Engineers

25. Quella, F.: Innovationsstrategien für umweltverträgliche Produkte. Umweltwirtschaftsforum 8(3), 47–51 (2000)

26. ISO/TR 14062:2002-11: Environmental Management—Integrating Environmental Aspects into Product Design and Development. ISO—International Organization for Standardization, Geneva/Switzerland

27. IEC/PAS 62814:2012-12: Dependability of Software Products Containing Reusable Components. Guidance for Functionality and Tests. Bureau Central de la Commission Electrotechnique Internationale, Geneva/Switzerland. ISBN 978-2-83220

28. Kocak, S.A., Miranskyy, A., Alptekin, G.I., Bener, A.B., Cialini, E.: The impact of improving software functionality on environmental sustainability. In: Hilty, L., Aebischer, B., Andersson, G., Lohmann, W. (eds.) ICT4S 2013: Proceedings of the First International Conference on Information and Communication Technologies for Sustainabilty, ETH Zurich, pp. 95–100. https://doi.org/10.3929/ethz-a-007337628

29. Naumann, S., Dick, M., Kern, E., Johann, T.: The GREENSOFT model: a reference model for green and sustainable software and its engineering. Sustain. Comput. Inf. Syst. I, 294–304 (2011)

30. Murugesan, S.: Making IT Green. IT Professional. IEEE Comput. Soc. **12**(2), 4–5 (2010)
31. Schall, D., Hoefner, V., Kern, M.: Towards an enhanced benchmark advocating energy-efficient systems. In: "Performance Evaluation and Benchmarking", Third TPC Technology Conference TPCTC 2011, 8 Sept. 2011, Seattle WA, USA, Lecture Notes in Computer Science, pp. 31–45. Springer, Berlin, Heidelberg (2011)
32. van Bokhoven, F., Bloem, J.: Pilot result monitoring energy usage by software. In: Hilty, L., Aebischer, B., Andersson, G., Lohmann, W. (eds.) ICT4S 2013: Proceedings of the First International Conference on Information and Communication Technologis for Sustainabilty, ETH Zurich, pp. 108–110 (2013). https://doi.org/10.3929/ethz-a-007337628
33. Waste framework directive: Directive 2008/98/EC of the European Parliament and of the Council of 19 November 2008 on waste and repealing certain Directives. Off. J. Eur. Union **51**, 3–30 (2008). ISSN 1725-2555
34. End-of-life vehicle directive: Directive 2000/53/EC of the European Parliament and of the Council of end-of-life vehicles of 18 September 2000. Off. J. Eur. Union **43**, 34–43 (2000). ISSN 0378-6978
35. RoHS2 Directive: Directive 2011/65/EU of the European Parliament and of the Council of 8 June 2011 on the restriction of the use of certain hazardous substances in electrical and electronic equipment. Off. J. Eur. Union **54**, 88–110 (2011). ISSN 1725-2555. RoHS3 directive: Directive of the European Parliament and of the Council of 15 November 2017 on the restriction of the use of certain hazardous substances in electrical and electronic products. Off. J. Eur. Union 8–11 (2017)
36. Packaging and packaging waste directive: European Parliament and Council Directive 94/62/EC of December 1994 on packaging and packaging waste. Off. J. Eur. Union **37**, 10–23 (1994). ISSN 0378-6978
37. General product safety directive (GPSD): Directive 2001/956/EC of the European Parliament and of the Council of 3 December 2001 on general product safety. Off. J. Eur. Union **45**, 4–17 (2002)
38. EMC directive: Directive relating to electromagnetic compatibility (EMC); Directive 2014/30/EU of 26 February 2014 on the harmonization of the laws of the Member States relating to electromagnetic compatibility and repealing Directive 89/336/EEC Text with EEA relevance. Off. J. Eur. Union 79–106 (2014)
39. Battery directive: Directive 2006/66/EC of the European Parliament and of the Council of 6 September 2006 on batteries and accumulators and waste batteries and accumulators repealing Directive 91/157/EEC. Off. J. Eur. Union **49**, 1–14 (2006). ISSN 1725-2555
40. Low voltage directive: Directive 2014/85/EC on the harmonization of the laws of Member States relating to electrical equipment designed for use within certain voltage limits. Off. J. Eur. Union 357–374 (2014)
41. Machinery directive: Directive 2006/42/EC on machinery and amending directive 95/16/EC (recast). Off. J. Eur. Union **49** 24–86 (2006). ISSN 1725-2555
42. REACH (Registration, Evaluation, Authorization of Chemicals) regulation: Regulation (EG) Nr. 1907/2006 of the European Parliament and of the Council of 18 December 2006 concerning the registration, evaluation, authorization and restriction of chemicals (REACH), establishing a European chemicals agency, amending directive 1999/45/EG and repealing Council regulation(EEC) Nr. 793/93 and Commission regulation (EC) Nr. 1488/94 as well as Council directive 76/769/EEC and Commission directives 91/155/EEC, 93/67/EEC, 93/105/EC and 2000/21/EC of 30 December 2006. Off. J. Eur. Union **49**, 1 (2006). ISSN 1725-2555
43. Energy: Eco-Design—Legislation—European Commission. DG Energy, Brussels/Belgium. http://ec.europa.eu/energy/efficiency/ecodesign/legislation_en.htm
44. Sicon Process: Sicon GmbH Hilchenbach/Germany. www.sicontechnology.com
45. Safe Automotive Software Architecture (SAFE): Deliverable D3.2.4.b: Definition of COTS Integration for SW and HW, The SAFE consortium 2014, ITEA2 project-10039
46. VDI 2343 Sheet 4:2012-01 Recycling of Electrical and Electronic Products—Marketing. VDI Guideline. VDI—Association of German Engineers, Düsseldorf/Germany

47. Smith, R.: A Component-Based Layered Abstraction Model for Software Portability Across Autonomous Mobile Robots, Thesis. Queensland University, Australia (2005)
48. ISO 14021:2016 Environmental Labels and Declarations—Self Declared Environmental Claims (Type II Environmental Labeling). ISO—International Organization of Standardization, Geneva /Switzerland
49. ISO 14025:2006 Environmental labels and declarations—Type III Environmental Declarations—Principles and Procedures. ISO—International Organization of Standardization, Geneva/Switzerland
50. LAGA Mitteilung M 31a: Anforderung an die Entsorgung von Elektro- und Elektronik-Altgeräten, Bund Länder-Arbeitsgemeinschaft Abfall (LAGA), 23.01. 2017; Ministerium für Umwelt, Klima und Energiewirtschaft Baden-Württemberg, Kernerplatz 9,70182 Stuttgart (only in German)
51. Project ReBorn: Innovative Reuse of Modular Knowledge Based Devices and Technologies for Old, Renewed and New Factories; Project No. 609223, Start: 09/2013, project co-funded by the European Commission within the Seventh Framework, c/o Steinbeis-Europa-Zentrum der Steinbeis Innovation GmbH, Dr. Patricia Wolny,Erbprinzenstraße 4-12, 76133 Karlsruhe
52. Lorenz, T., Fröhlich, P., Bertau, M.: Seltene Erden aus Leuchtstoffrückständen. Nachrichten aus der Chemie **63**, 984–986 (2015)
53. Hummel, O.: Thesis: Semantic Component Retrieval in Software Engineering. University of Mannnheim (2008)
54. Hoffmann, M.: Thesis: Konstruktive Zuverlässigkeit – Eine Methode für zuverlässige Systemsoftware auf unzuverlässiger Hardware. University of Erlangen-Nürnberg (2016)
55. ISO 14001:2015-11 Environmental Management Systems—Requirements with Guidance for Use. ISO—International Standardization Organization, Geneva/Switzerland
56. ISO/IEC 24700: 2005-07 Quality and Performance of Office Equipment that Contains Reused Components. ISO—International Standardization Organization, Geneva/Switzerland
57. ISO 9001:2015-09 Quality Management Systems—Requirements. ISO—International Standardization Organization, Geneva/Switzerland
58. Eiband, M., Eveleigh, T.J., Holzer, T.H., Sarkani, S.: Legacy Systems: Making the Right Choice. Defense ARJ **20**(2), 154–173 (2013)
59. Galligan, A., Maher, P., Ospina, J.: Cleaner Greener Production Programme Phase 4 (CGPP4), Final report, 2008-CP-4-S2. Iameco 2—Low Carbon, Resource Efficiency and Long Life in PC Design, Multimedia Computer Systems Ltd. (MicroPro), Revised 23.05.11, 98 Nutgrove Avenue, Rathfarnham, Dublin 14 (www.micropro.ie and www.iameco.com)
60. https://ec.europa.eu/growth/single-market/ce-marking_en
61. CSR-Reporting of large companies (directive of the EU): Directive 2014/95/EU of the European Parliament and of the Council of 22 October 2014 Amending directive 2013/34/EU as regards disclosure of non-financial and diversity information by certain large undertakings and groups. Off. J. Eur. Union **330**, 1–9
62. ISO 26000: 2010 Guidance on Social Responsibility. ISO—International Standardization Organization, Geneva/Switzerland
63. Dodd Frank Act: Securities and Exchange Commission, 17 CFR Parts 240 and 249b [Release No. 34-67716; File No. S7-40-10], Effective Date: 13 Nov 2012. Compliance Date: Issuers must comply with the final rule for the calendar year beginning 1 Jan 2013 with the first reports due 31 May 2014.
64. Proposal for a regulation of the European Parliament and of the Council setting up a Union system for supply chain due diligence self-certification of responsible importers of tin, tantalum and tungsten, their ores, and gold originating in conflict-affected and high-risk areas. Brussels, COM (2014) 111 final
65. Wimmer,W., Lee, K.M., Quella, F., Polak, J.: Ecodesign—The Competitive Advantage. Springer, Dordrecht (2010). ISBN 978-90-481-9126-0
66. ISO 8887-1:2017 Technical Product Documentation—Design for Manufacturing, Assembling, Disassembling and End-of-Life Processing – Part 1: General Concepts and Requirements. International Organization for Standardization, Geneva, Switzerland

67. BSI 8887-220:2010-03 Design for Manufacture, Disassembly and End-of-Life Processing (MADE), Design for Manufacture, Assembly, Disassembly and End-of-Life Processing (MADE). The Process of Remanufacture. Specification. BSI London/UK

68. ANSI/RIC 001.12016: Specifications for the Process of Remanufacturing. http://www.rem ancouncil.org/files/5fdeCD/RIC2015-Specifications-for-the-Process-of-Remanufacturing-Draft-11-25-151.pdf, Remanufacturing Industries Council 1335 Jefferson Road #20157, Rochester, NY 14602-0157

69. IEC 24700:2005-07-01 Quality and Performance of Office Equipment that Contains Reused Components. Bureau Centrale de la Commission Electrotechnique Internationale, Geneva/Switzerland

70. http://susproc.jrc.ec.europa.eu/Washing_maschines_and_washer_dryers/docs/Prepstudy_-WASH_20150601_FINAL_v2.pdf

71. Vautier, M., Philippot, O.: Is software ecodesign a solution to reduce the environmental impact of electronic equipment. In: EGG 2016 Proceedings, Fraunhofer IZM, Berlin, pp. 1–8 (2016). ISBN 978-3-00-053763-9

72. Bustani, A., Sahni, S., Gutowski, T., Graves, S.: Appliance Remanufacturing and Energy Savings, Environmentally Benign Manufacturing Laboratory, 28 Jan 2010, MITI-1-a-2010, Sloan School of Management: https://web.unit.edu/ebm/www/Publications/Mitei-1-a-2010.pdf

73. Schischke, K., Proske, M., Nissen, N.F., Lang, K.-D.: Modular products: smartphone design from circular economy perspective. In: EGG Conference Proceedings, Fraunhofer IZM Berlin, pp. 1–8 (2016). ISBN 978-3-00-053763-9

74. Berwald, A., Tinnetti, B., Stobbe, L., Nissen, N., Zedel, H.: Ecodesign with extended product scope on the Example of Enterprise Servers. In: EGG Proceedings, 7–9 Sept 2016, Fraunhofer IZM Berlin, pp. 1–7 (2016). ISBN 978-3-00-053763-9

75. European Remanufacturing Network. https://www.remanufacturing.eu/ Market Study Worldwide all Product Groups

76. Goodship, V., Stevels, A.: WEEE Directive Handbook. Woodhead Publishing, Philadephia (2012). ISBN 978-0-85709-089-8

77. Luttrop, C., Lagerstedt, J.: Ecodesign and the ten golden rules: generic advice for merging environmental aspects into product development. J. Clean. Prod. **14**, 1396–1408 (2006)

78. Fair mobile phone: see https://www.fairphone.com

79. Tomra: Recycling of Household Plastic Waste. https://www.tomra.com/de-de/sorting/recycling

80. United Nations, Department of Economic and Social Affairs Disability. https://www.un.org/development/desa/disabilities/about-us/sustainable-development-goals-sdgs-and-disability.html

81. Sustainable Development Goals, ZVEI Wegweiser für nachhaltige Entwicklung in der Elektroindustrie (Guide only in German), www.ZVEI.org. ZVEI—Zentralverband Elektrotechnik- und Elektronikindustrie e. V. Abteilung Umweltschutzpolitik Lyoner Straße 9 60528 Frankfurt am Main

82. Research and Markets "Global Refurbished Medical Devices Markets 2019–2023". https://www.researchandmarkets.com/research/pv65s6/global?w=12

83. Refrigerators, Commission Regulation (EU) 2019/2019 of 1 October 2019 Laying Down Ecodesign Requirements for Refrigerating Appliances Pursuant to Directive 2009/125/EC of the European Parliament and of the Council and repealing Commission Regulation (EC) No 643/2009, OJ 315, pp. 187–208

84. Televisions, Commission Regulation (EC) No 642/2009 of 22 July 2009 Implementing Directive 2005/32/EC of the European Parliament and of the Council with Regard to Ecodesign Requirements for Televisions OJ L 191, pp. 42–52

85. Refrigerating Appliances with a Direct Sales Function, Commission Regulation (EU) 2019/2024 of 1 October 2019 Laying Down Ecodesign Requirements for Refrigerating Appliances with a Direct Sales Function Pursuant to Directive 2009/125/EC of the European Parliament and of the Council OJ L 315, pp. 313–334

86. Welding Equipment, Commission Regulation (EU) 2019/1784 of 1 October 2019 Laying Down Ecodesign Requirements for Welding Equipment Pursuant to Directive 2009/125/EC of the European Parliament and of the Council OJ 272, pp. 121–135
87. Servers, Commission Regulation (EU) 2019/424 of 15 March 2019 Laying Down Ecodesign Requirements for Servers and Data Storage Products Pursuant to Directive 2009/125/EC of the European Parliament and of the Council and amending Commission Regulation (EU) No 617/2013 OJ 74, pp. 46–66
88. Household Washing Machine and Washer Dryers, Commission Regulation (EU) 2019 /2023 of 1 October 2019 Laying Down Ecodesign Requirements for Household Washing Machines and Household Washer-Dryers Pursuant to Directive 2009/125/EC of the European Parliament and of the Council amending Commission Regulation (EC) No 1015/2010, OJ 315, pp. 285–312
89. Household Dishwashers, Commission Regulation (EU) 2019/2022 of 1 October 2019 Laying Down Ecodesign Requirements for Household Dishwashers Pursuant to Directive 2009/125/EC of the European Parliament and of the Council amending Commission Regulation (EC) No 1275/2008 and repealing Commission Regulation (EU) No 1016/2010, OJ 315, pp. 267–284
90. Report from the Commisision to the European parliament, The Council, The European Economic and Social Committee and the Committee of the Regions on the Implementation of the Circular Economy, Brussels, 4.3.2019 COM (2019) 190 final
91. Electronic displays, Commission (EU) 2019/2021of 1 October 2019 Laying Down Ecodesign Requirements for Electronic Displays Pursuant to Directive 2009/125/EC of the European Parliament and of the Council, Amending Commission Regulation (EC) No 1275/2008 and Repealing Commission Regulation (EC) No 642/2009, OJ L 315, pp. 241–266
92. Matsumoto, M., Ijomah, W.: Remanufacturing. In: Lee, K.-M., Kauffmann, J. (eds.) Handbook of Sustainable Engineering, pp. 389–408. Springer, Dordrecht/Netherlands (2013). ISBN 978-1-402-08940-4

Subject Index—Part I

© The Editor(s) (if applicable) and The Author(s), under exclusive license to Springer Nature Switzerland AG 2021
F. Belli and F. Quella, *A Holistic View of Software and Hardware Reuse*, Studies in Systems, Decision and Control 315, https://doi.org/10.1007/978-3-030-72261-6

357

Subject Index—Part II